AF790746

Solution-Processed Solar Cells

Materials and device engineering

Online at: https://doi.org/10.1088/978-0-7503-3255-2

IOP Series in Renewable and Sustainable Power

The IOP Series in Renewable and Sustainable Power aims to bring together topics relating to renewable energy, from generation to transmission, storage, integration, and use patterns, with a particular focus on systems-level and interdisciplinary discussions. It is intended to provide a state-of-the-art resource for all researchers involved in the power conversation.

Series Editor
Professor David Elliott
Open University, UK

About the Editor
David Elliott is Emeritus Professor of Technology Policy at the Open University, where he developed courses and research on technological innovation, focusing on renewable energy policy. Since retirement, he has continued to write extensively on that topic, including a series of books for IOP Publishing and a weekly blog post for *Physics World* (physicsworld.com/author/david-elliott)

About the Series
Renewable and sustainable energy systems offer the potential for long-term solutions to the world's growing energy needs, operating at a broad array of scales and technology levels. The IOP Series in Renewable and Sustainable Power aims to bring together topics relating to renewable energy, from generation to transmission, storage, integration, and use patterns, with a particular focus on systems-level and interdisciplinary discussions. It is intended to provide a state-of-the-art resource for all researchers involved in the power conversation.

We welcome proposals in all areas of renewable energy including (but not limited to) wind power, wave power, tidal power, hydroelectric power, PV/solar power, geothermal power, bioenergy, heating, grid balancing and integration, energy storage, energy efficiency, carbon capture, fuel cells, power to gas, electric/green transport, and energy saving and efficiency.

Authors are encouraged to take advantage of electronic publication through the use of colour, animations, video, data files, and interactive elements, all of which provide opportunities to enhance the reader experience.

A list of recently published and forthcoming titles published in this series can be found here: https://iopscience.iop.org/bookListInfo/iop-series-in-renewable-and-sustainable-power#series.

Solution-Processed Solar Cells

Materials and device engineering

Richard A Taylor

Department of Chemistry, University of the West Indies, St. Augustine Campus Trinidad, Tobago

and

Faculty of Science and Technology, North Carolina Agricultural and Technical State University, Greensboro, NC-27411, USA

Karthik Ramasamy

Los Alamos National Laboratory, P.O. Box 1663, Los Alamos, NM-87545, USA

IOP Publishing, Bristol, UK

© IOP Publishing Ltd 2025. All rights, including for text and data mining (TDM), artificial intelligence (AI) training, and similar technologies, are reserved.

This book is available under the terms of the IOP-Standard Books License

No part of this publication may be reproduced, stored in a retrieval system, subjected to any form of TDM or used for the training of any AI systems or similar technologies, or transmitted in any form or by any means, electronic, mechanical, photocopying, recording or otherwise, without the prior permission of the publisher, or as expressly permitted by law or under terms agreed with the appropriate rights organization. Certain types of copying may be permitted in accordance with the terms of licences issued by the Copyright Licensing Agency, the Copyright Clearance Centre and other reproduction rights organizations.

Permission to make use of IOP Publishing content other than as set out above may be sought at permissions@ioppublishing.org.

Richard A Taylor and Karthik Ramasamy have asserted their right to be identified as the authors of this work in accordance with sections 77 and 78 of the Copyright, Designs and Patents Act 1988.

ISBN 978-0-7503-3255-2 (ebook)
ISBN 978-0-7503-3253-8 (print)
ISBN 978-0-7503-3256-9 (myPrint)
ISBN 978-0-7503-3254-5 (mobi)

DOI 10.1088/978-0-7503-3255-2

Version: 20250801

IOP ebooks

British Library Cataloguing-in-Publication Data: A catalogue record for this book is available from the British Library.

Published by IOP Publishing, wholly owned by The Institute of Physics, London

IOP Publishing, No.2 The Distillery, Glassfields, Avon Street, Bristol, BS2 0GR, UK

US Office: IOP Publishing, Inc., 190 North Independence Mall West, Suite 601, Philadelphia, PA 19106, USA

Richard Taylor dedicates this book to his late parents, Gilbert Taylor and Audriana Taylor. Karthik Ramasamy dedicates this book to his late father Ramasamy P. Both Richard and Karthik also dedicate this book to the late Professor Paul O'Brien, OBE, who was supervisor and mentor, respectively.

Contents

Preface xii

Acknowledgments xiv

Author biographies xv

Solution-processed solar cells: materials and device engineering xvi

1 Energy and the solar photovoltaics landscape: an overview **1-1**

1.1 The global renewable energy and photovoltaics sector 1-1
1.2 Developments in solar PV cell technologies 1-4
1.3 The next-generation photovoltaics market sector 1-6
1.3.1 Global copper-based chalcogenide solar cell market 1-6
1.3.2 Global quantum dot solar cell market 1-6
1.3.3 Global dye-sensitized solar cell market 1-7
1.3.4 Global perovskite solar cell market 1-7
1.3.5 Global organic solar cell market 1-7
References 1-8

2 Introduction to solar cell devices **2-1**

2.1 Overview of solar cell device architectures 2-2
2.1.1 Evolution of solar cells and efficiencies 2-2
2.1.2 Copper-based chalcogenide solar cells 2-4
2.1.3 Colloidal quantum dot solar cells 2-5
2.1.4 Dye-sensitized solar cells 2-6
2.1.5 Perovskite solar cells 2-7
2.1.6 Organic/polymer solar cells 2-8
2.2 Basics of device performance characteristics 2-8
2.2.1 Current–voltage (*I*–*V*) characteristics 2-8
2.2.2 Incident-photon-to-current conversion efficiency (IPCE) 2-11
2.2.3 Shockley–Queisser limit 2-12
2.3 Photoconductivity and charge transport properties 2-13
2.3.1 Energy loss factors 2-13
2.3.2 Charge separation and recombination 2-15
2.3.3 Charge collection efficiency 2-17
2.4 Strategies for defect and interface engineering 2-18
2.4.1 Defect and interface engineering in dye-sensitized solar cells 2-19
2.4.2 Defect and interface engineering in perovskite solar cells 2-20

2.4.3 Defect and interface engineering in organic solar cells 2-21
References 2-22

3 Solution-processed fabrication methods 3-1

3.1 Types of processing methods 3-1
3.1.1 Spin coating 3-2
3.1.2 Drop casting 3-3
3.1.3 Blade coating 3-5
3.1.4 Spray coating 3-7
3.1.5 Printing technology 3-9
3.1.6 Spray pyrolysis 3-10
3.1.7 Chemical bath deposition 3-12
3.1.8 Successive ionic layer adsorption and reaction (SILAR) 3-13
3.1.9 Electrodeposition 3-15
3.1.10 Electrochemical polymerization 3-16
3.2 Strategies towards scaling up 3-18
3.2.1 The challenge with solution processing methods 3-18
3.2.2 Select strategies for scaling up 3-18
3.2.3 Scaling up by roll-to-roll fabrication 3-20
3.3 Conclusion and prospects 3-22
References 3-22

4 Copper-based chalcogenide solar cells 4-1

4.1 Structure and properties of copper-based chalcogenide materials 4-1
4.2 Solar cell architecture 4-3
4.2.1 Back contact metal layer 4-3
4.2.2 *p*-type absorber layer 4-4
4.2.3 *n*-type buffer layer 4-4
4.2.4 Transparent conducting oxide (TCO) 4-4
4.3 Solution processing of Cu-based chalcogenide materials 4-5
4.3.1 Solution processing of CIGS absorbers 4-5
4.3.2 Nanocrystals ink-based approach for CIGS 4-19
4.3.3 Solution processing of CZTS absorbers 4-23
4.3.4 Nanocrystals ink-based approach for CZTS 4-32
4.4 Cell degradation and failure diagnostics 4-33
4.4.1 CIGS cell degradation and failure diagnostics 4-34
4.4.2 CZTS cell degradation and failure diagnostics 4-34

4.5	Conclusion, challenges and future prospects	4-35
	References	4-36
5	**Colloidal quantum dot solar cells**	**5-1**
5.1	Structure and chemical properties of CQDs	5-2
	5.1.1 Classification of CQDs	5-2
	5.1.2 Categories of CQDs	5-5
	5.1.3 Optical and electronic properties	5-11
5.2	Strategies for engineering colloidal quantum dots	5-17
	5.2.1 Synthesis: nucleation and growth control	5-17
	5.2.2 Synthetic methods	5-19
	5.2.3 Strategies for doping colloidal nanoparticles	5-20
	5.2.4 Surface passivation of colloidal nanoparticles	5-21
5.3	Cell architecture, device physics and performance	5-23
	5.3.1 Schottky CQD solar cell	5-23
	5.3.2 Depleted heterojunction CQD solar cell	5-25
	5.3.3 Bulk heterojunction CQD solar cell	5-26
	5.3.4 CQD-sensitized solar cell	5-27
	5.3.5 Tandem/multijunction CQD solar cell	5-28
	5.3.6 Tandem perovskite CQD solar cell	5-29
	5.3.7 CQD-organic hybrid solar cell	5-30
5.4	Cell degradation and failure analysis	5-32
	5.4.1 Evaluation of degradation mechanisms and analysis	5-32
5.5	Conclusion, challenges and future prospects	5-37
	References	5-38
6	**Dye-sensitized solar cells**	**6-1**
6.1	Working principles of dye-sensitized solar cells	6-1
6.2	Structure and properties of dye sensitizer materials	6-3
	6.2.1 Ruthenium-based dyes	6-4
	6.2.2 Porphyrin-based dyes	6-6
	6.2.3 Organic dyes	6-9
	6.2.4 Quantum dot sensitizers	6-11
6.3	Photoanode development	6-12
	6.3.1 Binary metal oxides	6-14
	6.3.2 Graphene-based materials	6-18
	6.3.3 Doped and hybrid materials	6-19

6.4 Electrolyte development 6-22
6.4.1 Liquid electrolytes 6-23
6.4.2 Quasi-solid electrolytes 6-23
6.4.3 Solid-state electrolytes 6-24
6.5 Counter electrode development 6-27
6.5.1 Metal CE materials 6-28
6.5.2 Carbon CE materials 6-29
6.5.3 Polymer CE materials 6-31
6.5.4 Transition metal compound CEs 6-33
6.5.5 Composite/hybrid CE materials 6-34
6.6 Transparent and flexible conducting substrates 6-35
6.7 Cell architecture, device physics and performance 6-36
6.7.1 Device modelling (*first principles*) 6-37
6.7.2 Grätzel cell 6-38
6.7.3 Solid-state DSSCs 6-39
6.7.4 *p*-Type DSSCs and *p*-type solid-state DSSCs 6-41
6.7.5 Quantum dot-sensitized solar cells 6-43
6.7.6 Perovskite-sensitized solar cells 6-45
6.7.7 Tandem DSSCs 6-46
6.8 Cell degradation and failure 6-48
6.8.1 Cell stability 6-48
6.8.2 Failure diagnostics 6-50
6.9 Conclusion, challenges and future prospects 6-51
References 6-52

7 Perovskite-based solar cells 7-1

7.1 Structure and properties of perovskites 7-1
7.1.1 Defect structure 7-2
7.2 Solution preparation of perovskites 7-3
7.2.1 One-step and step-by-step processes 7-4
7.2.2 Perovskites through intermediate complexes 7-11
7.3 Cell architecture, device physics and performance 7-12
7.3.1 Electron transport layer 7-12
7.3.2 Hole transport layer 7-17
7.3.3 Lead (Pb)-free perovskite solar cells 7-19
7.3.4 Tandem solar cells 7-21

7.4 Cell degradation and failure diagnostics 7-24
7.5 Conclusion, challenges and future prospects 7-26
References 7-27

8 Organic and polymer-based cells 8-1
8.1 Working principles of organic and polymer-based solar cells 8-2
8.2 Structures and properties of photoactive materials 8-3
8.2.1 Acceptor materials 8-4
8.2.2 Donor materials 8-10
8.2.3 Small-molecule donors 8-13
8.3 Cell architecture, device physics and performance 8-15
8.3.1 Single-layer OSCs 8-15
8.3.2 Bilayer OSCs 8-16
8.3.3 Bulk heterojunction OSCs 8-18
8.3.4 Pseudo-bilayer OSCs 8-20
8.3.5 Tandem/multijunction OSCs 8-21
8.4 Cell degradation and failure 8-23
8.4.1 Cell stability 8-23
8.5 Conclusion, challenges and future prospects 8-32
References 8-33

Preface

The utilization of abundant solar energy employing efficient photovoltaics (PV) is a particularly attractive option for meeting global energy demand, which has seen accelerated growth. This growth is in response to the rapid industrialization of human society over the past century with concomitant incessant carbon-based energy consumption and endless negative environmental impact. Solar photovoltaics which directly harvest and transform sunlight into electricity, represent a form of renewable energy technology now at the center of global energy security strategies. Presently, single crystalline silicon wafer cells dominate the market share of commercially available solar cells and has to date the highest efficiency of 26.5%. However, the vacuum-based solar cell fabrication approach particularly for traditional silicon-based solar cell technologies makes these technologies prohibitively expensive and limited in their applications. Thus, there has been tremendous development in cost-effective solution-based solar cells fabrication leading to alternative and new/emerging technologies like dye-sensitized solar cells, quantum dots-sensitized solar cells, nano-ink and perovskite solar cells with impressive efficiency over 25%. Solution-processed solar cells are fabricated by depositing thin films of materials from solutions, typically involving layer-by-layer deposition. This approach is cost-effective and scalable, making it attractive for large-scale solar cell production. Therefore, these emerging technologies are expected to result in more wide-scale application of PV technologies across the spectrum of sectors of residential, commercial, and utility-scale applications—such as in heating, agriculture, transportation, and everyday devices.

This book aims to present a comprehensive but focused outline of this vastly growing area with reference to fundamental chemical and physical aspects of various solution-processed solar cells including different components of solar cells stacks, their characteristics and use, and working mechanisms with self-explanatory illustrations. It will thoroughly examine the state-of-the-art of these different solution processing approaches in advancing each technology, highlighting challenges related to cell optimization with perspectives aiming towards PhD students and researchers in the field. Importantly, the discussions are predicated on fundamental chemistry and physics concepts of the various types of materials and how their properties are exploited for these technologies. The discussions not only present a comprehensive account of these emerging PV technologies but highlight the current device challenges and prospects, the opportunities for further advancement and their potential for applications.

Though there are a number of review articles on these types of solar cell devices, currently there is no book that collectively and comprehensively covers these topics. Therefore, this book serves as an overall introduction to early-stage researchers in these fields, and seeks to establish and intertwine fundamental concepts with practical experimental approaches and outcomes. It provides the reader with an insightful approach to understanding the tools and chemical approaches that a materials scientist uses to manipulate the structures and properties of materials towards developing these emerging PV technologies while considering the structural, synthetic and processing protocols. Overall, the practical accounts will be useful for the young experimentalist, and professors will find this as an effective tool for reference material in teaching at the advanced stages. Overall, this book uniquely interfaces the chemical considerations including engineering with the device physics of a range of emerging solution-processed PV technologies.

The book has two main sections: (1) introductory concepts of photovoltaics, the field and approaches to solution processing of solar cells; and (2) materials and device engineering and characteristics of different classes of solution-processed solar cell types.

In chapter 1 we present an overview of the PV sector with a focus on the market driving forces leading to the development of the PV sector. Chapter 2 focuses on the fundamental principles of solar cells, a brief description of the types of solution-processed devices, their performance characteristics, photoconductive and charge transport characteristics. Chapter 3 outlines approaches for solution processing of solar cells, highlighting specific examples from published literature and issues related to scaling up for large-area device fabrication.

In the second section, we highlight the chemical and physical principles that are characteristic for different device types prepared via solution-processed methods. In particular, chapter 4 focuses on copper-based chalcogenide solar cells with the premise of materials structure and properties, device architectures and engineering, solution processing of different components and ends with a discussion on cell degradation and failure. Chapter 5 explores colloidal quantum dot solar cells, starting with the structure and chemical properties of colloidal quantum dots, their related optical and electronic properties, strategies for engineering colloidal quantum dots and controlling their properties. The influence of device physics and performance based on the characteristics of colloidal quantum dots and device engineering and their relation to cell degradation mechanisms are also explored. Chapter 6 outlines aspects of dye-sensitized solar cells, their cell architectures, performance and device modeling. There is a survey of the types of dye-sensitized molecules, their structures and chemical properties and materials development of other cell components such as the photoanode, electrode and electrolyte. The chapter ends with an assessment of cell degradation mechanisms and failure. Chapter 7 deals with the highly promising perovskite-based solar cells. It is underpinned with perovskite structures, their defect chemistry, properties and their preparation. There is a survey of device architectures achieved through solution processing and the chapter concludes with an evaluation of cell degradation and failure which has been a tremendous limitation for these types of devices. Finally, chapter 8 focuses on organic/polymer solar cells. It is premised on an extensive survey of these types of solar absorbers, their molecular structural donor and acceptor features and their influence on photoactivity. The prominent and promising types of device architectures are featured, and their performance referenced, including in terms of the degradation mechanisms that lead to device failure.

Acknowledgments

The authors would like to thank all those who made this book possible. Special thanks to Richard Taylor's graduate student, Kimberly Weston for helping with citations, and former student, Dr Shanna-Kay Ming, PhD for making some content available through her thesis work. We are very grateful for the team at IOP, starting with Caroline Mitchell for initiating us to conduct the writing of this book and Mia Foulkes for guiding us to its completion. We also thank our respective families, friends and colleagues who have encouraged and supported us.

Author biographies

Richard A Taylor

Richard A Taylor is Senior Lecturer (Associate Professor) in materials chemistry at The University of the West Indies (UWI), St. Augustine Campus, Trinidad and Tobago. He holds a PhD in Chemistry from The UWI Mona Campus, Jamaica. He has been Research Scientist at the North Carolina A&T State University, visiting scholar at the Schools of Chemistry and of Materials, University of Manchester, UK and visiting scientist at the National Synchrotron Light Source-II and Center for Functional Nanomaterials, Brookhaven National Laboratory, US. His research focus is on optoelectronic materials including, chalcogenide semiconductor thin films and nanomaterials (quantum dots), 2D materials, novel metal–organic liquid crystals, and luminescent metal–organic frameworks.

Karthik Ramasamy

Karthik Ramasamy is a scientist at Los Alamos National Laboratory (LANL). His current work is focused on developing and studying materials for energy and catalytic applications. Prior to rejoining LANL, he was the Vice President of Materials at UbiQD, Inc, where he led research and development and manufacturing of quantum dots (QD) and nanocomposites and helped them become the second largest QD manufacturer in North America. He received his PhD in chemistry from the University of Manchester, UK, in 2010. He has authored/co-authored 70 publications and written 9 book chapters and spoken at more than 50 international meetings. He has 12 granted, published and pending patents. His work has been cited more than 5000 times with an h-index of 37. He is a Fellow of the Royal Society of Chemistry (FRSC) and an associate editor of Frontiers in Nanotechnology. He is a member of American Chemical Society (ACS) and the American Association for the Advancement of Science (AAAS).

Solution-processed solar cells: materials and device engineering

Solution-processed solar cells are a rapidly developing emerging or next generation class of solar cell technology in which the active materials are processed from a liquid solution using simple means. This approach of fabricating solar cells involves depositing materials onto thin, transparent and flexible substrates using techniques such as spin coating, inkjet printing, or roll-to-roll processing. The main advantages of solution-processed solar cells are their potential ease of fabrication and low production cost, potential for large-scale production, flexibility in substrate materials, and versatility in application. However, they also face challenges related to long-term stability, efficiency, and scalability that need to be addressed for widespread commercial use. With continuous advances in research and development, these types of solar cells are expected to be relatively competitive to conventional silicon-based solar cells with more wide-scale application. This book surveys the main types of this emerging generation of photovoltaic devices including: metal chalcogenide thin film solar cells, organic solar cells (OSCs), perovskite solar cells (PSCs), quantum dot solar cells (QDSCs) and dye-sensitized solar cells (DSSCs). The discussion will span materials properties and their development, device architectures, engineering processing of devices and their photophysics, electronic and electrical attributes.

IOP Publishing

Solution-Processed Solar Cells
Materials and device engineering
Richard A Taylor and Karthik Ramasamy

Chapter 1

Energy and the solar photovoltaics landscape: an overview

1.1 The global renewable energy and photovoltaics sector

The increase in global population and concomitant energy consumption led to energy shortages within the latter period of the 20th century which has continued into the 21st century and it is a critical global issue. Currently, the main sources of the world's energy are derived from fossil fuels such as coal, petroleum and natural gas. The demand for and use of fossil-based energy have tremendously impacted global greenhouse gas emissions implicated in climate change. This along with concerns about the health effects of air pollution, energy security and access, as well as volatile and unpredictable oil prices in recent decades, have driven the production and utilization of alternative, low-carbon technology and renewable energies. The Global Energy Review 2023 published by the International Energy Agency (IEA) reports that global fossil fuel consumption is projected to peak in 2030, as shown in figure 1.1 [1]. This is against the backdrop of increased investment and use of renewable, low-emissions electricity and fuels alongside energy efficiency improvements. Additionally, according to the U.S. Energy Information Administration (EIA), in 2023, about 4.18 trillion kWh of electricity were generated at utility-scale electricity generation facilities in the United States [2]. About 60% of this electricity generation was from fossil fuels—coal, natural gas, petroleum, and other gases with about 19% from nuclear energy and about 21% from renewable energy sources. This is in relation to petroleum and natural gas energy consumption amounting to 74%, renewables amounting to 9% and nuclear also at 9% (figure 1.2) [3]. Furthermore, the EIA reports that in the United States, energy-related carbon dioxide (CO_2) emissions decreased slightly in 2023 compared to 2022 [4]. Although emissions decreased across many economic sectors, more than 80% of U.S. energy-related CO_2 emissions reductions in 2023 occurred in the electric power sector with the reductions due to reduced coal-fired electricity generation. This change in the energy

doi:10.1088/978-0-7503-3255-2ch1 © IOP Publishing Ltd 2025. All rights, including for text and data mining (TDM), artificial intelligence (AI) training, and similar technologies, are reserved.

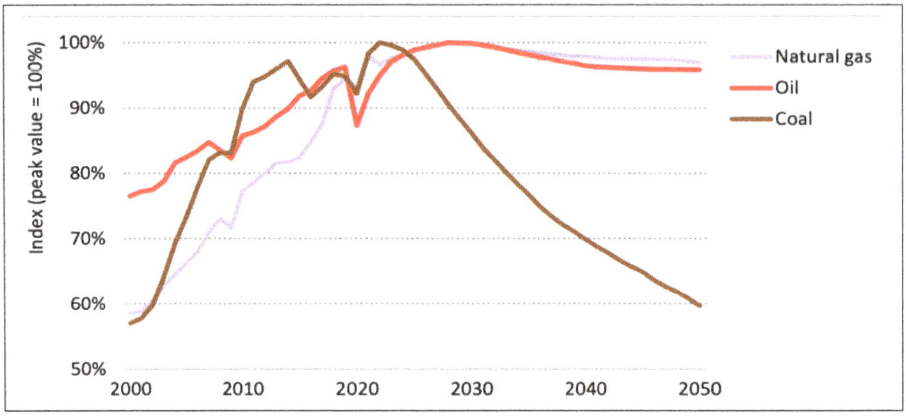

Figure 1.1. Fossil fuel consumption by fuel, 2000–50. Reprinted with permission from [1]. Copyright 2022 International Energy Agency. CC By 4.0.

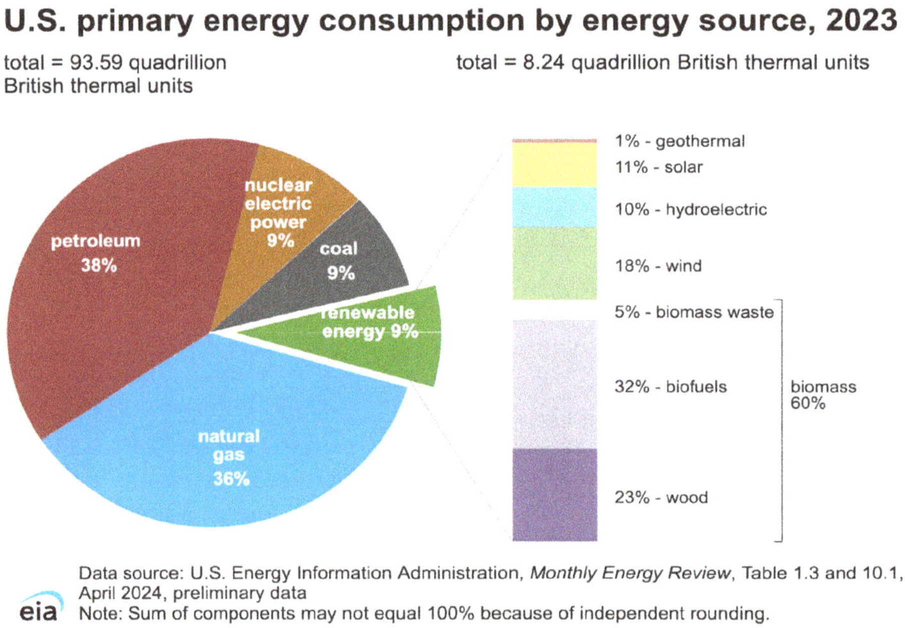

Figure 1.2. US electricity generation by source. Reprinted from [3] U.S. Energy Information Administration (July 2024).

source mix away from coal, which has the highest carbon intensity among fossil fuels, decreased electric power sector CO_2 emissions by 7% relative to 2022. Moreover, globally, it is projected that CO_2 emissions in 2030 in the power sector are projected to be 50% lower than today, largely due to government incentives such as tax credits that accelerate the deployment of solar photovoltaics (PV) and wind energy [1].

Amongst the renewable forms of energy sources, PV has been one of the pioneering and most promising technologies over the last several decades with the fastest deployment. The field of PV has been rapidly expanding since its emergence in the 1950s with the fabrication of monocrystalline silicon wafer cells. What is encouraging is that new solar capacity added between now and 2030 will account for 80% of the growth in renewable power globally, according to the Renewables 2024 Report from the IEA [5]. The report states that this growth is fuelled by declining costs, shorter permitting timelines and widespread social acceptance. Additionally, cost competitiveness and government policy support are expected to stimulate the growth of distributed applications among residential and commercial consumers as more households and companies seek to reduce their electricity expenditure. Because of the demand for PV electricity, there have been increasing production and installed capacity especially in recent years. For example, according to The Global Energy Review 2021, the total installed capacity of solar PV reached 570 GW globally by the end of 2018, representing the second-largest renewable electricity source after wind and was projected to increase by 149 GW in 2022 [6]. Overall, solar PV and wind contribute to two-thirds of the growth in renewables with solar PV electricity generation expected to rise by almost 3300 TWh in 2030. However, notwithstanding this expected growth, the Global Energy Review reports that solar PV manufacturers are scaling back investment plans due to a deepening supply excess and record-low prices [5]. In particular, global solar manufacturing capacity is expected to reach over 1100 GW by the end of 2024, outpacing projected PV demand. This oversupply has led to negative net margins for integrated solar PV manufacturers in 2024 and resulting in the cancellation of about 300 GW of polysilicon and 200 GW of wafer manufacturing capacity projects, valued at approximately US$ 25 billion.

The growth of the PV sector worldwide can be attributed to the high demand for renewable energy and the declining cost/kilowatt-hour of silicon-based cells. Fundamentally, the sector's growth is sustained by increased research and development in materials' properties, cell architectures and device characteristics, reduction in costs of processing/fabrication technology, along with government support initiatives, for example, through funding to develop technology, and incentives for deployment and use [7]. The significant interest and rapid research were primarily influenced by the oil embargo (oil crisis) in the 1970s which provided the impetus for discovery of alternative forms of energy [8]. Today the global solar PV industry is supported by thousands of companies employing several hundred thousand workers. According to Fortune Business Insights, the global solar power market size was valued at US$ 253.69 billion in 2023 and projected at US$ 273 billion in 2024 and to value US$ 436.36 billion by 2032, exhibiting a compound annual growth rate of 6% during the forecast period [9]. In particular, North America dominated the solar power industry with a market share of 41.30% in 2023. Additionally, the PV market in the U.S. is projected to grow significantly, reaching an estimated value of US$ 103.96 billion by 2032, driven by the need to combat climate change through renewable energy sources reinforced by government tax credit and feed-in-tariff programs.

1.2 Developments in solar PV cell technologies

Fundamentally, solar energy is an attractive energy source since it is the most ubiquitous and abundant form available to the Earth, and solar power is projected to become the most significant source of energy by 2050 [3]. The developments in the global PV sector are mainly because solar energy has several key advantages over other renewable energy technologies, including its global distribution, lack of hazardous waste, and decentralized generation [4]. Solar PV cell devices are primarily composed of effective light harvesting, semiconductor/photo-absorber materials which convert incident solar photon energy directly to electrical energy. Arguments for the utilization of PV include:

1. The energy source is pervasive and limitless.
2. No noise pollution and physical degradation.
3. It is an environmentally friendly alternative to fossil fuels.
4. Fabricated devices generally have a long lifetime.
5. The power generated can range from megawatts to microwatts [7, 10].

The field of solar PV has involved much investigation that has developed suitable materials and fabrication methods for solar cell devices with improved efficiencies, as illustrated in the best research-cell efficiencies chart in figure 1.3 published by the U.S. based National Renewable Energy Laboratory (NREL) [11]. Devices included in the chart of the current state-of-the-art have efficiencies reported on a standardized basis that are confirmed by independent, recognized test labs such as NREL, National Institute of Advanced Industrial Science and Technology—AIST, Joint Research Centre-European Solar Test Installation—JRC-ESTI, and Fraunhofer Institute for Solar Energy Systems—Fraunhofer-ISE. The measurements for new entries must be with respect to Standard Test or Reporting Conditions as defined by the global reference spectrum for flat-plate devices, and the direct reference spectrum for concentrator devices, as listed in standards IEC 60 904-3 edition 2 or ASTM G173.

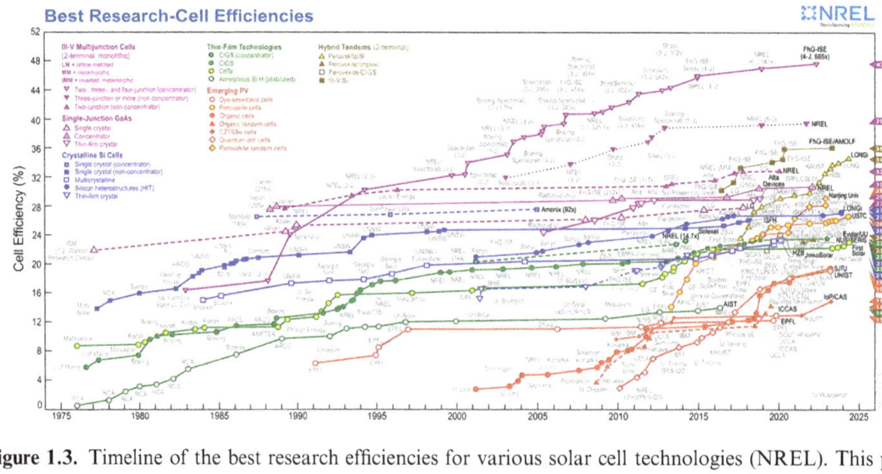

Figure 1.3. Timeline of the best research efficiencies for various solar cell technologies (NREL). This plot is courtesy of the National Renewable Energy Laboratory, Golden, CO [11].

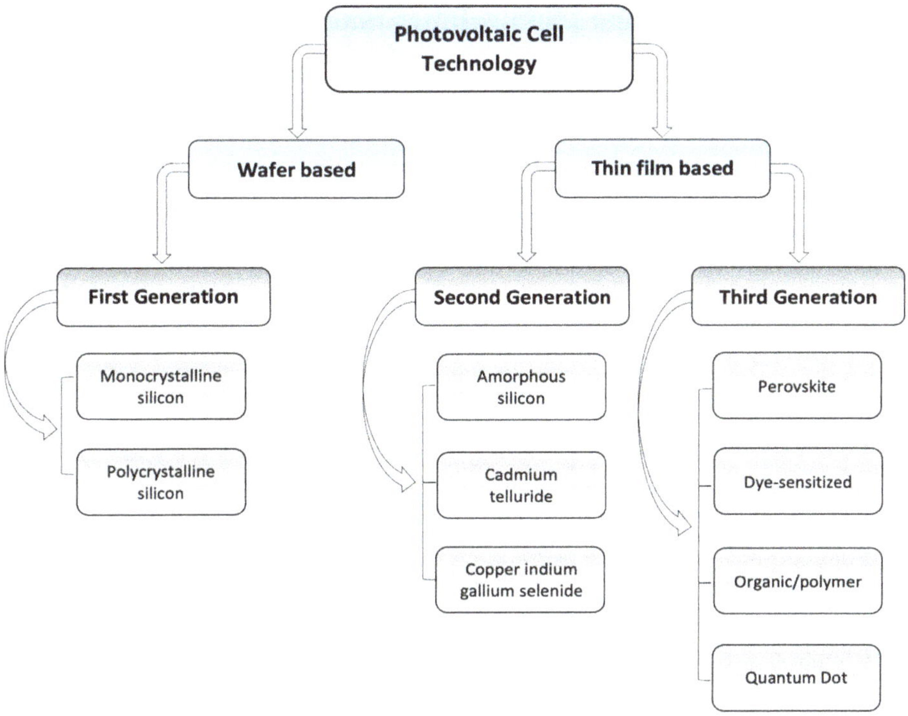

Figure 1.4. Classification of PV technology showing main types of devices.

Developments in the field have evolved over several categories of devices: (1) first generation (1G); (2) second generation (2G); and (3) third generation (3G), as shown in figure 1.4. 1G solar cells are based on crystalline or polycrystalline silicon wafers which, whilst producing high power conversion efficiencies (PCEs) above 20%, also have high production costs. The goal of reducing costs resulted in 2G solar cells commonly based on amorphous silicon, copper indium gallium selenide (CIGSe) and cadmium telluride (CdTe). However, their PCEs were generally poor, around 5%–15% compared to their 1G counterparts. Of these, CdTe, though very promising, has issues with toxicity of cadmium, whilst CIGSe has issues with market availability of indium. Additionally, these types of solar cells are fabricated using traditional vacuum-based methods which make them prohibitively expensive, especially when one considers their associated cost/kilowatt-hour. On the other hand, 3G solar cells described as emerging/next-generation solar cell technologies have become a prime focus due to impressive efficiencies from ~14% for dye-sensitized solar cells to the best-in-class perovskite solar cells at ~30% [11]. Additionally, their ease of fabrication using solution-based methods, adaptability towards wide-scale application in various forms, for example thin films, and projected lower cost/kilowatt-hour make these types of 3G cells, which also include quantum dot solar cells, nano-ink, polymer, organic and hybrid organic–inorganic solar cells, particularly attractive.

1.3 The next-generation photovoltaics market sector

Overall, the demand for electricity generation via renewable energy sources at low cost and the drive to achieve power grid parity is expected to boost the demand of third-generation solar cells. Generation of electricity using third-generation solar cells is inexpensive, as compared to that generated by first- and second-generation solar cells, primarily because of the cost of materials, simpler and inexpensive processing and improved efficiencies. They are also regarded as more environmentally friendly, more sustainable and can be deployed over a wider range of applications and end uses. The march towards their development requires various considerations inclusive of novel and better solar absorbers, easy and scalable device engineering, lower costs of production, high chemical and physical stability and long lifetime, and improved power conversion efficiencies, comparable to the conventional, market-dominated silicon-based devices.

1.3.1 Global copper-based chalcogenide solar cell market

The global copper-based chalcogenide solar cell market is mainly categorized into copper indium gallium selenide (CIGS) and copper indium selenide (CIS) thin film solar cells. This thin film solar cell market size was valued at approximately US$ 12 billion in 2024 and is expected to reach US$ 31.1 billion by 2033, growing at a compound annual growth rate (CAGR) of about 10% from 2025 to 2033 [12]. One challenge to this market was the COVID-19 pandemic which disrupted the global market growth, affecting production, supply chains, and venture deployment. The sector is also categorized based on application—residential, commercial, ground station and others—with integration into vehicles and wearables. A sizeable feature of the market is the CIGS skinny-film solar cell market which is due to the growing recognition and demand for flexible and lightweight solar panels. Unlike conventional rigid solar panels, bendy CIGS panels leverage the skinny and pliable nature of the CIGS semiconductor fabric, allowing them to be integrated into unconventional surfaces and applications. This is driven by expectations for solar solutions in sectors including transportation, transportable power devices, and constructing-integrated photovoltaics (BIPV).

1.3.2 Global quantum dot solar cell market

According to a Facts and Factors market research report [13], the global quantum dot solar cell market is expected to grow from US$ 758 million in 2020 to US$ 2.32 billion by 2026, at 20.5% CAGR during the forecast period of 2021–26. The global quantum dot solar cell market is segmented based on type of cell, whether quantum dot solar cells, quantum dot hybrid solar cells, or quantum dot nanowire solar cells. The market is also categorized based on absorber material, either cadmium-based and cadmium-free or is based on the end-user, including consumer, healthcare, military defence, commercial, and telecommunications. More specifically, the sector is categorized according to the application including single-junction cell, multi-junction cell, roofing tiles, windows, walls, and heat sensors.

1.3.3 Global dye-sensitized solar cell market

The global dye-sensitized solar cell market was valued at US$ 114.57 million in 2023 and is expected to reach US$ 296.68 million by 2031, with a projected CAGR of 12.63% during the forecast period of 2022–29, according to Data Bride Market Research [14]. The market is segmented by type of dye sensitizer, either natural or synthetic, or by application, whether portable charging, BIPV, building-applied PV (BAPV), automotive-integrated PV (AIPV), embedded electronics, outdoor advertising, solar chargers. Of these, portable charging application accounts for the largest segment of this market.

1.3.4 Global perovskite solar cell market

Of the emerging class of PV, halide perovskites are a promising lab-scale PV technology, achieving PCEs of over 25% in single-junction cells and over 29% in tandem cells. Notwithstanding the high-throughput potential of manufacturing using various processing techniques, significant technological barriers must be overcome before they can be commercially available for the power sector markets [15]. These include stability and durability, scalability, module efficiency, and manufacturing (yield, process control, etc). Additionally, their levelized cost of electricity (LCOE) must be competitive with that of incumbent technologies at the time for deployment. According to Data Bridge Market Research, the global perovskite solar cell market which was US$ 79.05 million in 2022, is expected to reach US$ 120.29 million by 2030, with a CAGR of 56.5% during the forecast period 2023–30 [16]. The global perovskite solar cell market is segmented based on device structure to include hybrid, flexible and multi-junction cells or by cell structure, either planar and mesoporous cells, product type (rigid perovskite solar cells, flexible perovskite solar cells), fabrication method (solution, vapour-deposition, vapour-assisted solution), application (smart glass, solar panel, perovskite in tandem solar cells, portable devices, utilities, BIPV), end use industries (manufacturing, energy, industrial automation, aerospace, consumer electronics), device type (hybrid, flexible and multi-junction cells). Of these, BIPV dominate the application segment of the global perovskite solar cell market due to their dual purpose—serving as the structure's outer layer and generating electricity for on-site use or export to the grid. BIPV systems can save on material and electricity costs, reduce pollution, and increase a building's architectural appeal. Also, whilst vacuum-based methods are the primary forms of fabrication, solution processing and hybrid solution-vacuum-based processing are becoming important, in order to control cost and device engineering.

1.3.5 Global organic solar cell market

There have been tremendous advances in the organic solar cells market which was valued at US$ 97.4 million in 2020 and is estimated to expand at a CAGR of 10.9% from 2023 to 2031, projected at over US$ 600 million by the end of 2030 [17]. The major factor driving the growth of the organic solar cell market is their rising popularity and growing awareness regarding their properties and their products, especially considering their use in flexible devices. The market is segmented by cell architecture such as bilayer membrane heterojunction, Schottky type, as examples, or by absorber material type such as polymers and small molecules, or by application such as BIPV, consumer electronics, wearable devices, automotive, military, among others. Of these, the BIPV segment is expected to have an upward trajectory during the forecast period of 2021–27. This is because, these cells can be fabricated using absorber materials across a range of energy and in various colours, and their ability to make efficient transparent devices as well as providing thermal insulation, among other environmental benefits. The use of organic cells in defence and military applications is projected to increase because of the lightweight and flexible characteristics, and for remote charging purposes.

References

[1] *World Energy Outlook* 2023 International Energy Agency 2024 https://iea.blob.core.windows.net/assets/86ede39e-4436-42d7-ba2a-edf61467e070/WorldEnergyOutlook2023.pdf

[2] *What is U.S. Electricity Generation by Energy Source?* U.S. Energy Information Administration 2024 https://eia.gov/tools/faqs/faq.php?id=427&t=3 (accessed 2024)

[3] *U.S. Energy Facts Explained.* U.S. Energy Information Administration 2024 https://eia.gov/energyexplained/us-energy-facts/(accessed 2024)

[4] *U.S. Energy-Related Carbon Dioxide Emissions, 2023* U.S. Energy Information Administration 2024 https://eia.gov/environment/emissions/carbon/pdf/2023_Emissions_Report.pdf

[5] *Renewables 2024—Analysis and Forecast to 2030* International Energy Administration 2024 https://iea.blob.core.windows.net/assets/45704c88-a7b0-4001-b319-c5fc45298e07/Renewables2024.pdf

[6] *Global Energy Review 2021* International Energy Administration 2021 https://iea.blob.core.windows.net/assets/d0031107-401d-4a2f-a48b-9eed19457335/GlobalEnergyReview2021.pdf

[7] Tyagi V V, Rahim N A A, Rahim N A and Selvaraj J A L 2013 Progress in solar PV technology: research and achievement *Renew. Sustain. Energy Rev.* **20** 443–61

[8] Dharmadasa I M 2012 *Advances in Thin Film Solar Cells* (Singapore: Pan Stanford Publishing Ple. Ltd)

[9] *Renewables—Solar Power Market.* Fortune Business Insights 2024 https://fortunebusinessinsights.com/enquiry/request-sample-pdf/solar-power-market-100764 (accessed 2024)

[10] Trykozko R 1997 Principles of solar energy conversion of solar energy *12th Annual of School of Optoelectronics: Photovoltaics—Solar Cells and Infrared Detectors (Warsaw, Poland)*

[11] *NREL Best Research-Cell Efficiency Chart.* National Renewable Energy Laboratory 2024 https://nrel.gov/pv/interactive-cell-efficiency.html

[12] *CIGS Thin Film Solar Cell Market Size, Share, Growth, and Industry Analysis, by Type (Copper Indium Gallium Selenide (CIGS) Solar Cells, Copper Indium Selenide (CIS) Solar Cell) by Application (Residential, Commercial, Ground Station, Others), and Regional Forecast to 2033.* Business Research Insights 2025 https://businessresearchinsights.com/market-reports/cigs-thin-film-solar-cell-market-122190 (accessed 2025)

[13] *Quantum Dot Solar Cell Market Size, Share Global Analysis Report, 2021—2026.* Facts and Factors Research 2024 https://fnfresearch.com/quantum-dot-solar-cell-market (accessed 2024)

[14] *Global Dye-Sensitized Solar Cell (Dssc) Market—Industry Trends and Forecast to 2031.* Data Bridge Market Research 2024 https://databridgemarketresearch.com/reports/global-dye-sensitized-solar-cell-dssc-market(accessed 2024)

[15] Siegler T D, Dawson A, Lobaccaro P, Ung D, Beck M E, Nilsen G and Tinker L L 2022 The path to perovskite commercialization: a perspective from the United States Solar Energy Technologies Office *ACS Energy Lett.* **7** 1728–34

[16] *Global Perovskite Solar Cell Market—Industry Trends and Forecast to 2030.* Data Bridge Market Research 2024 https://databridgemarketresearch.com/reports/global-perovskite-solar-cell-market#:~:text=Data%20Bridge%20Market%20Research%20analyses,the%20forecast%20period%202023%2D2030. (accessed 2024)

[17] *Global Organic Solar Cell (OPV) Market—Industry Trends and Forecast to 2030.* Data Bridge Market Research 2024 https://databridgemarketresearch.com/reports/global-organic-solar-cell-opv-market (accessed 2024)

IOP Publishing

Solution-Processed Solar Cells
Materials and device engineering
Richard A Taylor and Karthik Ramasamy

Chapter 2

Introduction to solar cell devices

Solar cells, also known as photovoltaic (PV) cells, are devices that convert solar light energy directly into electricity using the PV effect. They are typically made from semiconductor materials like silicon or photoactive molecules such as organic polymers, which absorb photons the from sunlight and generate free electrons. These photogenerated electrons are allowed to flow through an external circuit based on an electric field established across a junction, producing an electric current. This current is extracted through conductive metal contact electrodes and can then be used to power your home and the rest of the electric grid. Solar cells were developed as critical systems for electrical power production due to greater power consumption and population increase. However, these days, these applications lend themselves across the spectrum of sectors of residential, commercial, and utility-scale applications—such as in transportation, transportable power devices, and constructing-integrated photovoltaics (BIPV), heating, agriculture, and everyday devices. The power conversion efficiency of solar cells is one of the most important parameters in directly converting light to electricity. This is measured as the amount of electrical power from the cell compared to the energy from the light shining on it, which indicates how effective the cell is at converting energy from one form to the other. The amount of electricity produced from PV cells depends on characteristics such as intensity and wavelengths of the light available and multiple other performance attributes of the cell such as charge transport and extraction. The PV sector has developed in order to produce devices that are more efficient, cost effective and sustainable. This has included the traditional silicon-based devices to more emerging devices such as inorganic thin film, perovskite, quantum dot, organic, and systems such as multi-junctions and solar concentrators. These developments have been possible through materials development, device engineering and performance assessment, facilitated by funding and incentives.

doi:10.1088/978-0-7503-3255-2ch2

© IOP Publishing Ltd 2025. All rights, including for text and data mining (TDM), artificial intelligence (AI) training, and similar technologies, are reserved.

2.1 Overview of solar cell device architectures

2.1.1 Evolution of solar cells and efficiencies

The first solar PV device was fabricated in 1839 when Edmund Becquerel submerged brass electrodes in liquid cuprous oxide and observed an electrical property when the electrodes were exposed to sunlight [1]. Decades later in 1873, Willoughby Smith, W G Adams and R T Day discovered the PV properties of selenium, while Charles Fritts discovered the PV characteristics of gold plated amorphous silicon [2, 3]. It wasn't until the mid-20th century that efficiencies of solar cells surpassed 1% achieved in 1954 by Chapin and co-workers [3] with the utilization of a monocrystalline silicon-based cell which had an efficiency of 6%.

The traditional single *p–n* junction solar cells comprising *p*-type and *n*-type semiconductor layers have a theoretical efficiency limit of 30%, the *Shockley–Queisser (SQ) limit*, determined by a detailed balance model developed by Shockley and Queisser in 1960 [4]. The energy loss processes which contribute to more than half the energy being lost or unutilized are illustrated in figure 2.1. Lattice thermalization and lack of absorption of sub-bandgap photons are the main contributors to this energy loss [5]. Due to non-absorption of sub-bandgap energies, a significant portion of the solar spectrum is not utilized, whilst thermal losses are a result of the material's inability to utilize the excess energy from carriers generated from high energy photons, and thus rids the system of this thermal energy through lattice vibrations. Consequently, various approaches have been employed to optimize cell efficiency, including adapting the semiconductor absorber material to better utlize the energy from the solar spectrum through altering the structure and chemical composition, including defect chemistry. More extensively, the development of PV devices utilizing other absorber materials such as perovskites, dye-sensitizers, quantum dots and organic molecules is now widely pursued to overcome the efficiency limitations.

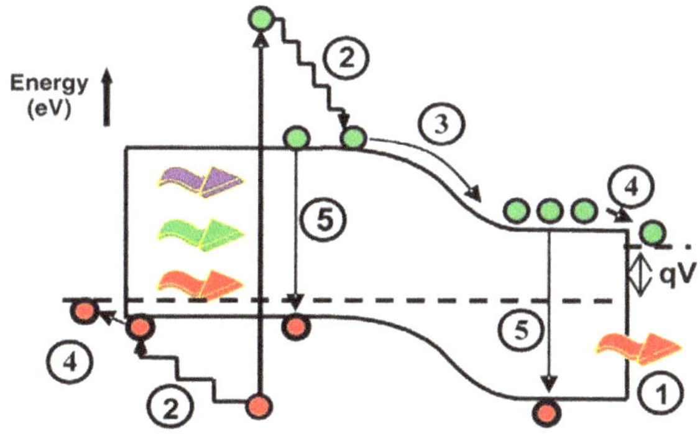

Figure 2.1. PV cell energy loss processes: (1) non-absorption of below bandgap photons; (2) lattice thermalization or hot carrier losses; (3) *p–n* junction voltage losses; (4) contact voltage losses; and (5) recombination (band to band or parasitic). Reprinted with permission from [5]. Copyright 2007 Elsevier Ltd. CC BY-NC 3.0.

The single *p–n* junction solar cell is a type of *First-generation* solar cell fabricated in response to the need for alternative forms of electrical energy. These are comprised of micron thick wafers of monocrystalline (*m*-Si) and polycrystalline silicon (*p*-Si) semiconductors. Overall, the device characteristics influence efficiency of photon energy conversion. For example, if the absorber semiconductor is too thick, there is a long carrier diffusion length and a higher probability for charge carrier recombination in the photoexcited semiconductor. Since the thickness of the Si wafer is ~100 μm, pristine *m*-Si is for maximum carrier diffusion and collection leading to an appreciable power conversion efficiciency (PCE) [6, 7]. However, the production of monocrystalline cells is expensive, with 20–40% of the cost associated with processing, and is therefore a major hindrance to wide-scale use [8]. Consequently, polycrystalline cells were developed as a means to reduce cost. Though production cost was lower, the efficiency of theses devices was drastically less than monoscrystalline cells. Nonetheless, these cells with the two highest PCEs of 26.1 and 24.4% for single- and polycrystalline cells, respectively, dominate the commercialized PV sector with about 80% market share [9–12].

In contrast to the silicon crystalline based cells, *Second-generation* (thin film) cells use less material and have fewer manufacturing requirements resulting in lower cost for PV units. Thin films generally have high absorption coefficients and are around 35–260 nm thinner than crystalline wafers, permitting enhanced versatility and economic viability. As a result, thin film cells can be deposited onto flexible substrates which increase portability, broadening the scope for use in hand-held, mobile technology, 'green' or hybrid car manufacturing as well as other exciting applications [10, 13]. The typical single-junction thin film solar cell architecture of *n*-type cadmium sulfide (CdS) and *p*-type and cadmium telluride (CdTe) is shown in figure 2.2. Absorber materials incorporated in commercially viable thin film cells include metal chalcogenides such as copper indium gallium selenide (CIGS), copper indium selenide (CIS) and CdTe/CdS as well as amorphous silicon and polycrystalline silicon.

Today, PV technology has advanced towards the development of new, state-of-the-art solar cell architectures—*Third-generation* solar cells. For these, the aim is to optimize efficiency, possibly surpassing the SQ limit, lower cost and increase stability, make environmentally friendly, improve material versatility, especially with respect to their compatibility with flexible and printable substrates—thin film architecture for wider scale application [11]. As seen from the National Renewable Energy Laboratory (NREL) chart 2024 [12] (figure 1.3—chapter 1), the best research-cell efficiencies for these emerging PV devices in some cases rival the *c*-Si single *p–n* junction thin film solar cells. These emerging thin film solar cells fabricated via non-vacuum, solution-processed methods are particularly attractive relative to vacuum-based fabrication due to their simplicity, low cost, use of versatile materials, promising device performance, and have the potential to revolutionize the PV industry. In particular, solution-processed solar cells have attracted tremendous attention due to their potential for scalability, favourable performance-to-weight ratio, easy manufacturing with low environmental impact, as well as short energy payback periods [13]. The following sections present an overview of each of these

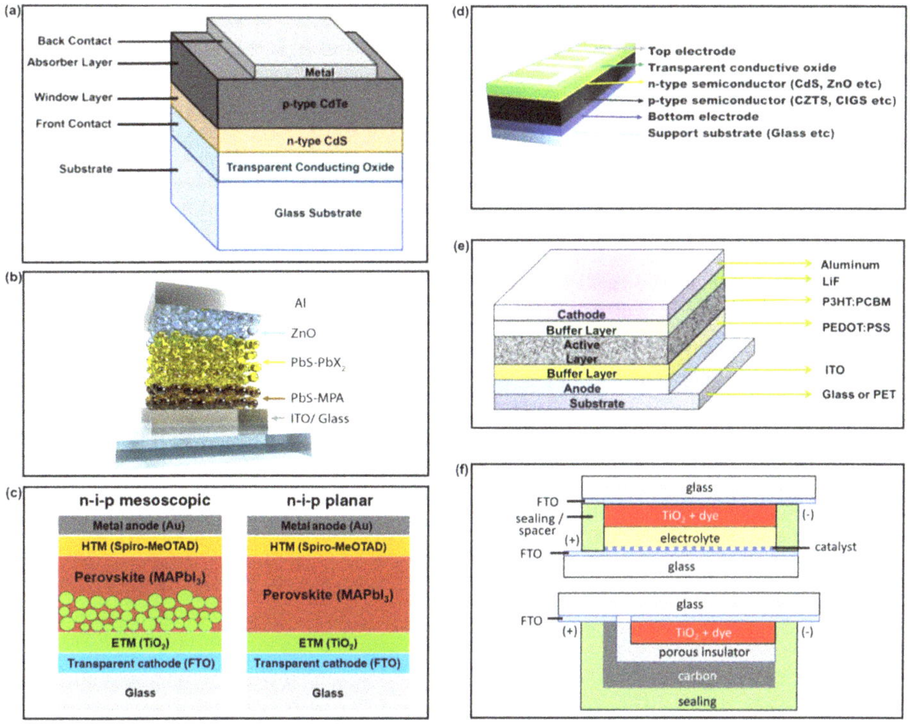

Figure 2.2. Types of solar cell architectures. (a) A conventional thin film *p–n* junction. Adapted with permission from [14], copyright 2022 Elsevier. (b) *p–n* junction colloidal quantum dot cell. Adapted with permission from [15], copyright 2021 Elsevier. (c) mesoscopic and planar perovskite cell. Adapted from [16], CC BY 4.0. (d) copper chalcogenide thin film cell. Adapted with permission from [17], copyright 2018 Wiley-VCH Verlag GmbH & Co. KGaA. (e) Conventional organic cell. Adapted with permission from [18], copyright 2016 Elsevier. (f) Sandwhich and monolithic dye sensitzed cell. Adapted from [19], CC BY 3.0.

types of emerging, solution-processed solar cells with each covered extensively in ensuing chapters.

2.1.2 Copper-based chalcogenide solar cells

Thin film PV technology offers an attractive alternative to *c*-Si solar cells for applications that require light weight and flexible substrates. Of these, thin-film copper-based chalcogenides such as $CuInS_2/Se_2$, $Cu(In,Ga)(S,Se)_2$ (CIGSSe) and $Cu_2ZnSn(S,Se)_4$ (CZTSSe) hold tremendous promise because of the relatively environmentally benign and abundant constituent elements, high absorption coefficient and comparatively narrow bandgaps, spectrally requisite for solar energy absorption. Among all types of thin film solar cells, copper indium gallium selenide (CIGSSe)-based cells attained an efficiency of 23.6% in 2024 [12], which is quite close to the maximum for *c*-Si technology. Atmospheric solution-based deposition of these materials offers an attractive alternative to traditional vacuum deposition methods due to use of less raw materials, lower production costs, large-area production throughput, better compatibility with flexible substrates, and scalable roll-to-roll fabrication processes. Overall, these solution-based thin-film metal chalcogenides offer a promising path for the mass production of low-cost solar cells prepared at low temperatures.

Along with CIGSSe, chalcopyrite $CuInS_2$ (CIS) and kesterite Cu_2ZnSnS_4 (CZTS) are the main representatives in the family of copper chalcogenide materials extensively studied as thin film PV absorbers. Both chalcopyrite and kesterite structures are derived from the zinc-blende structure of zinc sulfide (ZnS), while the chalcopyrite CIS is the predecessor of kesterite CZTS. The standard hetero-junction configuration of these thin film solar cells comprises several layers, including a glass substrate, metal back contact (primarily molybdenum, Mo), *p*-type absorber layer, *n*-type buffer layer, transparent conducting oxide (TCO) layer and a top electrode, as shown in figure 2.2. In most cases, the *p–n* junction interface is formed from CIGSSe/CZTSSe *p*-type absorber and doped CdS *n*-type layer. Photogenerated electrons and holes are separated by the built-in electric field, which are then collected by the TCO and Mo back contacts, respectively.

2.1.3 Colloidal quantum dot solar cells

Quantum dots are three-dimensionally confined semiconductor nanocrystals (1–10 nm) usually synthesized using standard inert conditions from organometallic/inorganic precursors in organic or aqueous media. Colloidal quantum dots (CQDs) are normally surface passivated by organic ligands that stabilize their growth, maintain colloidal dispersion and generally isolate them from their electronic environment. The unique control over these CQDs' size, ligand chemistry and post-synthesis annealing conditions has enabled many recent advances in the properties and performance of solution-processed solar cells, with achievement of high efficiency at low cost [20]. Quantum dot-based solar cells offer the advantage of tunable bandgap, which can adapt the solar cell to the solar spectrum to utilize more energy, producing extra charge carriers, increasing conductivity and photoconversion [6]. Notably, there are several types of device architectures for quantum dot solar cells, but all of these involve the quantum dots as mono/multilayer films in which the charge carriers must travel significant distances to be extracted, and is therefore a disadvantage. Additionally, they can be processed to create junctions (multiple) on inexpensive substrates such as plastics, glass or metal sheets and can easily be combined with organic polymers and dyes. It is envisaged that quantum dot solar cells can surpass the Shockley–Queisser PCE limit prescribed for conventional *p–n* junction solar cells [20]. This will depend on charge carrier diffusion length of the depletion region, charge carrier extraction efficiency, carrier concentration, doping and trapping density. In addition to the quantum-confinement particle size effects, several strategies have been employed to optimize diffusion mechanisms for enhanced charge carrier extraction and mobility in quantum dot films in these devices. For example, increasing the doping in an *n*-type layer that forms a depletion region in an adjacent *p*-type light-absorbing colloidal quantum dot layer enables a deeper depletion region [20]. Overall, CQDs offer the benefits of facile solution processing alongside size dependent tunable bandgaps, multi-exciton generation, tailored towards adaptability across a range of architectures. In that respect, they have been incorporated in CQD-sensitized solar cells, Schottky junction CQD cells, depleted heterojunction CQD cells, depleted bulk heterojunction CQD cells, bulk—nano heterojunction CQD cells, and multijunction (tandem) CQD solar cells with recorded PCEs upwards of approximately 19.2% (figure 1.3) [12].

2.1.4 Dye-sensitized solar cells

Dye-sensitized solar cells (DSCs), also known as the *Grätzel* cells, introduced by Grätzel and co-workers [21] are solution-processed PV devices consisting of a thin layer of *n*-type semiconducting metal oxide, mainly TiO_2 nanoparticles, as an electron transport material sensitized by a metal–organic/organic dye molecule. The dye molecules are responsible for light absorption, charge separation and injection of charge carriers into the semiconductor, which only plays the role of the electron transport layer (ETL). In the conventional Grätzel cell, a redox electrolyte is responsible for regeneration of the oxidized dye and charge transport between electrodes. DSC devices can be fabricated using low cost procedures such as inkjet or screen printing, enabling preparation of large-area devices on flexible substrates.

According to NREL (figure 1.3), the record efficiency of DSCs stands at 13.0% under full sun illumination [12]. Notably, while the efficiency of several classes of PV technology has been steadily increasing, those associated with DSCs have plateaued which has stymied their commercialization. This is primarily attributed to issues of leakage of the liquid electrolyte, its corrosive nature and the negative environmental impact. Furthermore, a primary undesirable process that reduces device performance involves charge recombination of the injected electrons with the oxidized sensitizer or with the oxidized state of the redox couple, the result of which makes the absorbed photon responsible for the injected electron useless for the production of electrical potential.

To overcome these limitations, solid-state DSCs (ssDSCs) in which the liquid electrolyte is replaced by a solid material have been developed in recent years owing to solution-processed methods [22]. In addition to the ssDSCs, other architectures include, quasi-solid-state DSCs, quantum dot DSCs and tandem DSCs, with the latter projected to be more competitive. The theoretical efficiency limit for a typical dye-sensitized solar cell is around 30%, whilst for two junction tandem solar cells, it is ~42%. However, fabrication of tandem devices is extremely complicated which may significantly limit their utilization. DSCs have become a promising alternative third-generation PV technology because it is likely a sustainable technology with simple fabrication including solution processing, use of non-toxic materials, and it is an adaptable technology with the use of flexible materials amenable for various applications including solar windows, indoor applications and consumer electronics.

2.1.5 Perovskite solar cells

Perovskite solar cells (PSCs) are the most rapidly developing of the emerging thin film solar cells in which most of their layers are solution-processed under mild conditions. Perovskite is a calcium titanium oxide mineral of chemical formula $CaTiO_3$, and compounds of similar crystal structure, i.e., ABX_3 are called perovskites. The first perovskite semiconductor, caesium lead halide ($CsPbX_3$, X = Cl, Br or I) was introduced by Moller in 1957, where it was categorized as a semiconductor due to its high photoconductive properties [23]. The most recent efficiencies reported by NREL (figure 1.3) has perovskite cells at 29.5% which is impressive since Kojima and co-workers in 2009 [24] reported methylammonium lead iodide, $MAPBI_3$ ($CH_3NH_3PbI_3$) and methylammonium lead bromide, $MAPbBr_3$ ($CH_3NH_3PbBr_3$) as perovskite sensitizers in a liquid electrolyte-based DSC with PCEs of 3.81% and 3.13%, respectively. Later, having overcome the common stability issue with the perovskite, Kim and co-workers [25] developed a solid-state, at that time, highly efficient (PCE of 9.7%) and stable perovskite solar cell in 2012. The performance of these cells is predicated mainly on perovskites' tunable direct bandgap, high absorption coefficient, good charge carrier transport, high PCE and improved stability. However, the field has advanced tremendously with recent devices composed of earth abundant materials that can be processed from solution at low temperatures. This alongside efforts into the design of novel device architectures, careful control of the morphology of functional layers and optimized interfacial engineering are at the basis of enhanced PCEs for PSCs. However, current challenges to be overcome with these solar cells relate to the instability of the perovskite lattice at ambient temperature and device lifetime, as well as controlling the perovskite structure by solution-processed methods, including scaling up for commercialization.

A simple and conventional configuration of a PSC consists of halide perovskite nanoparticles as light harvesters, as shown in figure 2.2. Upon absorption of solar photons, the excited electrons are injected into the interfaced electron transporting layer (ETL) embedded in mesoscopic TiO_2, and the excited holes migrate to the hole transporting layer (HTL]. The active and buffer layers are sandwiched between a transparent conductive oxide (TCO) electrode coated onto a glass substrate or a flexible transparent layer and a metal back contact electrode. Overall, the configuration of PSCs is generally classified based on the orientation of the ETL or HTL layers in relation to incident photons, either as a regular (*n–i–p*) or inverted (*p–i–n*) heterojunctions. These two classes are subdivided into mesoscopic and planar structures incorporating either mesoporous or planar layers, respectively. Of the variety of architectures, the most commonly studied include the mesoscopic *n–i–p* configuration, the planar *n–i–p* configuration, the planar *p–i–n* configuration, the mesoscopic *p–i–n* configuration, the ETL-free configuration and the HTL-free configuration.

2.1.6 Organic/polymer solar cells

An organic solar cell (OSC) is a type of device that uses conductive small organic molecules or polymers for photo-absorption and conductivity. The first promising organic solar cell was a donor–acceptor bilayer architecture with a PCE of 1% reported by Tang and co-workers in 1986 [26]. The two layer heterojunction sandwiched between two electrodes consisted of *p*-type copper phthalocyanine complex (donor) and *n*-type perylenediimide derivative (acceptor). Currently, thin film bilayers of donor- and acceptor-type organic semiconductors form the core of heterojunction organic PV cells. According to the best research-cell efficiency chart provided by the National Renewable Energy Laboratory (NREL), the PCE of OSCs has shown a rapid increase in the past few years, with the state-of-the-art OSCs yielding a certified highest efficiency of 19.2% for organic tandem cells [12]. This type of organic tandem cell along with other thin film bilayers of donor- and acceptor-type organic semiconductors form the core of heterojunction organic PV cells with improved efficiency over time. Such remarkable improvements in efficiency has been possible with engineering new materials of variable bandgaps, improvements in the solubility of materials, advances in device architecture, including control of molecular ordering in thin films, addition of new buffer layers and adoption of novel approaches for fabrication, including solution processing.

Critically, improvements in devices have focussed on the stability and the degradation of the organic solar cells during operation, for example with materials that undergo structural changes after reaction with the ambient atmosphere. Almost all organic PV cells have a planar layered structure, where the organic active layer/s is/are sandwiched between two different electrodes, typically a transparent conducting oxide (TCO)/anode of indium tin oxide (ITO) and the other, opaque, aluminium back contact/cathode. There are three main categories of device architecture developed for OSCs, namely, single-layer, bilayer and bulk heterojunction cells. Bulk heterojunctions (BHJs) consist of a larger nanostructured layer of interpenetrating donor–acceptor interface of photoactive molecules embedded in a conjugated polymer matrix. In these, the larger donor–acceptor interface on the order of 100 nm increases the diffusion length of the excitons, enabling more effective charge separation, limiting recombination losses. A typical single-junction organic BHJ device shown in figure 2.2 consists of layers of different materials, such as a transparent bottom electrode of ITO, a hole transport layer, the active donor polymer/acceptor molecule layer and a top metallic electrode layer. In contrast, a tandem junction BHJ cell consists of multiple stacks of the single-junction BHJ, each having different combinations of donor–acceptor active layers with interlayers serving to match the charge transport across both cells.

2.2 Basics of device performance characteristics

2.2.1 Current–voltage (*I*–*V*) characteristics

PCE, the fraction of the incident photons converted to electricity, is the main parameter used to determine the photoconversion performance of a solar cell. The efficiency parameter, η is defined as the ratio of energy (power) output from the solar cell to input energy from the sun and is calculated accordingly:

$$\eta = \frac{P_{max}}{P_{in}} = \frac{J_{max}\, V_{max}}{P_{in}} = \frac{J_{sc}\, V_{oc} FF}{P_{in}} \tag{2.1}$$

Here, P_{max} is the maximum power, P_{in} is the input power, V_{max} is the maximum voltage, J_{max} is the maximum current density, V_{oc} is the open-circuit voltage, J_{sc} is the current density, and FF is the fill factor.

Efficiency is a key metric in the development of PV systems and its increase is critical in reducing the cost per kilowatt-hour of PV electricity [27]. Importantly, the efficiency depends on the spectrum and intensity of the incident sunlight, and the temperature of the solar cell. Therefore, conditions under which efficiency is measured must be carefully controlled in assessing the performance of devices. Consequently, terrestrial solar cells are measured under Air Mass 1.5 Global (AM 1.5G) illumination (100 mW cm^{-2}) and at a temperature of 25 °C. The abbreviation, AM 1.5G is the air mass coefficient that describes the terrestrial spectral irradiance distribution of the solar spectrum on earth, as shown in figure 2.3, according to the American Society for Testing and Materials (ASTM International Standard). A scan of the current–voltage (*I–V*) relationship associated with a *p–n* junction diode or any solar cell configuration allows for the computation of cell efficiency. For a solar cell, the maximum power, P_{max} is attained when the product of current density, J and cell voltage, V is maximum. However, since J_{max} and V_{max} are not easily determined in the *I–V* scan, they are not commonly used to evaluate solar cell performance.

A typical *I–V* plot as shown in figure 2.4 of a solar cell is the superposition of the *I–V* curve of the solar cell in the dark with the photogenerated current from which power can be extracted. Without illumination, the cell has the characteristics of a large diode and when illuminated, the *I–V* curve shifts into the lower quadrant of the plot and power is generated. The extent of illumination and generation of charge carriers determine the level of shift and power output, and the maximum power output occurs at maximum voltage and current. The essential parameters attributed to the performance of the solar cell are determined from the *I–V* curve, as illustrated in figure 2.4.

The short-circuit current, I_{sc} is the maximum current measured when the negative and positive terminals directly connected with each other are short-circuited to give zero voltage, producing maximum electrical output. The I_{sc} is due to the generation and collection of both photogenerated charge carriers, and since the current production also depends upon the surface area of the cell, the maximum current density, J_{sc} is primarily used. This short-circuit current density is the ratio of short-circuit current and cell surface area; equation (2.2).

$$J_{sc} = \frac{I_{sc}}{A} \tag{2.2}$$

The short-circuit current also depends on other factors including:

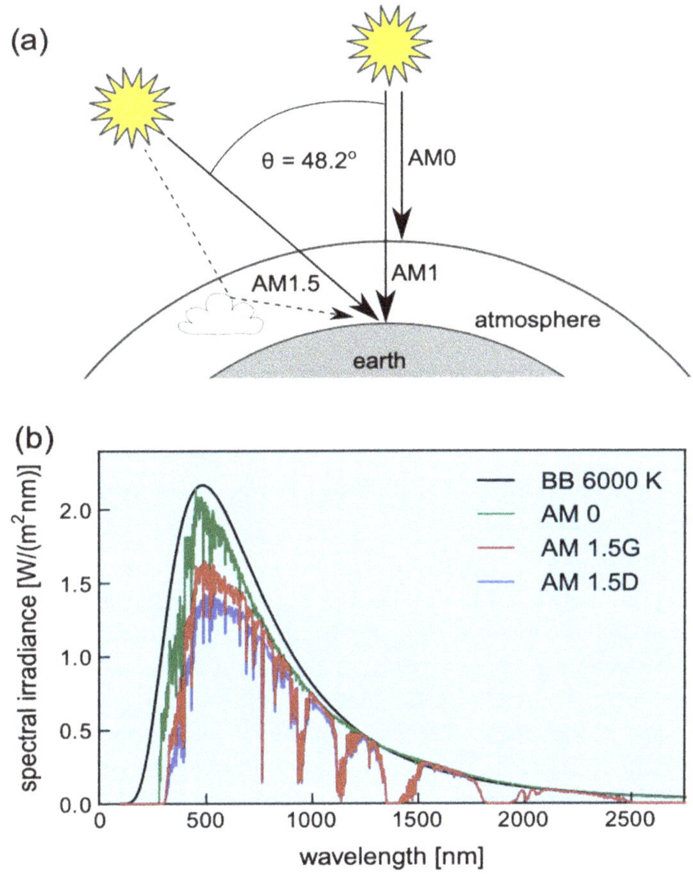

Figure 2.3. (a) Schematic representation of the spectral irradiance outside the earth's atmosphere (AM 0) and on the earth's surface for direct sunlight shown by a solid arrow (AM 1.5D) and the direct sunlight together with the scattered contribution from atmosphere (solid and dashed arrow) integrated over a hemisphere (AM 1.5G). (b) Spectral irradiance according to ASTM G173-03 in comparison to the spectrum used by Shockley and Queisser of a black body with a surface temperature of 6000 K (BB 6000 K). Reprinted with permission from [28], copyright 2016 Elsevier.

1. The number of photons (i.e., a measure of the power of the incident light source) and light intensity.
2. The spectrum of the incident light, standardized to the AM 1.5G spectrum, designed for flat plate PV modules having an integrated power of 100 mW cm^{-2}.
3. The absorption and reflection properties of the solar cell, related to the threshold bandgap, E_g of the photo-absorbers.
4. The charge carrier collection probability of the solar cell, which depends primarily on the charge carrier extraction, including the diffusion length of carriers.

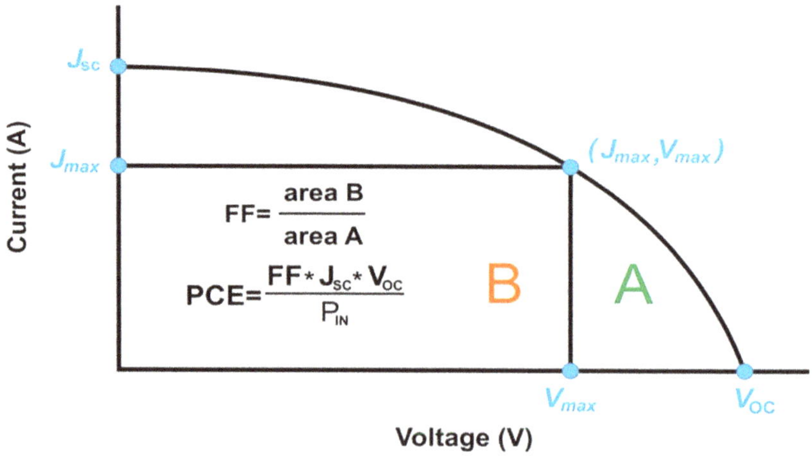

Figure 2.4. *I–V* curve to evaluate solar cell performance. Reprinted with permission from [29], CC BY 4.0.

The open-circuit voltage, V_{oc}, is the maximum voltage attained due to zero current when the cell is not connected to any load. It corresponds to the amount of forward bias, i.e. voltage delivered across the junction due to its bias with the photocurrent. V_{oc} not only depends on the photocurrent but also on the saturation current and charge carrier recombination in the device. As such, V_{oc} depends on carrier concentration which increases with temperature for most thin film cells, and in an ideal device is limited by radiative recombination of charge carriers and bandgap of the photoexcited semiconductor.

Overall, the measure of the quality of a cell is determined through the fill factor, *FF*, which is defined as the maximum power from an actual solar cell to the maximum power from an ideal solar cell. That is, it is the ratio of the maximum power from the solar cell to the product of V_{oc} and I_{sc}. *FF*s are usually reported with PCEs of solar cell devices as a measure of their performance.

2.2.2 Incident-photon-to-current conversion efficiency (IPCE)

In addition to the PCE, defined through the efficiency parameter, η, measurements of incident photon-to-current conversion efficiency (IPCE), also called quantum efficiency (QE) are important in understanding solar cell performance. Such measurements are used to correlate the discrete efficiency of the cell as a function of wavelength with the short-circuit current measurements of cells under AM 1.5G illumination. In effect, IPCE is a measure of the ratio of the photocurrent versus the rate of incident photons as a function of photon wavelength. If all photons of a certain wavelength are absorbed and the resulting majority and minority carriers are collected (extracted) from the cell, then the QE at that particular wavelength is unity. However, this is not achieved since there are a number of energy losses associated with the cell, as shown previously. Critically, IPCE or QE can be used to determine other important characteristics of the cell such as the carrier diffusion length, a measure of charge migration (*vide infra*). Since QE is affected by carrier recombination mechanisms, it can be viewed as the charge collection probability due to the generation profile of a single wavelength, integrated over the device thickness and normalized to the number of incident photons. Also, if the cell's QE is integrated over the entire solar spectrum, the amount of current that the cell produces when exposed to sunlight can be evaluated. In effect, IPCE measurements can be used to estimate the short-circuit current density, J_{sc} at AM 1.5 illumination according to:

$$J_{sc} = e \int \mathrm{IPCE}(\lambda)\phi_{sun}(\lambda)d\lambda \qquad (2.3)$$

Here, ϕ_{sun} is the spectral flux distribution of sunlight, λ is the wavelength of photon, and e is the elementary charge. Since the energy of a photon is inversely proportional to its wavelength, QE is often measured over a range of different wavelengths to characterize a device's efficiency at each photon energy level. Accordingly, there are two ways to measure the QE of a solar cell:

1. *External quantum efficiency (EQE)*, which is the ratio of the number of charge carriers collected by the solar cell, to the number of incident photons of a given energy, including those that have been lost via transmission and reflection.
2. *Internal quantum efficiency (IQE)*, which is the ratio of the number of charge carriers collected by the solar cell, to the number of incident photons of a given energy that are absorbed by the cell, and is larger than the EQE.

In effect, since QE depends on the rate of collection of charge carriers, a low value indicates that the active layer of the solar cell is unable to make good use of the photons, most likely due to poor carrier collection efficiency reflected through short diffusion length, low carrier concentration, recombination losses, poor charge separation and interfacial losses as primary factors.

2.2.3 Shockley–Queisser limit

The SQ model for efficiency proposed for *p–n* junction inorganic solar cells is the established reference for all solar cells. Also known as the *detailed balance limit*, the SQ limit is the maximum theoretical efficiency of a solar cell using a single *p–n* junction to collect power from the cell. This model was initially proposed by William Shockley and Hans-Joachim Queisser in 1961, estimating a maximum efficiency of 30% at 1.1 eV for a standard silicon *p–n* junction [4]. The proposed detailed balance principle correlates photon absorption and emission of a semiconductor in which it is assumed that every microscopic process must have the same rate as its inverse process in thermal equilibrium and in the case of solar cells encompasses the optical, thermal and electronic components of the photoconduction process. More precisely, the model in its simplest and most common context makes several fundamental assumptions:

1. there is maximum absorptivity of photons at the bandgap of the photo-absorber;
2. only one electron–hole pair per absorbed photon is collected at short-circuit;

3. heat extraction from the carrier system is such that the carrier temperature equals the cell and ambient temperatures;
4. electron–hole recombination involves radiative emission; and
5. there are no Ohmic losses (internal resistance to the flow of charge carriers) since the contacts are perfectly selective for each carrier.

Accordingly, the primary losses responsible for the SQ limit are attributed to: (1) below bandgap energy not absorbed; (2) thermalization of 'hot' carriers; (3) electron–hole recombination; and (4) isothermal dissipation loss during carrier collection, all collectively accounting for >55% of the total absorbed solar energy.

Several of the next-generation solar cell devices are proposed to exceed the established SQ limit since they are overcoming some of the energy losses associated with the *p–n* junction assumed under the detailed balance model. The concept of third-generation PV is to significantly increase device efficiencies whilst still using thin film processes and abundant non-toxic materials. This can be achieved by circumventing the SQ limit for single bandgap devices, using multiple energy threshold approaches. Such approaches can be realised either by incorporating multiple energy levels in tandem or intermediate band devices or by modifying the incident spectrum on a cell by converting either high energy or low energy photons to photons more suited to the cell bandgap. Overall, these methods have advantages and disadvantages and are at various stages of development.

2.3 Photoconductivity and charge transport properties

2.3.1 Energy loss factors

Since photoconversion efficiency is a direct determinant for reducing the cost per kilowatt-hour of PV electricity, evaluating its influential factors is paramount. Invariably, the main energy loss factors impacting solar cell efficiency are correlated with three primary physical mechanisms involved in the photo-electrical energy conversion process. The first is the *absorption* of solar photons by the absorber material which is dependent on the absorption coefficient, α, a function of incident photon energy, E. The second is the *transport* of photogenerated separated free charge carriers towards the electrical contacts, measured in terms of carrier mobilities of electrons, μ_n and holes, μ_p and is dependent on the properties of the absorber materials and its interfaces. The third involves the *recombination* of charge carriers, measured as recombination rates expressed as lifetime, τ. Invariably, the higher the value of these quantities, the higher the efficiency of the device, and therefore, their product, $\alpha\mu_n\mu_p\tau$ would represent a general figure of merit for solar cell efficiency [30]. In effect, charge carrier mobility is a measure of charge carrier collection efficiency at the electrodes and is the product of mobility and lifetime for the absorber/photoanode diffusion length. Invariably, low mobility of carriers can be compensated by a long lifetime and *vice versa*. However, diffusion length, L which is related to the absorber/photoanode thickness must be adapted to their absorption coefficient. In effect, in all solar cells, a certain thickness of the absorber layer is required to effectively absorb all solar photons and this is dependent on its absorption coefficient and light scattering properties. L is considered as the average distance that excited carriers travel before recombination and depends on the lifetime and mobility of the carriers by the diffusivity according to:

$$L = \sqrt{D\tau} = \sqrt{\frac{\mu\tau k\mathrm{T}}{e}} \tag{2.4}$$

Here, D is the diffusivity (a measure of how quickly a group of particles fill a space), τ is the lifetime in seconds, k is the Boltzmann's constant, T is the temperature and e is the elemental charge of the negative carrier. Therefore, higher diffusion lengths are indicative of absorber layers with longer lifetimes and is an important quality to consider for absorber materials and their devices.

As an important aspect in optimizing PV cell efficiency, analyzing loss factors is a complex and variable process that requires useful models depending on the type of cell, absorber and interfacial materials comprising the cell. One useful approach proposed by Ehrler and co-workers [27], which could be generalized involves optimizing light and carrier management towards increasing efficiency as functions of photocurrent and recombination losses, respectively. Certainly, this is a generalized approach and there are other models that must be considered depending on the type of cells and the factors that most influence the performance characteristics. In the model proposed by Ehrler, as shown in figure 2.5, for a particular material/cell, the extent of photocurrent losses (j) is computed as the ratio of the short-circuit current of the record cell, J_{sc} to the maximum possible short-circuit current, J_{SQ} for that material and cell architecture, as calculated from the detailed balance limit [27]. A low ratio of these values ($j = J_{sc}/J_{SQ}$), indicates that better light management must be applied to improve the solar cell and is dependent on optimized light penetration through the photoactive components of the cell and reduction of light absorption in

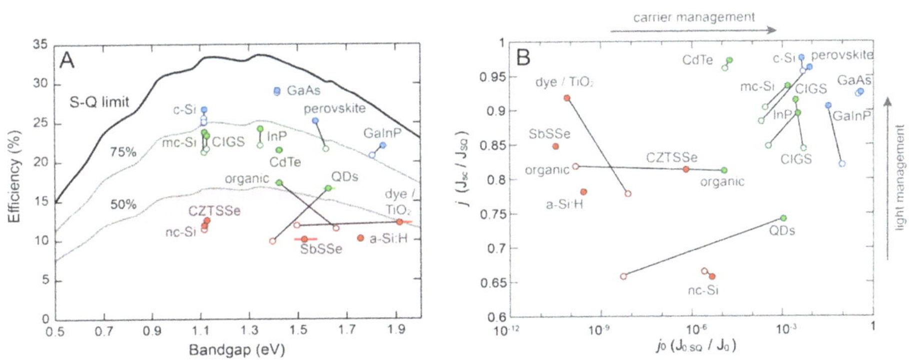

Figure 2.5. (a) Record efficiency of solar cells of different materials against their bandgap, in comparison to the SQ limit (top solid line). (b) Current density relative to the maximum possible current density, under standardized AM 1.5 illumination conditions, versus minimum dark recombination current density relative to the recombination current derived for the record cells in panel A. The open symbols show the record efficiency in April 2016, the solid symbols show the numbers in July 2020. Reprinted with permission from [27], copyright 2020 American Chemical Society.

inactive regions of the cell. Recombination losses, j_0 are computed as the ratio of the lowest possible recombination current at the absorber's bandgap, $J_{0,\text{sQ}}$, calculated from detailed balance, to the dark recombination current derived from record solar cell data, J_0. A low ratio of j_0 ($J_{0,\text{sQ}}/J_0$) indicates that better charge carrier management by the reduction of bulk, interfacial and surface recombination is required to improve the solar cell performance. It is to note that for this model, the analysis is performed for an idealized cell and it doesn't account for factors such as series or shunt losses in the cell. However, detailed treatments can only be made based on the cell architecture accounting for factors that influence α, μ_n, μ_p and τ.

Overall, in evaluating the energy loss factors affecting cell efficiency, the recombination event must be de-coupled from charge separation towards increased charge collection efficiency. There are two main factors to achieve the required charge collection efficiency. The first is to exploit or improve the kinetics at interfaces, favouring the required forward process, and the second takes advantage of internal electrical fields caused by a built-in voltage and distribution of photo-generated charges. The second is dependent on the selectivity at the electrodes for charge carrier migration. However, the extent of these two factors depends on the mechanisms of charge transport for different cell architectures in terms of their materials, interfacial contacts and the electronic processes. For example, in a DSC, the electric field is of little relevance for charge collection at short-circuit, however, charge separation relies on interfacial kinetics, energy steps at interfaces and diffusion driven transport [31, 32].

2.3.2 Charge separation and recombination

In the absorber material, photogeneration creates two unbound charge carriers of opposite polarity which are extracted at the electrodes. Effective local charge separation avoids immediate, geminate recombination of these mobile electron–hole pairs. In all solar cells, a certain thickness of the absorber layer, *d*, is required to effectively absorb all solar photons. The desired range of values of *d* is determined by the absorption coefficient and light scattering properties of the device that increase the optical pathlength of weakly absorbed photons in the absorber. A useful criterion for classifying solar cell types based on charge carrier separation is the ratio between the width, *w* of the space charge region and *d*. It is important to note that *d* whilst related must not be assumed as the diffusion length, *L*, a measure of the mobility and lifetime diffusion of the charge carrier upon photoexcitation in the absorber layer.

The charge carrier lifetime is an important parameter that determines the performance of solar cells and is limited by the rate of electron–hole recombination, which can be radiative or non-radiative [33]. The kinetics of the decay in the concentration of photogenerated charge carriers (n) is often modelled using a third-order rate equation (2.5).

$$\frac{dn}{dt} = G - k_1 n - k_2 n^2 - k_3 n^3 \tag{2.5}$$

Here, the charge carrier generation rate (G) is offset by one-particle (k_1), two-particle (k_2) and three-particle (k_3) recombination processes (k being rate constant). The contribution from each to the total recombination is dependent on the carrier density and thus the illumination intensity. In a low carrier density regime ($n < 10^{15}$ cm^{-3}), a deep sub-bandgap electronic level of a semiconductor facilitates the capture of carriers and it is often the dominant non-radiative recombination mechanism. Recombination effectively becomes a first-order process (k_1) as the kinetics are usually limited by the capture of the minority carrier. At intermediate carrier densities, two-particle band-to-band recombination (k_2) processes occur radiatively. For indirect bandgap semiconductors, the values of k_2 are smaller than those for direct bandgap materials, which makes k_1 and k_3 processes more important. The k_3 process, Auger recombination, involves collision of a carrier with a second carrier of the same type which becomes excited to a higher energy state, whilst the other carrier recombines with a third carrier of the opposite charge. Typically, this recombination generally limits the carrier lifetime significantly only in a very high carrier density regime ($n > 10^{17}$ cm^{-3}).

Since the mechanisms of charge separation and recombination are unique to the solar cell architecture, herein these are discussed in terms of two different types as examples—the DSC which has molecular photo-absorber embedded in a semiconductor support (photoanode) and the conventional *p–n* junction cell, as the inorganic thin film prototypical reference. In homogeneous semiconductors such as inorganic thin films with relatively high dielectric permittivities and low exciton binding energies, local charge generation and separation are very efficient [31]. Thus, each photogenerated carrier rapidly forms part of the respective ensemble of free carriers in the conduction and valence bands (carrier thermalization) with the in-built electric field across the *p–n* junction providing different polarities for diffusion of the carriers in opposite directions towards extraction at the electrode interfaces. However, for other types of cells involving isolated absorber entities such as dye-sensitizers and quantum dots, initial charge separation is achieved by preferential injection kinetics to either electron or hole transport layers. In these, charge extraction competes with internal recombination of the electron–hole pair due to the interface energetics of various contact materials. Therefore, it is important to facilitate fast extraction by a good match of the energy levels between both materials involved and by ensuring the existence of downward energy pathways from the absorber to the carrier transport material. It is also required that the kinetic and energetic charge transfer pathways prevent the return of injected carriers to the original absorber molecule. Essentially, recombination of the primary electron–hole pair is usually not an issue in conventional DSCs, mainly because the molecular and electronic structures of the dye absorber and semiconductor substrate are coupled, favouring efficient charge separation and fast electron injection. However, the energy loss is mainly attributed to interfacial inefficiencies between the dye–semiconductor photoanode and the redox couple, and the photocurrent is dependent on the energy gap between the electron and hole transport materials in both electrodes. However, in some metal-free organic dyes, the extent of internal recombination is relevant, for example quantum dot sensitizers which display complex carrier confined dynamics, primarily surface defect state recombination which is influenced by quantum dot surface passivation. In these, the poor interfacial connectivity between quantum dots results in charge recombination before extraction.

2.3.3 Charge collection efficiency

Since charge carrier separation and recombination are uniquely dependent on contact interfaces in next-generation solar cell architectures, contact selectivity is therefore a critical criterion for good cell efficiency. In effect, contact selectivity for either carrier is paramount for establishing an electric field flux for one carrier and limiting the other and is dependent on the voltage gradient (built-in potential) between the quasi-Fermi states/electrochemical potentials between the electrodes within the cell [31]. Characteristically, selective contacts typically are *n*- or *p*-type materials, possessing the required kinetic properties to diffuse and extract only one carrier at its interface with the absorber by alignment of energy levels. In effect, selectivity is high if recombination of minority carriers is minimized and the series resistance across the interface is low, resulting in easy extraction of all majority carriers accompanied by no or low photovoltage leakage due to recombination of minority carriers. In contrast, poor selectivity is characterized by high series resistance and recombination current with high photovoltage leakage. Overall, materials with small work functions are selective contacts to electrons since their ohmic potential results in a small injection barrier for electrons due to good alignment with the conduction band minimum of the absorber layer. On the other hand, materials with a large work function may form an efficient hole extraction contact. Therefore, a major influence of built-in potential on the selectivity will depend on whether this potential is built up across the device junctions.

Since contact interfaces are possible routes for unwanted recombination, it is even more critical that their charge extraction kinetics supersede charge recombination rates to achieve good charge collection, as a measure of their selectivity. For example, in DSCs, charge carrier selectivity is achieved using materials and morphologies that provide an easy pathway for one carrier and a large barrier to the other [31]. Here, electrons injected from the dye are fully collected in the semiconductor layer since the photoanode is spatially and electrically separated from the counter electrode, and holes cannot be injected into the electron collecting transparent conducting layer due to the presence of a typical blocking layer. In cases where there is excellent charge separation and high mobilities, all resistive and recombination losses would therefore depend exclusively on the carrier concentrations at contact interfaces. Though mobility and lifetime are mainly influential on charge carrier collection, the pathways for improvement are material dependent. Therefore, a primary strategy to improve charge carrier collection is to improve the microstructure of the material through reducing the defect density or increasing the grain size, for example. Overall, a general approach prescribed for efficient charge collection is focussed on the absorber material in terms of its carrier mobility, lifetime and doping density. However, the interface characteristics between absorber and contact layers is relevant. Generally, solar cells that have poor collection efficiency will have poor conversion efficiency at the optimum thickness of the collector and for devices that have efficient charge collection, the mobility is less relevant and the conversion efficiency will remain constant irrespective of mobility at optimum collector thickness.

2.4 Strategies for defect and interface engineering

Control of electronic transport involving defects and interfaces is essential for realizing high efficiency solar cells upwards of the SQ limit. It is therefore important during solution-processed device fabrication to employ strategies based on materials properties to limit these effects since they may have detrimental impacts on electronic transport and charge carrier extraction. Important in these strategies is to understand the dynamics of electron transport arising from these effects and the device characteristics. Figure 2.6 illustrates the different types of defects implicated in absorber materials and electron/hole transporting layers. However, the influence of each on solar cell performance will vary depending on the device configuration and the fabrication processes. Defects can generally be classified as deep-level or shallow-level according to energy levels with respect to the conduction and valence bands. They are also classified into point defects (i.e., vacancies, interstitials, and antisites), defect pairs (i.e., Frenkel and Schottky), one-dimensional (1D) (i.e., dislocations), two-dimensional (2D) (i.e., surface and grain boundaries), three-dimensional (3D) [e.g., precipitates (small volumes with a different crystal structure), large voids, and second-phase domains] [34]. Whilst intrinsic and extrinsic localized or point defects in absorber materials can limit the transport of carriers in terms of their mobility and diffusion length, they can act as donor or acceptor states to effectively tune carrier concentrations and Fermi level for realizing *p*-type (hole-

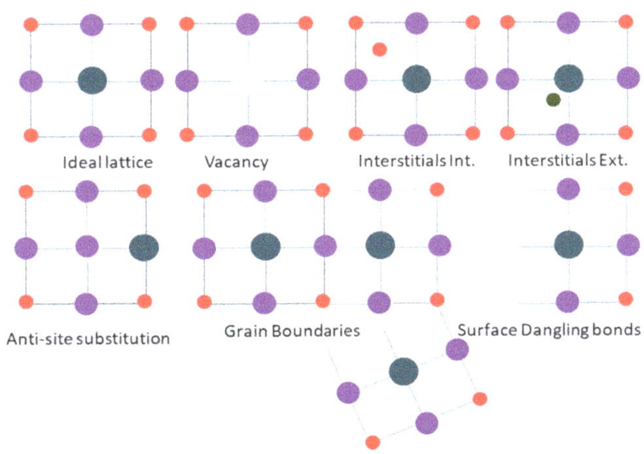

Figure 2.6. Types of defect in halide-based perovskite crystal lattices, in comparison to the ideal lattice of ABX_3 perovskite. Here red, dark cyan and purple circles represent A, B, and X, respectively. The olive circle represents an interstitial extrinsic impurity. Reprinted from [35], copyright IOP Publishing Ltd. All rights reserved.

conducting) and *n*-type (electron-conducting) materials. Though the defect chemistry of materials is widely studied, their influence on the charge carrier properties are specific to the type of solar cell. Essentially, optical transitions in absorbers involving defect states can enhance photocurrent generation through sub-bandgap absorption. However, deep-level defect states are generally detrimental since they are often responsible for carrier trapping through non-radiative recombination (NRR) mechanisms that decrease carrier extraction efficiency, thereby limiting solar cell efficiency. Overall, the nature of recombination mechanism (either radiative or non-radiative) in the absorber determines the open-circuit voltage and transport properties of the device. The different non-radiative processes include Shockley–Read–Hall (SRH) recombination, Auger recombination, surface, and interfacial recombination [35]. Primarily for interfaces, the main origins of NRR energy losses stem from interface defects, imperfect energy level alignment between layers, and interfacial reactions. Therefore, to address point defect and interface engineering in next-generation solar cells, a precise understanding of defects, interfaces and their processes is necessary to enable control of functionality. In these, photogenerated electrons and holes must have lifetimes sufficient for collection before recombination, as previously discussed. Here, we highlight representative cases of the dynamics and approaches to defects and interfacial engineering in select types of devices. Overall, the strategy for mitigating issues related to defects and interfacial dynamics, though specific to a cell type can be employed to other cells, primarily on the premise of optimizing electronic transport and minimizing NRR processes. Of these, defect passivation, energy level modulation, and suppression of interfacial inefficiencies among others are essential in improving cell performance.

2.4.1 Defect and interface engineering in dye-sensitized solar cells

In the case of DSSCs, competition between electron transport in the semiconductor and electron recombination in the semiconductor/dye/electrolyte interface is key for high performance because electrons in the semiconductor interface are easily recombined with the holes in electrolyte. Doping the photoanode oxides with cations is one strategy to improve electron transport, since this exerts a larger dipole moment and strong electronic coupling of the dye's lowest unoccupied molecular orbital (LUMO) that changes the interface energetics between it and the dye for faster electron transfer than the recombination. For example, doping TiO_2 with strontium ions realigns the conduction band edge, thereby increasing the efficiency of electron–hole separation at the interface, and thus reducing the electron/electrolyte recombination rate, thereby increasing V_{oc}. For this strategy, a wide range of other cation dopants has been used to improve charge injection in the photoanode [36]. Incorporating dopants and additive species is a critical strategy employed to alter the semiconductor–electrolyte interface to increase PV performance [19]. In such cases, nitrogen-heterocyclic compounds such as 4-tert-butylpyridine (tBP) and N-methylbenzimidazole (NMBI) are typically used to shift the Fermi level and inhibit electron–hole recombination, thus improving the V_{oc} in liquid cells. Likewise, additives in solid-state electrolytes and hole transport materials such as LiTFSI and tBP are used to alter TiO_2 energy levels and passivate its surface, reducing trapping energy states allowing for improved charge injection and reduced recombination processes at the TiO_2/HTM interface. Another strategy employed to improve the competition involves limiting electron recombination by reducing surface defects density that trap electrons within the semiconductor [37].

An important strategy involving interfacial engineering for improving device performance is to incorporate interface materials as essential components. In a typical DSC, the mesoporous oxide (e.g. TiO_2) layer offers sufficient surface area for dye loading and solar light absorption leading to efficient photon-to-electricity conversion. However, the mesoporous nature of the semiconductor can also be an effective interface with the liquid electrolyte facilitating electron recombination via the triodide species to the conducting substrate. In this context, the inclusion of a thin blocking/passivating layer in the interface of the conducting glass substrate and mesoporous layers is one strategy to deter this recombination, thereby improving the performance [36]. Also, application of a metal oxide as an interface modifier/blocking layer in the photoanode material is a common strategy to retard the back electron transport, since it forms a surface energy barrier, which retards the recombination.

2.4.2 Defect and interface engineering in perovskite solar cells

One of the key factors that have limited the efficiencies of perovskite solar cells is related to the defect density and NRR processes from deep-level states acting as charge carrier trap states in the photoactive layers despite the high defect tolerance of metal halide perovskites [38]. In particular, the SRH and Auger recombinations related to defects induced in bulk perovskites decrease carrier lifetimes and are mainly responsible for the V_{oc} losses. However, surface and interfacial defects highly influence charge transport in these devices [35]. Furthermore, despite the high defect tolerance of metal halide perovskites, the presence of deep-level states acting as charge carrier trap states and the resulting NRR losses have been implicated in limiting the PCE of perovskite solar cells. Several studies have shown that deposited perovskite layers host a combination of point and linear defects particularly arising from dangling bonds mainly located at grain boundaries or the surfaces [35, 38, 39]. For example, grain boundaries as sources of high defect densities were confirmed by the fact that the trap density in solution-processed polycrystalline $MAPbI_3$ perovskite films (10^{16}–10^{17} cm^{-3}) is much higher than that in perovskite single crystals (10^{9}–10^{10} cm^{-3}) [34, 35, 38]. Additionally, defects at grain boundaries and interfaces have been proven to be responsible for triggering the degradation of perovskite films since they provide charge accumulation sites as well as infiltration pathways of water vapour [38]. Overall, bulk charge NRR may originate from the perovskite light absorber layer, electron transport layer (ETL), hole transport layer (HTL), and electrodes [34]. Among them, NRR loss from transparent conductive oxide (TCO) and metal electrodes is negligible and may be ignored as compared to those from other layers. Because the defects and poor electrical properties of the ETLs and HTLs can lead to bulk charge NRR losses in charge transport layers (CTLs), many efforts have been devoted to developing and optimizing the CTLs to improve carrier transport and reduce defects.

In that regard, for perovskite cells, strategies for defects and interface engineering typically involve increasing the grain size, grain boundary and surface passivation, as well as interfacial modification, and improvements in the energy level alignment between layers to minimize NRR losses. For example, it has been shown that perovskite films with large grains exhibit lower trap state density and reduced recombination centers. Therefore, the film formation process should involve mechanisms that reduce the crystallization rate resulting in larger grains, and an effective method is solvent annealing, where the thermal annealing of perovskite films is processed under vapours of N,N-dimethylformamide (DMF) or dimethyl sulfoxide (DMSO) [38]. Additionally, there has been a range of strategies involved in the passivation of the surface of halide perovskite semiconductors. For example, a range of species have been used as dopants or passivating layers for surface and grain boundaries including, ions of guanidinium, n-butylammonium, alkyl-ammonium, and isobutylammonium as common examples. To passivate point defects with shallow energy levels, arising from bulk defects, as well as defects at grain boundaries, doping with small concentrations of ions have been effective. For example, passivating iodine interstitials and Pb–I antisites in perovskites, metal cations such as Cs^{+} ion is common [35]. On the other hand, anions are used to passivate lead and halide vacancies. Overall, Lewis acids are used to passivate defects which have free electrons such as Pb–I antisites or free iodide ions, whilst Lewis bases are used to passivate electron-deficient defects such as lead interstitials, the combination of which forms a Lewis acid–base pair that eliminates the deep-level defects associated with cation–anion pair (donor–acceptor) defects. The use of several types of large molecule Lewis acids and bases were shown to provide effective surface passivation in perovskite cells, reducing the number of defects and increasing carrier lifetimes. For example, the Lewis base poly(methyl methacrylate) (PMMA) has been proposed as a nucleation template to improve crystallization, planarizing and passivation. It has been used to improve the interfacial dynamics between the perovskite and the electron transport layer as well as between the perovskite and the hole transport layer.

2.4.3 Defect and interface engineering in organic solar cells

For organic solar cells, molecular assembly and thin film morphology control are key challenges since charge transport, separation and recombination are highly sensitive to morphological parameters of thin films across all length scales, which include paracrystalline size at the nanoscale, domain percolation at the mesoscale, and domain alignment and boundary distribution at the macroscale [40]. In BHJ configurations in particular, which typically consists of a network of electron-donating conjugated polymers and electron-accepting fullerenes, the solution deposition leads to multiple grain boundary configurations between randomly combined particles. Therefore, important factors that must be considered for surface and interface property control towards optimizing the morphology of the BHJ active layers include: (1) miscibility between donor and acceptor materials; (2) semi-crystalline characteristics of phase-separated domains; and (3) vertical concentration gradient, combined with substrate engineering. Overall, fabrication strategies must consider surface energy, surface morphology and surface dynamics, and the critical role of substrates in determining the morphology of these solvent-based films. For example, slow-drying solvent additives are important tools for manipulating phase separation of the active layer during deposition and crystallization. It has been observed that both thermal annealing and solvent vapour annealing produce thermodynamically stable layers with ordered and pure domains with interfacial mixed-phase to promote efficient charge carrier dissociation, resulting in optimal PV performance.

A critical attribute of BHJ organic cells is the influence of the interfacial layers. In these, an exciton is split into free carriers at the donor–acceptor interface and the free carriers are transported via percolated donor and acceptor pathways to the corresponding selective electrodes. Importantly, the energy level alignment at the electrode interfaces plays an essential role, where an ideal interface requires good Ohmic contact with minimum resistance and high charge selectivity to prevent charge carriers from migrating to the opposite electrodes. Therefore, the main functions of interface materials are to: (1) adjust the energy level alignment between the active layer and the electrodes; (2) form a selective contact for precisely each carrier; (3) tailor the built-in electric field/polarity of the device; (4) improve chemical/interfacial stability between the active layer and electrodes; and (5) act as an optical spacer to modulate light absorption [41, 42]. Overall, interface materials can be categorized as conducting, semiconducting and dipole layers. The choice, whether metal, metal oxides, organic molecules, *n*-type, *p*-type, self-assembled monolayers, carbon-based materials, organic–inorganic hybrids/composites or metal salts will be important based on their structural, conductive/electronic, morphological interfacial compatibility, and chemical/physical stability characteristics, the methods of deposition, and their impact on increased performance and device stability.

References

[1] Jager-Waldau A 2011 Progress in chalcopyrite compound semiconductor research for photovoltaic applications and transfer of results into actual solar cell production *Sol. Energy Mater. Sol. Cells* **95** 1509–17

[2] Mickey C D 1981 Chemical principles revisted *J. Chem. Educ.* **58** 418–23

[3] Fraas L M and Partain L D 2010 *Solar Cells and their Applications* (New York: Wiley)

[4] Shockley W and Queisser H J 1961 Detailed balance limit of efficiency of *p-n* junction solar cells *J. Appl. Phys.* **32** 510–9

[5] Conibeer G 2007 Third-generation photovoltaics *Mater. Today* **10** 42–50

[6] Taylor R A and Ramasamy K 2017 Colloidal quantum dots solar cells *Nanoscience* **vol 4** (London: The Royal Society of Chemistry) pp 142–68

[7] Coughlan C, Ibáñez M, Dobrozhan O, Singh A, Cabot A and Ryan K M 2017 Compound copper chalcogenide nanocrystals *Chem. Rev.* **117** 5865–6109

[8] Afzaal M 2005 O'Brien, P. recent developments in II–VI and III–VI semicomductors and their applications in solar cells *J. Mater. Chem.* **16** 1597–602
[9] Green M, Emery K, Hishikawa Y and Warta W 2014 Solar cell efficiency tables (version 36) *Prog. Photovolt.: Res. Appl.* **18** 346–52
[10] El Chaar L, lamont L A and El Zein N 2011 Review of photovoltaic technologies *Renew. Sustain. Energy Rev.* **15** 2165–75
[11] Green M A, Hishikawa Y, Dunlop E D, Levi D H, Hohl-Ebinger J, Yoshita M and Ho-Baillie A W Y 2019 Solar cell efficiency tables (version 53) *Prog. Photovoltaics Res. Appl.* **27** 3–12
[12] NREL Best Research-Cell Efficiency Chart 2024 National Renewable Energy Laboratory https://nrel.gov/pv/interactive-cell-efficiency.html
[13] Tyagi V V, Rahim N A A, Rahim N A and Selvaraj J A L 2013 Progress in solar PV technology: research and achievement *Renew. Sustain. Energy Rev.* **20** 443–61
[14] Rahman K S 2022 Cadmium telluride (CdTe) thin film solar cells *Comprehensive Guide on Organic and Inorganic Solar Cells* ed M Akhtaruzzaman and V Selvanathan (New York: Academic) pp 65–83 ch 3.1
[15] Goossens V M, Sukharevska N V, Dirin D N, Kovalenko M V and Loi M A 2021 Scalable fabrication of efficient p-n junction lead sulfide quantum dot solar cells *Cell Rep. Phys. Sci.* **2** 100655
[16] Hussain I, Tran H P, Jaksik J, Moore J, Islam N and Uddin M J 2018 Functional materials, device architecture, and flexibility of perovskite solar cell *Emergent Mater.* **1** 133–54
[17] Li W, Tan J M R, Leow S W, Lie S, Magdassi S and Wong L H 2018 Recent progress in solution-processed copper-chalcogenide thin-film solar cells *Energy Technol.* **6** 46–59
[18] Wang Q, Xie Y, Soltani-Kordshuli F and Eslamian M 2016 Progress in emerging solution-processed thin film solar cells—part I: polymer solar cells *Renew. Sustain. Energy Rev.* **56** 347–61
[19] Muñoz-García A B *et al* 2021 Dye-sensitized solar cells strike back *Chem. Soc. Rev.* **50** 12450–550
[20] Kramer I J and Sargent E H 2014 The architecture of colloidal quantum dot solar cells: materials to devices *Chem. Rev.* **114** 863–82
[21] O'Regan B and Grätzel M 1991 A low-cost, high-efficiency solar cell based on dye-sensitized colloidal TiO2 films *Nature* **353** 737–40
[22] Benesperi I, Michaels H and Freitag M 2018 The researcher's guide to solid-state dye-sensitized solar cells *J. Mater. Chem.* C **6** 11903–42
[23] Habibi M, Zabihi F, Ahmadian-Yazdi M R and Eslamian M 2016 Progress in emerging solution-processed thin film solar cells—part II: perovskite solar cells *Renew. Sustain. Energy Rev.* **62** 1012–31
[24] Kojima A, Teshima K, Shirai Y and Miyasaka T 2009 Organometal halide perovskites as visible-light sensitizers for photovoltaic cells *J. Am. Chem. Soc.* **131** 6050–1
[25] Kim H-S *et al* 2012 Lead iodide perovskite sensitized All-solid-state submicron thin film mesoscopic solar cell with efficiency exceeding 9% *Sci. Rep.* **2** 591
[26] Tang C W 1986 Two-layer organic photovoltaic cell *Appl. Phys. Lett.* **48** 183–5
[27] Ehrler B, Alarcón-Lladó E, Tabernig S W, Veeken T, Garnett E C and Polman A 2020 Photovoltaics reaching for the Shockley–Queisser limit *ACS Energy Lett.* **5** 3029–33
[28] Rühle S 2016 Tabulated values of the Shockley–Queisser limit for single junction solar cells *Sol. Energy* **130** 139–47

[29] Sharma K, Sharma V and Sharma S S 2018 Dye-sensitized solar cells: fundamentals and current status *Nanoscale Res. Lett.* **13** 381

[30] Kaienburg P, Krückemeier L, Lübke D, Nelson J, Rau U and Kirchartz T 2020 How solar cell efficiency is governed by the $\alpha\mu\tau$ product *Phys. Rev. Res.* **2** 023109

[31] Kirchartz T, Bisquert J, Mora-Sero I and Garcia-Belmonte G 2015 Classification of solar cells according to mechanisms of charge separation and charge collection *Phys. Chem. Chem. Phys.* **17** 4007–14

[32] Bisquert J, Cahen D, Hodes G, Rühle S and Zaban A 2004 Physical chemical principles of photovoltaic conversion with nanoparticulate, mesoporous dye-sensitized solar cells *J. Phys. Chem.* B **108** 8106–18

[33] Park J S, Kim S, Xie Z and Walsh A 2018 Point defect engineering in thin-film solar cells *Nat. Rev. Mater.* **3** 194–210

[34] Chen J and Park N-G 2020 Materials and methods for interface engineering toward stable and efficient perovskite solar cells *ACS Energy Lett.* **5** 2742–86

[35] Singh S, Laxmi and Kabra D 2020 Defects in halide perovskite semiconductors: impact on photo-physics and solar cell performance *J. Phys.* D **53** 503003

[36] Sengupta D, Das P, Mondal B and Mukherjee K 2016 Effects of doping, morphology and film-thickness of photo-anode materials for dye sensitized solar cell application—a review *Renew. Sustain. Energy Rev.* **60** 356–76

[37] Lou Y-Y, Yuan S, Zhao Y, Wang Z-Y and Shi L-Y 2013 Influence of defect density on the ZnO nanostructures of dye-sensitized solar cells *Adv. Manuf.* **1** 340–5

[38] Wang F, Bai S, Tress W, Hagfeldt A and Gao F 2018 Defects engineering for high-performance perovskite solar cells *npj Flex. Electron.* **2** 22

[39] Ran C, Xu J, Gao W, Huang C and Dou S 2018 Defects in metal triiodide perovskite materials towards high-performance solar cells: origin, impact, characterization, and engineering *Chem. Soc. Rev.* **47** 4581–610

[40] Zhang H, Li Y, Zhang X, Zhang Y and Zhou H 2020 Role of interface properties in organic solar cells: from substrate engineering to bulk-heterojunction interfacial morphology *Mater. Chem. Front.* **4** 2863–80

[41] Yin Z, Wei J and Zheng Q 2016 Interfacial materials for organic solar cells: recent advances and perspectives *Adv. Sci.* **3** 1500362

[42] Steim R, Kogler F R and Brabec C J 2010 Interface materials for organic solar cells *J. Mater. Chem.* **20** 2499–512

Chapter 3

Solution-processed fabrication methods

Solution-processed solar cells are a promising type of photovoltaic (PV) technology that offers low-cost manufacturing and flexibility, especially for flexible and wearable applications. These solar cells utilize solution-processable materials for their active layers and transparent electrodes, allowing for large-scale fabrication and cost-effective production. Typical solution-processed fabrication methods involve using liquid precursor/material solutions to deposit thin films of materials onto substrates, offering advantages like cost-effectiveness and adaptability to various substrates, for example, transparent and flexible substrates, enabling the creation of flexible and wearable solar cells. Solution processing of solar cells offers several advantages compared to traditional wafer or thin film scale solar cells. The easy and reliable processing can be scaled up for large-area production, making them suitable for mass manufacturing. Also, the use of non-toxic and earth-abundant elements and materials in solution-processed solar cells can reduce environmental impact compared to traditional solar cells. Overall, solution processing allows for the use of a wide range of materials for solar cells including organic and polymer semiconductors, perovskites, and quantum dots, making solution-processed solar cells attractive alternatives to conventional solar cells. This chapter explores some of these methods and highlights challenges and strategies for scaling up.

3.1 Types of processing methods

Solution processing of solar cells and their components typically involves a sequential approach with a number of steps for depositing thin films of materials. Usually, the substrate, often glass or flexible materials like polyethylene terephthalate (PET), is cleaned using treatments like UV-ozone or specific solvents to enhance film adhesion and eliminate impurities that will affect film nucleation, growth and properties. Following this, the active layer material, like a perovskite compound or a blend of organic polymers, or an electrode component, is dissolved in a

doi:10.1088/978-0-7503-3255-2ch3

© IOP Publishing Ltd 2025. All rights, including for text and data mining (TDM), artificial intelligence (AI) training, and similar technologies, are reserved.

suitable solvent or solvent mixture to form a precursor solution. The precursor solution is deposited onto the cleaned substrate to prepare thin films using various techniques, such as spin coating, doctor blading, slot die coating, inkjet printing, screen printing, electrodeposition, among others. Following deposition, the thin film is dried under ambient conditions or annealed at specific temperatures to promote crystallization, grain growth, and improve film quality. The following outlines some of the main solution processing techniques used to prepare thin films components of various kinds of solar cells.

3.1.1 Spin coating

Spin coating is a simple solution-based method for the fabrication of thin films of materials onto substrates with thickness in the range of nm to μm and is the most widely used method for preparing thin films from solution. Spin coating is carried out using a simple spin coating machine, where a small amount of coating material in liquid form is drop casted to the center of the substrate which is then spun at high speeds. Spin coating consists of several stages: (1) fluid dispense, where the solution is cast on top of a planar substrate which can be spun either before or after drop casting; (2) spin up, where the substrate is spun at a predetermined speed; (3) stable fluid outflow for spreading the film; (4) spin off; and (5) the evaporation of the solvent to leave behind dry thin film (figure 3.1) [1]. The solution polarity, its viscosity, and substrate surface energies, spinning speed, and spinning duration, are

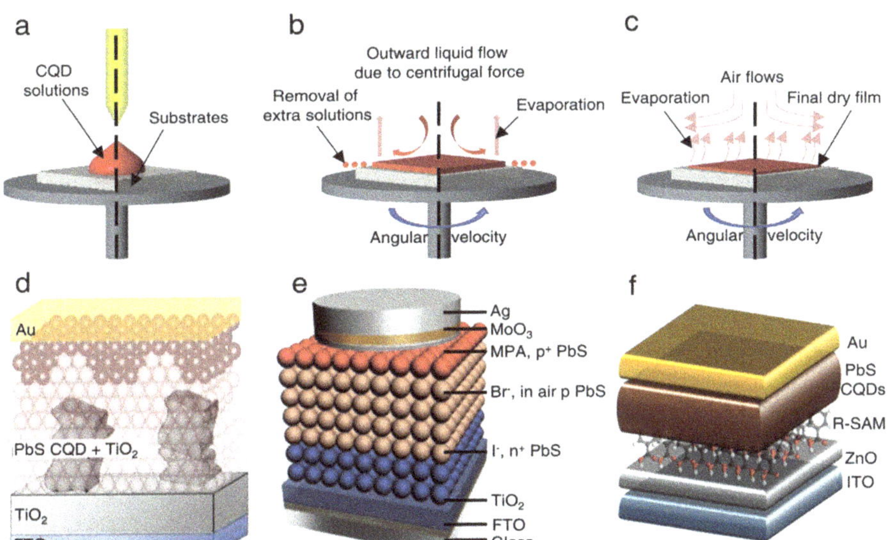

Figure 3.1. (a–c) Schematic illustration of the sequential spin coating process. (d) Schematic drawing of the cross-section of depleted bulk heterojunction PbS QD solar cells. (e) Schematic of inverted quantum junction PbS QD solar cells. (f) Schematic of a prototypical PbS QD device structure including a SAM between the ZnO and the PbS CQD film. Adapted with permission from [1] John Wiley & Sons. Copyright 2022 Wiley-VCH GmbH.

the key parameters of the spin coating process which determine the thin film's physical (structural and morphological), electrical, and thermal properties [2]. In general, the thickness of a spin-coated film is proportional to the inverse of the square root of spin speed. The precise thickness of a film will depend on the material concentration and solvent evaporation rate (which in turn depends upon the solvent viscosity, vapour pressure, temperature and local humidity).

Spin coating techniques are remarkably easy to use, versatile, inexpensive and highly effective for reproducibly depositing a uniform thin film. As a result, the technique has been extensively used to fabricate solution-processed organic and inorganic electronic devices including solar cells [3]. For example, recently, Eguchi and co-workers [4] reported performance optimization of perovskite solar cells of architecture, FTO/SnO_2/$CsAPbI_3$/OAI/spiro-OMeTAD/Au with an automated spin coating system and artificial intelligence technologies. In preparing the device, a layer of SnO_2 (4%, diluted with DI water) was deposited onto UV-ozone treated FTO substrates via spin coating at 3000 rpm for 20 s and annealed in ambient air at 150 °C for 30 min. The perovskite precursor solution which was prepared according to a Bayesian optimization method was spin-coated using the following sequence: static time 20 s, slope 2 s, and spinning at 1000 rpm for 10 s; slope 2 s, 6000 rpm for 30 s, and slope 3 s. During the spin coating step, 0.5 ml of diethyl ether was dropped onto the precursor using an automatic solution-dropping device after applying the maximum rotational speed for 14 s. After spin coating, the substrates were annealed at 70 °C–150 °C for 5 min, cooled to room temperature and spin coated with a 39 mM OAI/IPA solution at 4000 rpm for 20 s. The OAI layer was annealed at 100 °C for 5 min. The solution for the hole transport layer was prepared by dissolving 100 mg of spiro-OMeTAD, 8.7 mg of LiTFSI, 9.3 mg of CoTFSI, and 39.0 ml of *t*BP in 1.279 ml of chlorobenzene and spin coated at 4000 rpm for 20 s and the layer was annealed at 65 °C for 10 min. Finally, Au backing electrode was thermally deposited as the top layer of the cell.

Whilst spin coating is convenient and simple for fabricating lab-scale devices, it has some drawbacks. For example, primarily, most of the solution being spun off the substrate leads to substantial material waste and is therefore not practical for achieving uniform large-area films [5]. Furthermore, the drying kinetics of spin-coated solutions is much faster than other conventional industrial coating methods due to the combination of centrifugal force and gas flow caused by the spinning substrate, making the direct translation of some spin-coated formulations to large-scale coating methods extremely difficult.

3.1.2 Drop casting

Drop casting is a solution-processing technique used to create thin films of materials, including those used in solar cells. It is a simpler method than spin coating and involves adding a drop of a solution containing the desired material onto a substrate, after which the solvent evaporates, leaving behind a solid film. This method is particularly useful for screening various formulations for solar cell fabrication, and because it is simpler is more aligned with industry-relevant coating methods [5].

The self-spreading of the solution in the drop casting method is caused by the unbalanced surface tension of solution/substrate, solution/air, and air/substrate interfaces, enabling film preparation without assistance of a depositing apparatus. The method is practical since it conserves time and materials and it is therefore more industry relevant. The fluid dynamics and drying kinetics of films from a given precursor formulation is critical for the film formation behaviour. The drying kinetics affects the quality of the film in terms of its crystallinity and morphology which is dependent on solvent vapour pressure, temperature and humidity.

It has been reported that for perovskite films made under 70% humidity, the pinholes in the films increase, leading to significantly decreased PCE [6]. This approach contrasts the typical approach where perovskite solar cells are fabricated under highly controlled atmospheric conditions in order to maintain consistent and high performance levels with only trace levels of oxygen and water. Therefore, to investigate the effect of humidity on solar cell performance, Ding and Zuo [7] reported the use of the drop casting method to fabricate efficient perovskite solar cells under high humidity. In their method, as illustrated in figures 3.2(a), 4 ml $MAPbI_3$ solution was dropped onto the centre of a 2.5 × 2.5 cm substrate on a 60 °C hot plate and the solution spread spontaneously forming a round, uniform, wet, transparent film in 25 s. After being transferred to a 150 °C hot plate, the film colour turned yellow, then to black in 2 s, indicating the formation of $MAPbI_3$ crystalline film. Structural data confirmed that the crystals on the films made at low temperature (60 °C) and high temperature (>120 °C) were (110) and (200) oriented, respectively, which led to quite different film morphology. In their study, compared with spin-coated films, the drop casted films showed much better tolerance to humidity with $MAPbI_3$ solar cells made under 88% humidity delivering a PCE of 18.17%.

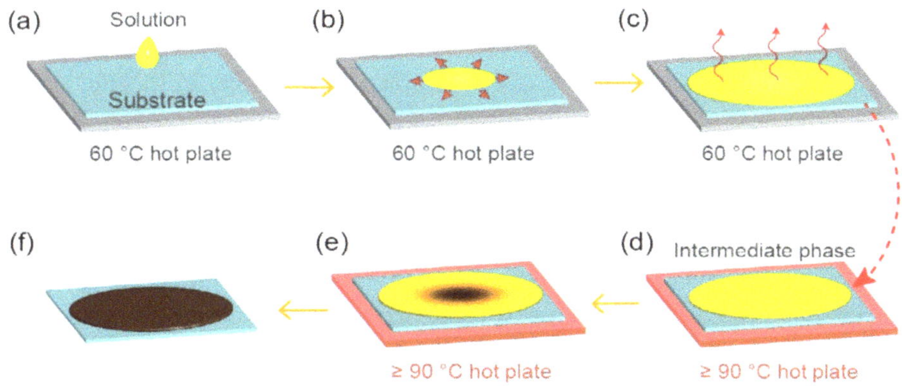

Figure 3.2. Illustration of the drop casting method for preparing $MAPbI_3$ film. (a) The solution is dropped onto the center of a substrate on a 60 °C hot plate. (b) The solution spreads on the substrate. (c) The solution stops spreading, resulting in a round wet film. (d) The substrate is transferred to a hot plate above 90 °C. (e) The film colour turns black, indicating the formation of $MAPbI_3$. (f) Round black $MAPbI_3$ film forms. Reprinted with permission from [7] John Wiley & Sons. Copyright 2021, Wiley-VCH GmbH.

In another case, Gao and co-workers [8] reported the fabrication of perovskite solar cells via solution-processing methods including drop casting. In their method, patterned indium tin oxide (ITO)-coated glass substrate (25 mm × 25 mm) was cleaned in a detergent, deionized water, acetone and isopropanol sequentially by ultrasonication and then treated with UV-ozone for 15 min. Then, m-PEDOT:PSS was spin coated at 5000 rpm for 30 s onto the ITO substrate, then heated on a hot plate at 150 °C for 10 min in air. After cooling to room temperature, the substrate was placed on a hot plate at 60 °C for 2 min, after which 6 μl of the perovskite precursor solutions (0.3 M Pb^{2+}) was deposited onto the center of the substrate. The solution spread on the substrate to form a circular wet film. For films prepared by drying under ambient conditions, no additional treatment was applied. For films prepared by nitrogen blow-drying, a stream of nitrogen was applied to the film when it started to dry resulting in a dark brown film. After nitrogen or air blow-drying, the substrate was heated at 100 °C for 2 min. After cooling to room temperature, 70 μl $PC_{61}BM$ in chloroform (10 mg ml^{-1}) was spin coated onto the perovskite layer at 1000 rpm for 30 s. Then, polyethylenimine ethoxylated (PEIE) (0.05% w/w in isopropanol) was spin coated onto the $PC_{61}BM$ layer at 4000 rpm for 30 s. Finally, 100 nm Ag was evaporated through a shadow mask to give an active area of 0.1 cm^2. The authors were able to demonstrate that film morphology, phase purity, and crystal orientation of the hybrid quasi-2D/3D perovskite films were improved significantly by applying a simple nitrogen blow-drying step following drop casting.

Overall, drop casting is a valuable technique for creating thin films for solar cells, offering advantages in terms of simplicity, scalability, and flexibility. By carefully controlling the process parameters, it is possible to produce high-quality films with desired properties for efficient solar devices.

3.1.3 Blade coating

Doctor blade coating, also known as blade coating, is a common technique used to apply a thin, uniform layer of an ink solution onto a substrate. The process involves using a blade to spread a coating of solution across a surface. In this method, the solvent plays a crucial role in determining the film's morphology, thickness, and crystallization, which can impact the solar cell performance [9]. This is because the solvent can affect the wetting behaviour, surface tension, and viscosity of the ink. By selecting solvents that provide excellent solubility and stability for the precursor materials, and also by controlling the solvent evaporation rate and temperature during the coating process, solvent engineering can be used to optimize the doctor blade coating process. Furthermore, solvent engineering can be used to enhance the process of film formation by modifying the viscosity of the ink through solvent mixing and enhancing the ink's spreading and wetting properties. Additionally, the quality of the film is influenced by the nature of the substrate surface, the speed of the blade, annealing temperature, and atmosphere.

One of the attractive advantages of the blade coating method is that it has minimal hardware costs and reduced material waste. It is also highly compatible with roll-to-roll (R2R) fabrication of large-area devices, and therefore has strong upscaling potential. However, the inherently slower film formation kinetics often result in unfavourable active layer microstructures, requiring empirical and material-inefficient optimization of solutions to reach the performance of spin-coated devices. As an alternative drying approach for optimized film formation, Hernandez–Sosa and co-workers [10] employed a gas-assisted blade-coating method for accelerating the film formation kinetics using a bulk heterojunction blend of the polymer donor, poly(3-hexylthiophene) (P3HT) and the non-fullerene acceptor, ITIC. Blade coating was performed under ambient nitrogen atmosphere using solutions and substrates preheated to 30 °C. P3HT:ITIC and PM6:ITIC films were blade coated from solutions in chlorobenzene (3.5 and 2.5 wt%, respectively, 1:1 wt/wt donor–acceptor ratio; $\nu = 7$ and 5 mm s^{-1}, respectively). The high boiling point of chlorobenzene and near-ambient deposition temperatures ensured slow film formation kinetics under ambient drying and gas-assisted drying. Solar cell devices were fabricated by annealing the active layers for 10 min under nitrogen atmosphere at 140 °C and thermal evaporation of a 10 nm MoO_3 hole transport layer and 100 nm Ag electrode.

Due to its simplicity, low-cost and high-throughput, the doctor blading deposition technique has been used for fabricating various thin films in PV devices including perovskite solar cells. For example, Yang and co-workers [11] reported large-area and high-performance perovskite, $CH_3NH_3PbI_3$ photodetectors fabricated via an *in situ* thermal-treatment doctor blading technique in ambient condition (humidity ~45%) as illustrated in figure 3.3. They showed that in comparison to spin coating deposition, the doctor blade deposited $CH_3NH_3PbI_3$ films had larger grain size, as well as good reliability and reproducibility in large area. These $CH_3NH_3PbI_3$ photodetectors exhibited high detectivity and high responsivity, as well as the fast

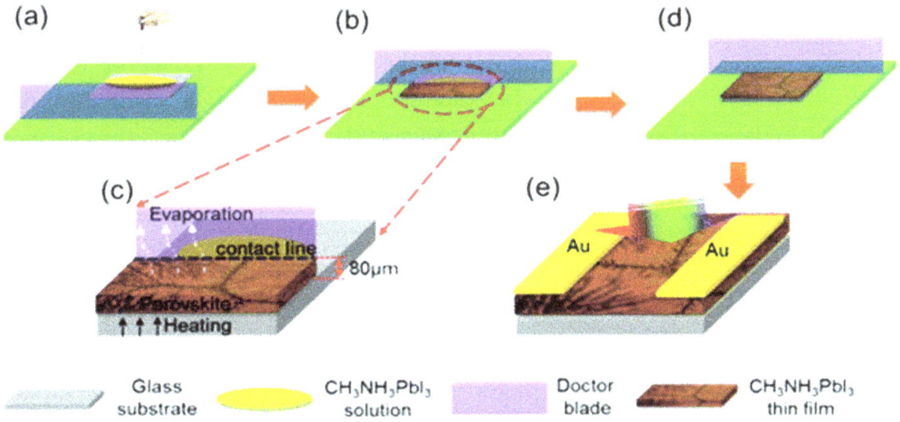

Figure 3.3. Schematic of *in situ* doctor blade coating technique for fabricating $CH_3NH_3PbI_3$ films. (a) $CH_3NH_3PbI_3$ solution is dropped before the doctor blade closes the substrate. (b) $CH_3NH_3PbI_3$ crystals nucleate and growth during doctor blade coating. (c) Magnified schematic in (b). (d) Large-area film growth with the moving of the doctor blade. (e) Schematic of a $CH_3NH_3PbI_3$ photodetector fabricated via doctor blade coating. The illumination light source irradiates from the front side of electrodes. Reprinted with permission from [11], copyright 2017, Elsevier.

response time, indicating their potential for large-scale and high-performance devices.

3.1.4 Spray coating

Spray coating was developed as an alternative to spin coating and involves applying a coating material in a liquid state by atomizing it into a fine spray and applying it to a substrate surface. This technique is widely used in various industries for applications like thin film coating, solar cell fabrication, including organic and perovskite devices. In comparison to other scalable techniques, spray deposition has two main advantages. One is that it is much faster because the spray head can move across a substrate at more than 5 metres per minute. The process involves several steps [12]. Firstly, an ink solution of the desired material is sheared into a mist of micron-sized droplets as the spray head moves over a surface. This is achieved in several ways, including via a nozzle, ultrasonically through piezoelectric transducers or as an electrospray. Secondly, the spray mist is guided to a substrate surface by a shaping gas. This results in coalescence of droplets forming a uniform wet film on the substrate. Finally, the wet film dries as solvent evaporates, leaving behind a solid thin film of desired thickness. This method requires effective control over the film-drying process based on the type of solvent and the conditions of drying. Therefore, optimizing surface temperature, solution flow rate, solvent choice, and spray-head height/speed are essential for high-quality films. Importantly, spray coating is an attractive technique for high volume manufacturing, since it allows large areas to be coated at high speed with only minimal loss of coating ink.

Spray coating was used as early as 2004 to fabricate hybrid organic–inorganic perovskite-like materials and there have been various types of solar cells fabricated via spray coating, including organic and perovskite solar cells. For example, Arumugam and co-workers [13] fabricated fully spray coated organic solar cells at low temperature on woven polyester cotton fabric for wearable energy harvesting applications. As shown in figure 3.4(a), the fabrication of the solar cell involved two deposition stages for the interface and one functional layer. Following screen printing of the interface layer on the fabric substrate, the bottom electrode comprising a flexible thin layer of silver nanowires (AgNWs) was spray coated first with a distance of 15 cm from the spray nozzle to the substrate and a differential pressure inlet/outlet of 0.3 bar. The spray coated AgNW layer was annealed at 130 °C for 5 min to obtain a AgNW film with thickness of ~100 nm. The zinc oxide nanoparticle (ZnO-NP) dispersion was then spray coated on top of the AgNW bottom electrode and annealed at 60 °C for 10 min to obtain a thin layer. Afterwards, the absorber layer of P3HT:ICBA was spray coated onto the top of the ZnO-NP layer. The deposited layers were subsequently annealed in argon at room temperature up to 135 °C for an hour. Then, the hole transport layer, PEDOT:PSS was spray coated and annealed at 100 °C for 5 min. To complete the device fabrication, a semi-transparent AgNW electrode was spray coated on top of the PEDOT:PSS layer.

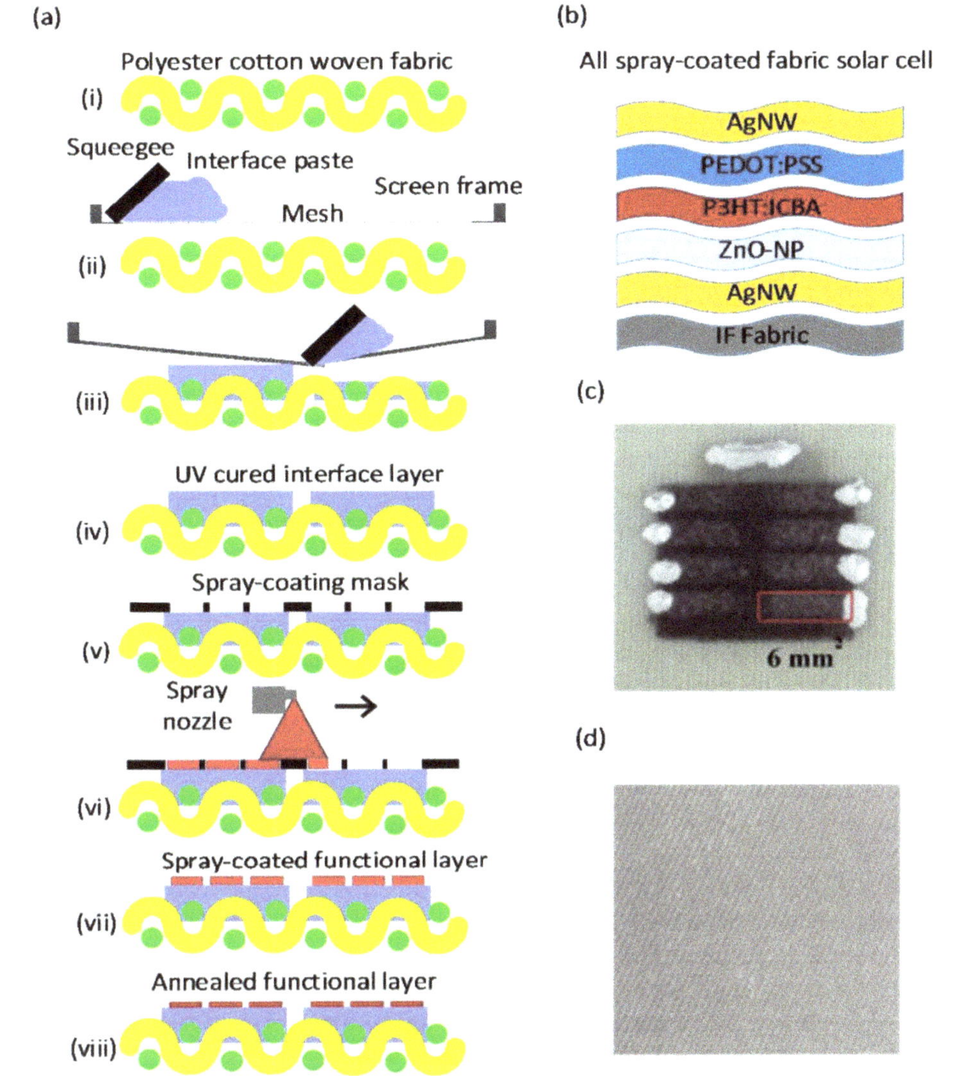

Figure 3.4. (a) Cross-sectional view of the fabrication process of spray coated fabric solar cells. (b) Device structure of a fully solution-processed spray coated fabric substrate. (c) The plan view of an optimized fabric solar cell. (d) The plan rear view of fabric solar cells. Reprinted with permission from reference [13]. Copyright 2016, Royal Society of Chemistry.

The first use of spray-coated perovskites in solar cells was reported by Barrows and co-workers [14] in 2014. The process developed was relatively simple in which a solution comprising a 3:1 ratio of methylammonium iodide to lead chloride (dissolved in dimethylformamide) was spray cast under ambient conditions using an ultrasonic spray coater to create a planar heterojunction device architecture—*p*-type/intrinsic/*n*-type (*p*–*i*–*n*) of the structure ITO/PEDOT:PSS/perovskite/PCBM/Ca/Al. They were able to show the effect of temperature of the substrate during spray coating, the volatility of the solvent and the post deposition anneal on the efficiency of the solar cells. Importantly, they demonstrated that maximum device efficiency was correlated with the creation of dense films having a surface coverage above 85%.

3.1.5 Printing technology

Printing of solar cells is a fabrication approach involving the deposition of cell components onto a substrate using printing methods like inkjet, screen or aerosol jet printing from highly concentrated ink formulations. This approach allows for flexible and potentially low-cost production of solar cells, especially for materials like perovskites, which can be solution-processed. In the case of inkjet printing, the operating principle is based on the ejection of inks in the form of droplets from a nozzle to give controlled properties and their precise deposition and fixation on a target substrate. The two most common approaches that have been employed in inkjet printing for the generation of ink droplets are: (1) continuous inkjet printing (CIP) and (2) drop-on-demand (DOD) inkjet printing [15]. CIP involves an ejection of a continuous flow of fluid through a nozzle, which then separates under the forces of surface tension. However, DOD inkjet printing enables the generation of a single drop based on requirements and reduces material waste. For printing, the physiochemical parameters of inks such as dispersibility, concentration, and rheological properties are essential in obtaining high-quality thin films with the desired chemical, physical, thermal, and electrical properties [2].

As a solution-processed approach, printing technologies have been employed to prepare a range of solar cells. For example, Lin and co-workers [16] used the inkjet printing technique (figure 3.5) for the deposition of $Cu_2ZnSn(S_xSe_{1-x})_4$ (CZTSSe) thin films from Cu–Zn–Sn–S molecular inks. The stable ink with suitable viscosity was prepared by dissolving Cu, Sn, and Zn metal salt precursors, thiourea and different amounts of NaF (varied from 0 to 1.12 M) in dimethyl sulfoxide. NaF was added to verify the influence of sodium ion in ink formation and CZTSSe device properties. The ink was filtered by using an 800 nm polytetrafluoroethylene filter before loading into the ink container of the printer—this being necessary to prevent impurities and inhomogeneities in the film as well as blockages during the printing process. The formulated ink was printed onto molybdenum-coated (800 nm) soda lime glass substrates using a PiXDRO LP50 printer at 60 °C. The as-deposited CZTS precursor thin films were preheated on a hot plate at 300 °C for 2 min to remove the residual solvent. For desired thickness, the inkjet printing and pre-heating steps were repeated four times. The last

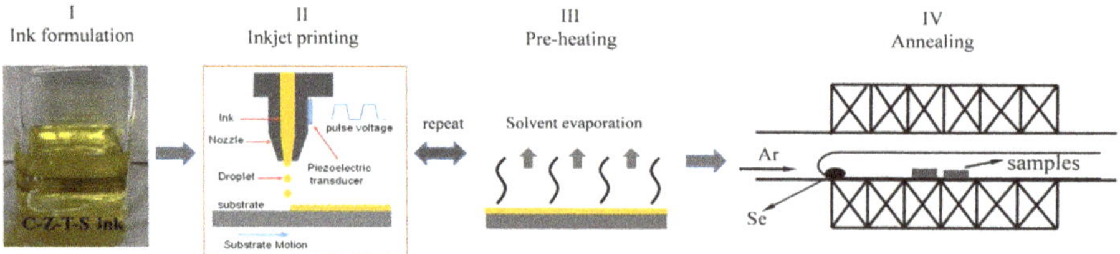

Figure 3.5. Schematic of the inkjet printing formation process of CZTSSe thin films. Reprinted with permission from [16]. Copyright 2018, Elsevier.

step involved annealing of the precursor films in a quartz tube furnace under a selenium atmosphere at 560 °C for 20 min to allow the formation and crystal growth of CZTSSe thin film absorbers which were soaked in 5 vol% HCl at 75 °C for 10 min and etched by 10% KCN for 3 min. Solar cells were fabricated by chemical bath deposition (CBD) (another solution-processed method to be discussed later) of a 50 nm CdS buffer layer at 75 °C for 7 min, sputtering of 50 nm i-ZnO and 500 nm aluminum doped ZnO window layers. A Ni/Al grid on top of the solar cell was deposited by evaporation using a shadow mask.

Of the printing methods, screen printing is a contact printing method that can risk damaging fragile substrates due to applied contact forces [17]. In contrast, inkjet printing is a non-contact, additive digital lithography method that allows for the precise deposition of layers while reducing material waste compared to screen printing. However, one of its main limitations is the need for low-viscosity inks. Furthermore, aerosol printing faces scalability challenges due to limitations in multi-nozzle systems. However, electrostatic inkjet (ESJET) printing is an alternative additive and non-contact digital lithography printing technology in which an applied potential generates an electrostatic force on the surface of the liquid ink which is emitted from the apex.

3.1.6 Spray pyrolysis

Spray pyrolysis is a method for depositing thin films and synthesizing nanoparticles by spraying a solution of chemical precursor compounds onto a heated surface, where they undergo pyrolysis (decomposition) to form the desired material. This technique is used for a preparing various kinds of solution-processed solar cells. During spray pyrolysis, aerosols are formed by atomization of a precursor solution with a nebulizer which is carried to the furnace by the carrier gas and heated at a given temperature. Each droplet acts as a microreactor and undergoes several chemical and physical changes including: (1) solvent evaporation from the droplet surface; (2) precipitation of the solute; (3) thermolysis; (4) porous particle formation; and (5) solid particle formation and sintering [18]. There are several factors involved in the spray pyrolysis process which could be easily controlled to develop tunable porous architectures such as the compositions of precursor solution and reaction parameters. The precursor solution properties including solubility, concentration, density, viscosity, and surface tension directly influence the size of the microreactor or atomized droplet. Similarly, the reaction parameters of temperature, pyrolysis time, and atmosphere can influence the precipitation of droplets.

Additionally, spray pyrolysis is a cost-effective method that can be easily transferred from a laboratory setting to industrial manufacturing. There are various types of spray pyrolysis techniques, including aerosol spray pyrolysis, flame spray pyrolysis (FSP), ultrasonic spray pyrolysis (USP) and electrostatic spray deposition (ESD) [19]. These processes have unique parameters and requirements, with some using a flame as the source of heat and reactants and others using ultrasonic waves to generate the spray. Electrostatic spray deposition (ESD) is a process that uses charged particles to deposit thin films onto a substrate.

In one study reported by Shahiduzzaman and co-workers [20], spray pyrolysis was featured in the fabrication of a multilayer material system to prepare an efficient front contact for perovskite single-junction and perovskite/perovskite tandem solar cells. As a first step, a high-quality, reproducible, and scalable TiO_2 compact electron transport layer (ETL) was prepared by spray pyrolysis deposition, which was a vital part of the front contact. Here, the transparent conducting fluorine-doped tin oxide FTO-patterned glass substrates were cleaned with soap solution, distilled water, acetone, ethyl alcohol, and again distilled water and further cleaned by UV-ozone treatment. A compact TiO_2 layer was deposited onto the FTO-patterned glass via spray pyrolysis at 450 °C from a precursor solution of titanium diisopropoxide bis(acetylacetonate) in isopropanol. As-deposited thin films were left at 450 °C for 30 min then allowed to cool to room temperature. Subsequently, the perovskite film was prepared by spin coating the precursor solution and annealing at 100 °C. Finally, 100 nm thick gold (Au) electrodes were deposited on top of the hole transport layer (HTL) to complete the device fabrication.

In another study, Ikeda and co-workers [21] prepared Cu_2ZnSnS_4 (CZTS) thin film solar cells by facile spray pyrolysis of the thin film absorber on an Mo-coated glass substrate of a precursor film from an aqueous solution containing $Cu(NO_3)_2$, $Zn(NO_3)_2$, $Sn(CH_3SO_3)_2$ and thiourea at 380 °C. The as-deposited film was placed in an evacuated borosilicate glass ampoule together with 20 mg sulfur powder and annealed at 580 °C–600 °C for 10–50 min. Subsequent deposition on the CZTS absorber film, of a CdS buffer layer was done by CBD. An ITO/ZnO bilayer was then deposited on the top of the CdS layer by radio frequency (RF) magnetron sputtering to form a device with a structure of ITO/ZnO/CdS/CZTS/Mo/glass. Also, Tajabadi and co-workers [22] used spray pyrolysis (figure 3.6) to deposit $CuInSe_2$ absorber materials as thin films onto transparent FTO, FTO/nickel oxide (NiO_x) and FTO/molybdenum oxide (MoO_3) substrates for bifacial PV devices. They demonstrated that the crystallinity, carrier concentration, mobility, and energy levels of the CISe layers were strongly dependent on the chemical nature of the substrate accessed via spray pyrolysis.

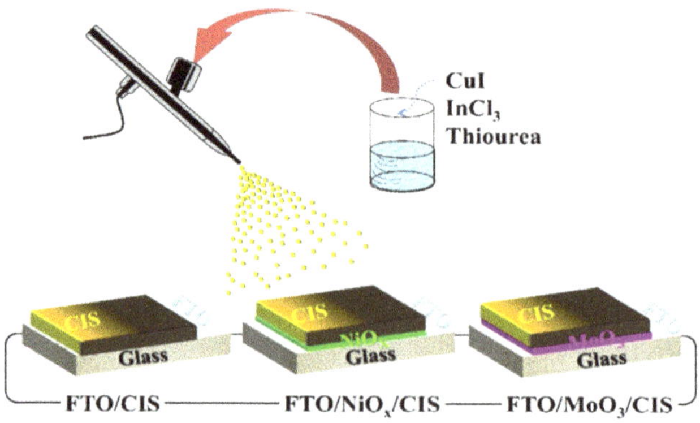

Figure 3.6. Schematic of the spray deposition method of $CuInS_2$ thin films. Reprinted from [22] CC BY 4.0.

3.1.7 Chemical bath deposition

CBD is a method for forming thin films of inorganic compounds such as metal chalcogenide and perovskites on substrates by immersion in a solution containing precursors, often aqueous solutions containing metal ions and a source of other ions (such as sulfide, selenide, or hydroxide) at controlled temperatures. This process utilizes chemical reactions to precipitate the desired material onto the substrate surface. That is, the film is typically formed through heterogeneous nucleation, where the ions in the solution deposit or adsorb onto the substrate's surface with subsequent growth of small particles that eventually form a thin film. Once the film is removed from the precursor solution it is usually rinsed and annealed under specific conditions.

One of the key features of CBD is its ability to easily modulate the morphology, structural characteristics, and optical properties of nanostructures and thin film topography through the manipulation of various parameters, including precursor concentration, pH, growth duration, and temperature. Accordingly, Bruno and co-workers [23] reported a CBD approach for obtaining zinc oxide nanostructures in acidic pH environments using hexamethyleneteramine (HMTA), $Zn(NO_3)_2{\cdot}6H_2O$, and aqueous NH_4F solutions. Through meticulous control of process parameters, such as the order of reactant addition, it was possible to obtain zinc oxide nanostructures with diverse morphologies such as microflowers, nanoparticle–nanowire, starlike structures, and belts. As another example, Aida and co-workers [24] reported nanocrystalline zinc sulfide, ZnS thin films deposited on glass substrates by the CBD technique. There, they demonstrated that the pH of the solution influenced the films properties. Structural and morphological characterizations revealed that as-grown ZnS films were amorphous, while thermal annealing at 550 °C, yielded nanocrystalline cubic structure ZnS.

In a recent study, Najm and co-workers [25] proposed a new technique based on CBD, in which a bi-functional linker molecule was used for selectively attaching pre-synthesized metal chalcogenide quantum dots (e.g., CdS, CdSe) to oxide surfaces (e.g., TiO_2 or ZnO). Here, they achieved high-quality CdS thin films by combining the Linker-Assisted CBD (LACBD) technology with doping through the CBD process. The approach consisted of two experimental steps. In the first, CdS was synthesized using several approaches, including standard CdS thin films, MPA-sensitized CdS, and Ag-doped CdS. In this method, CdS thin films were deposited onto ultrasonically cleaned and degreased soda lime glass substrates through CBD from a solution of thiourea (0.002 M) and cadmium sulfate (0.002 M) at a specific pH. For preparing Ag-doped CdS thin films, $AgNO_3$ (0.8 ml of 0.01 M stock solution) was slowly added to the $CdSO_4$ solution. MPA films were prepared in the following manner in which CdS synthesis was initiated after adding $AgNO_3$ in the $CdSO_4$ solution and then, MPA (0.1 M) was added to the mixture, 20 min after the initiation of the complete reaction. In the second stage, the CdS synthesis was initiated after adding $AgNO_3$ in the $CdSO_4$ solution. Then, MPA (0.1 M) was added to the mixture, 20 min after the initiation of the complete reaction. The following mechanism was proposed for the formation of CdS by the CBD method.

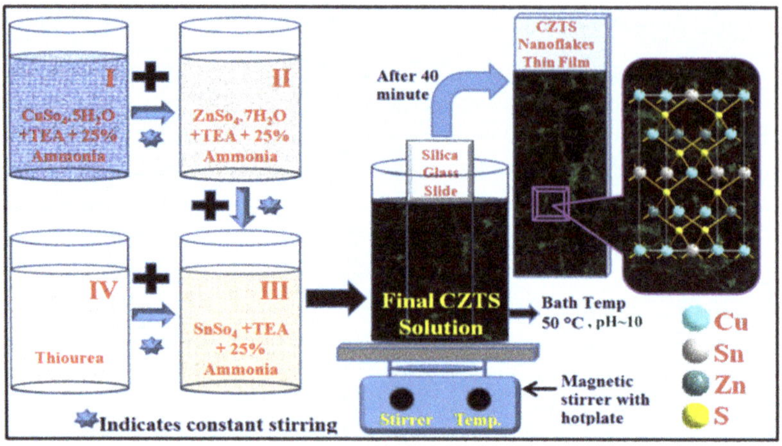

Figure 3.7. Schematic representative of CBD method of CZTS thin film. Reprinted with permission from reference [26]. Copyright 2018, Springer Nature.

$$CdSO_4 + 2NH_4OH \rightarrow Cd(OH)_2 + (NH_4)_2SO_4 \quad \text{Decomposition of cadmium salt}$$

$$Cd(OH)_2 + 4NH_4OH \rightarrow Cd(NH_3)_4^{2+} + 2OH^- + 4H_2O \quad \text{Intermediate complex}$$

$$(NH_2)_2CS + 2OH^- \rightarrow CH_2N_2 + 2H_2O + S^{2-} \quad \text{Decomposition of thiourea}$$

$$Cd(NH_3)_4^{2+} + S^{2-} \rightarrow CdS + 4NH_3 \quad \text{Formation of CdS}$$

In a similar study, Sharma and co-workers [26] deployed a facile, one-step synthesis of non-toxic kesterite Cu_2ZnSnS_4 nanoflakes thin film by CBD for solar cell application, as illustrated in figure 3.7.

One of the issues with CBD is that there can be uncontrollable precipitation of bulk material, especially in depositing metal chalcogenide semiconducting thin films. This unnecessary formation of bulk precipitation results in loss of material. Accordingly, in order to avoid such unwanted precipitation, a modified CBD method was developed, known as successive ionic layer adsorption and reaction (SILAR) method [27].

3.1.8 Successive ionic layer adsorption and reaction (SILAR)

Successive ionic layer adsorption and reaction (SILAR) is a method used to grow thin films, particularly of semiconducting materials, for applications like solar cells. A version of CBD, it is a layer-by-layer deposition technique that involves alternating immersion of a substrate into separate cation and anion precursor solutions. SILAR is known for being cost-effective, simple to operate, and suitable for large-area deposition at room temperature. Following immersion of the substrate, there is surface adsorption of cations from the solution, then with subsequent immersion into the anionic solution, adsorption of anions and reaction to form a thin layer of the desired material, usually upon annealing the layered precursor. The cycle of immersion, adsorption and reaction is repeated several times to increase film thickness. SILAR offers several advantages, including its low-temperature operation, suitability for various substrates and the ability to deposit thin multilayer structures. Additionally, it offers an extremely easy way to deposit doped thin films, it does not require high-quality target and/or substrates, and it does not require a vacuum [27]. Importantly, the deposition rate and film thickness can be easily controlled over a wide range through deposition cycles. It is also relatively inexpensive and can be used for large-area deposition without the need for complex instrumentation.

The SILAR method was first reported in 1985 by Ristov and co-workers [28] and further developed to primarily deposit thin films such as ZnS, CdZnS and CdS [27]. Its versatility lends itself to deposition of a range of inorganic materials films but it has been extensively used to deposit a wide range of binary, ternary and quaternary chalcogenides and composite films. For example, Liu and co-workers [29] fabricated flexible solar cells with Ag/AZO/i-ZnO/CdS/CZTS/Mo foil architecture involving SILAR as the first step, as shown in figure 3.8. The CZTS absorber layers were prepared by sulfurizing two stacked precursors with different stacking sequence: Cu_2SnS_x/ZnS/Mo (sample 1) and ZnS/Cu_2SnS_x/Mo (sample 2) via the SILAR method. This was accomplished by immersing the ultrasonically cleaned soda lime glass substrate into separate cation and anion precursor solutions for adsorption and reaction, and then rinsed with deionized water after each immersion to remove excess ions and avoid homogeneous precipitation. This process of immersion, adsorption and reaction was repeated to acquire the desired thickness and suitable compositions of thin films. Similarly, ZnS films were deposited from separate solutions of 0.5 M $ZnCl_2$ (cation) and 0.05 M Na_2S (anion); 0.5 M NH_4F was added to the cation solution to avoid hydrolysis of Sn^{2+}. Cu_2SnS_x (CTS) films were deposited from solutions of 0.02 M $SnSO_4$ (cation) and 0.005 M $CuSO_4$ (cation) and 0.05 M Na_2S (anion). The adsorption and reaction times were 20 and 30 s, respectively, and the rinsing time was 20 s. For the CTS/ZnS/Mo stacked precursor, 200 cycles of SILAR were implemented to deposit a ZnS film on the substrate and then a CTS thin film was deposited on the ZnS film with 100 cycles of SILAR (about 300 nm). To prepare ZnS/CTS/Mo precursor, the procedure was different only in deposition sequence. These two stacked precursors were then

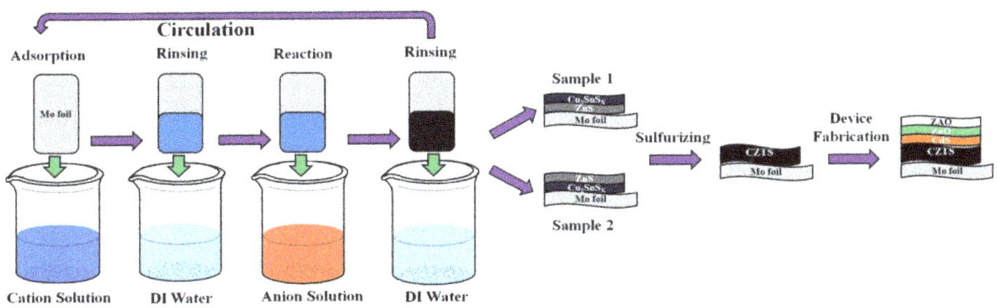

Figure 3.8. Schematic illustration of the fabrication of CZTS solar cells with SILAR as the first step. Reprinted with permission from reference [29]. Copyright 2014, Royal Society of Chemistry.

sulfurized at 500 °C for 30 min in an atmosphere from elemental sulfur in nitrogen. The CdS buffer layers (60–80 nm) were prepared by conventional CBD from aqueous solution of $CdSO_4$ (0.003 M), thiourea (0.3 M) and ammonia (0.28 M) at 80 °C for 8 min and the ZnO window (80 nm) and ZnO:Al (AZO) (400 nm) layers were subsequently deposited by RF magnetron sputtering.

3.1.9 Electrodeposition

Electrodeposition is a method used to create thin films, including those used in solar cells, where a material is deposited onto a substrate using an electric current. This technique offers advantages like large-scale production, uniform deposition, and cost-effectiveness. It has been particularly useful for preparing inorganic, quantum dot, polymer and perovskite solar cells. Electrodeposition is as simple electrochemical process which typically employs a three-electrode system consisting of a counter electrode (CE), a working electrode (WE), and a reference electrode (RE) [30]. Once a voltage is applied across the electrodes, an electrochemical process involving redox and transport enables material deposition onto the electrode surface. A key advantage of electrodeposition is that uniform deposition can be achieved based on parameters such as the cell potential, current density, and electrolyte composition. Electrodeposition has been shown to provide improved benefits in several solar cell configurations. It is one of the most selective processes because deposition only occurs at positions on a substrate where the substrate conductivity is highest.

For example, electrodeposition of $CuIn_xGa_{(1-x)}Se_2$ (CIGS) solar cells was performed from a precursor bath containing $CuCl_2$, $InCl_3$, H_2SeO_3, and $GaCl_3$ involving an electrolyte species, LiCl [31]. Electrodeposited CIGS layer (0.2 mm) was prepared at room temperature by applying a constant current density of 0.9 mA cm^{-2} for 10 min, followed by electrodeposition of the second Cu layer at room temperature from a $CuCl_2$ solution using a constant current density of 8.2 mA cm^{-2} for 3 min. The third layer of In was electrodeposited from a $InCl_3$ solution using a constant current density of 7.2 mA cm^{-2} for 6 min. The films were electrodeposited by employing a two-electrode cell at constant current mode in which the CE was platinum (Pt) gauze and the WE substrate was glass/Mo. The Mo film was about 1 mm thick and was deposited by dc sputtering. Solar cells from these films were completed by CBD of about 50 nm CdS, followed by RF sputtering of 50 nm of intrinsic ZnO and 350 nm of Al_2O_3-doped conducting ZnO. Bilayer Ni/Al top contacts were deposited in an electron-beam system. The final step in the fabrication sequence was the deposition of an antireflection coating (100 nm of MgF_2).

The first attempt to develop a perovskite layer using electrodeposition involved depositing a PbO layer as the initial material from an aqueous solution of 2 mm Pb $(NO_3)_2$ and 0.2 m H_2O_2 onto mesoporous TiO_2-coated FTO glass as WE, with a Pt wire as CE [30]. The resulting PbO layer was then exposed to iodine vapour to form a PbI_2 film and the conversion of PbI_2 into the $MAPbI_3$ perovskite was carried out using two different methods: solid–solid and solid–liquid interdiffusion reactions. For the solid–solid interdiffusion reaction, a solution of MAI (methylammonium iodide) in isopropanol (40 mg in 1 ml) was spin coated onto the PbI_2 film at

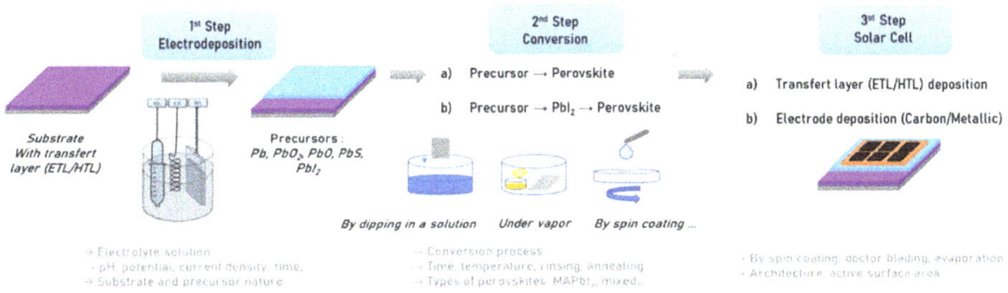

Figure 3.9. The electrodeposition process for perovskite solar cells. Reprinted from [30] John Wiley & Sons. Copyright 2023 The Authors. Advanced Materials Technologies published by Wiley-VCH GmbH .

6000 rpm for 30 s at room temperature, followed by drying at 100 °C for 1 h. In the case of the solid–liquid interdiffusion reaction, the PbI_2 film was immersed in an MAI solution (10 mg in 1 ml isopropanol) at room temperature for 5 min and dried at 70 °C for 10 min. Figure 3.9 outlines a schematic of the process of preparing a perovskite solar cell with the first step involving electrodeposition. The spin coating process resulted in a solar cell efficiency of 12.5%, while the immersion process yielded an efficiency of 9%.

Overall, the electrodeposition process has several advantages since it could provide: (a) high-quality film with very low capital investment; (b) a low-cost, high-rate process; (c) a large-area, continuous, multi-component, and low-temperature deposition method; (d) deposition of films on a variety of shapes and forms (wires, tapes, coils, and cylinders); (e) controlled deposition rates and effective material use (as high as 98%); (f) minimum waste generation since the solutions can be recycled.

3.1.10 Electrochemical polymerization

In the fabrication of organic polymer solar cells, the electroactive conducting polymer film is usually deposited via the process of electrochemical polymerization. Electro-polymerization is initiated by the oxidation of a monomer in an electrochemical cell, followed by the growth of the polymer film on the surface of the WE, which may be a carbonaceous, a metallic, or a conducting glass material. When applied to CEs in dye-sensitized solar cells or as hole transport materials in perovskite solar cells, electrochemical polymerization offers precise control over the polymer's properties, such as thickness and conductivity [32]. Overall, the process utilizes potentiostatic, galvanostatic, potential step and potential sweep methods. Therefore, factors such as applied potential, doping effect, electrolyte solution, direction of potential sweep, nature of the monomer, pH, solvent effect, temperature, and the voltammetric potential window determine the properties of the polymer films formed [33]. This process is typically carried out using the standard three-configuration electrodes electrochemical cell—the electrodes immersed in an electrolytic solution containing the monomer [34]. After the potential is applied, the monomer is electrochemically oxidized to form free radicals that initiate the polymerization process with subsequent deposition of the conducting polymer film on the WE surface. There are a number of factors that are critical in influencing the film's characteristics during electro-polymerization [35]. These include the nature of the electrolytic, temperature, pH, monomer concentration, and the stability of the formed cation-radical. Importantly, the deposited polymers can be obtained by anodic or cathodic reactions at the WE, the former being the most used. Additionally, control of the deposition conditions are important for the film morphology and thickness.

It has been reported that the electro-polymerization of porphyrins offers the advantage of controlling the film thickness, morphology and tuning redox properties by ionic doping and has been useful for PV donor/acceptor heterojunction devices. For example, Chauhan and co-workers [36] reported a method of electrochemical polymerization of tetra-(4-hydroxyphenyl) porphyrin for organic solar cells of the configuration, ITO/PEDOT:PSS/Poly-THPP/PCBM/Al. In their method, the poly-*tetrakis*(4-hydroxyphenyl)porphyrin (poly-THPP) films were deposited via cyclic voltammetry from a methanol electrolytic solution containing 1 mM THPP and 0.1 mM tetrabutylammonium perchlorate (TBAP) onto ITO electrode in a conventional one compartment electrochemical cell equipped with three electrodes (ITO, Pt foil and Ag/AgCl as working, counter and reference electrodes, respectively). The deposited film electrodes were rinsed with methanol and dried under nitrogen. The bilayer organic solar cell was fabricated by initially spin casting PEDOT:PSS layer (50 nm) on patterned ITO, and a bilayer device was formed by electropolymerizing THPP on PEDOT:PSS electrode and spin casting PCBM (26 mg ml^{-1} in 1,2-dichlorobenzene) on top of the poly-THPP electrode and finally evaporation of Al under vacuum (4×10^{-6} torr) for the back electrode. The authors demonstrated that tunability of the film structure, thickness and optoelectronic properties were dependent on the number of polymerization cycles.

Among the most used buffer layers for organic solar cells are organic conducting polymers like poly(3,4-ethylenedioxythiophene):poly(styrenesulfonate)—(PEDOT:PSS) with PEDOT layers being commonly deposited electrochemically, which is useful because the thickness, conductivity, and morphology can be easily controlled for optimizing performance of these devices. As an example, Del-Oso and co-workers [37] reported the electrochemical deposition of PEDOT films onto ITO electrodes for organic PV cells and demonstrated control of morphology, thickness, and electronic properties. The method involved electro-polymerization of the monomer ethylenedioxythiophene (EDOT) onto ITO electrodes in different dry organic electrolytic media, such as acetonitrile, acetonitrile–dichloromethane, and toluene–acetonitrile mixtures. It was found that electro-polymerization kinetics could be controlled by changing the polarity of the electrolytic media, and kinetics were slower for those with low polarity along with monomer concentration that controlled the morphology and thickness of the electropolymerized PEDOT films (E-PEDOT:ClO_4). The performance of the E-PEDOT:ClO_4 films were tested on ITO electrodes as anode buffer layer in organic solar cells with the configuration, ITO/E-PEDOT:ClO_4/P3HT:$PC_{61}BM$/Field's metal.

3.2 Strategies towards scaling up

3.2.1 The challenge with solution processing methods

Scaling up solution-processed solar cells, requires adapting deposition techniques to large-area production, optimizing film quality, and developing efficient modules. Strategies include scaling up precursor solutions, employing scalable deposition methods like blade coating, slot-die coating, and printing techniques, and scaling up charge-transport layers and electrodes. Much of the research in solution-processed PV architectures has been based on small devices fabricated via spin coating. Spin coating was established for batch fabrication of semiconductor manufacturing and is a robust, simple technique for producing uniform thin films. However, it is material consumptive and scalability is limited, creating a significant challenge to progress from the laboratory to large-scale manufacturing, and this has led to developments in other methods such as doctor blading, slot-die coating, spray coating, and inkjet printing [38]. With promising small-area devices of competitive efficiencies across the range of device architectures, efforts are required for scaling up to large-area devices with high efficiencies. Scaling up solution-processed solar cells involves adapting laboratory-scale fabrication methods to industrial-scale production, often through techniques like R2R manufacturing. This requires optimizing coating processes, improving material properties, and ensuring consistent film quality across large areas. While solution processing offers potential cost advantages, challenges remain in achieving high efficiencies and reproducibility in large-scale production. Critical is achieving uniform, high-quality thin films for high-efficiency solar cells and involves controlling film thickness, morphology, and uniformity through careful process optimization. Importantly, with large-area devices it will be essential to maintain consistent performance across large areas, ensuring reproducibility, and optimizing materials for large-scale production. In addition to a scalable deposition system that maintains function at high film quality, flexible electrodes that minimize series resistance must be developed, along with cell patterning approaches that minimize geometric losses, and optical designs that maximize efficiency for which significant efforts are required to achieve performances comparable to their lab-scale counterparts [38].

3.2.2 Select strategies for scaling up

The approach to upscaling methods of solution-processed solar cells is not trivial and a number of factors must be considered to achieve good device performance in larger area devices. When considering strategies towards upscaling solution-processed solar cells, it is important to note that active layer processing is generally thought to have a drastic impact on device performance when going from lab-scale processing to roll-to-roll processing [39]. This is because the morphology of active layers can change significantly when using a roll-to-roll friendly technique like slot-die coating due to challenges in controlling drying kinetics and solution flow as well as fabrication in air which commonly renders limitations to materials in terms of environmental stability. Solvent toxicity is another significant upscaling challenge, especially from the standpoint of environmental sustainability and health. For example, most organic solar cells are made using halogenated solvents such as chloroform (CF) and chlorobenzene (CB) and upscaling presents challenges. In upscaling, it is paramount to consider that the use of less toxic non-halogenated solvents instead of the commonly used halogenated ones which may have solubility and boiling point limitations and therefore requires additives which can affect the morphology of active layers which can impact device performance. Additionally, the impact of more sustainable solvents on charge carrier dynamics—charge carrier densities, lifetimes, and mobilities of active layers and interfaces is of strong consideration on the performance of upscaled large-area devices.

Strategies for scaling up, whilst generally applicable to all configuration of solar cells, requires specific considerations and approaches based on the device type. For perovskite solar cells, upscaling methods should consider the suitable storage of and engineering of precursors as well as the film deposition process [40]. This is because the precursor solution ratio and storage may have impacts on film quality, influencing both the nucleation and the crystal growth. Some strategies have involved introducing additives as stabilizers or mixing solvents to optimize the precursor ink viscosity as well as boiling point with impacts on the device performance. Also, the composition, stoichiometry, purity and stability play a part in the characteristics of the active layer and must be strategically engineered for optimized device performance. For example, it has been reported that the impurities (H_3PO_2, HI stabilizer) in synthesized MAI are of benefit to a spin-coated $MAPbI_3$ device, whereas these are detrimental to blade-coated ones [41]. In these devices the impurities tend to accumulate at the edges of perovskite grains, leading to a non-continuous film with micron wide gaps in between the grains. In contrast, with purified MAI, thick ($\sim$1 μm) and good coverage $MAPbI_3$ films with large area (2.5 cm × 2.5 cm) were achieved, resulting in 15.1% efficiency of small-area perovskite solar cells.

Other factors that limit the large-scale fabrication of perovskite solar cells include their thermomechanical instability, chemical instability, and rapid material degradation in the presence of oxygen, moisture, light, and heat [42]. In terms of thermomechanical stability, strategies for increasing the toughness of layers have been conducted, including, the use of a top carbon electrode—an inert hydrophobic layer that prevents moisture ingress, and/or the implementation of a scaffold, which provides mechanical support. It has been reported that scaffolding exhibits longer lasting stability, with various material classes from mesoporous metal oxides to natural clays as well as surface passivation, facile crystallization of the perovskite, and decreased defect concentration.

It has been reported that in the scaling up of high-quality large-area perovskite films, the low losses induced by the front contact sheet resistance and the interconnection dead-area and resistance are the main factors in deriving reproducible and reliable high-efficiency modules [43]. Furthermore, the inhomogeneity of layers increases when upscaling with solution processes and material compositions optimized for the small-area cells and defects at grain boundaries and vacancies are implicated. Therefore, the target is to realize optimized scalable processes to minimize losses and to reduce materials waste. To address this, Di Carlo and co-workers [43] employed a strategy to process a planar perovskite solar cell sub-module which was fully fabricated in ambient air by hybrid meniscus coating techniques assisted by air and a green antisolvent quenching method. To suppress non-radiative recombination losses, improve carrier extraction and control the perovskite layer growth on large surfaces, they utilized phenethylammonium iodide (PEAI) passivation and perovskite solvent addition strategies. They reported that the high homogeneous and reproducible layers delivered an efficiency of 16.13% (7% losses with respect to the small-area cell and zero losses with respect to the mini-modules) and a stability of more than 3000 h.

For colloidal quantum dot solar cells, spin coating is an ideal fabrication method that gives ample opportunity to experiment with different parameters for optimized performance, such as surface treatment, film thickness, and doping [1]. This can be considered a testing platform for translating these technical advances to large-area modules though there are some disadvantages of solar cell fabrication by spin coating to achieving this. One of the main challenges to scaling up with methods such as the layer-by-layer spin coating process is the ligand exchange that breaks the long-range ordering of quantum dots. An approach to circumvent this has been to modify the dispersion properties of quantum dot solutions such as particle weight fraction, solvent volatility, and solvent viscosity. Furthermore, approaches to prepare more pristine colloidal quantum dots effectively capped with insulating ligands are important, and their subsequent ligand exchange for shorter conductive ligands such as halides will be important for better dispersity in solvents and ink preparation [44]. With the development of the solution phase ligand exchange, colloidal quantum dot inks enable the direct fabrication of thick thin films with a one-step process. Such advances in ink technology will be critical in paving the way to fabricate high-performance large-area colloidal quantum dot solar cells with more advanced roll-to-roll compatible coating or printing techniques. Another challenge with scaling up colloidal quantum dot solar cells has to do with controlling film thickness and crystallinity. This is often associated with scaling up dip coating since it is difficult to precisely control film thickness when using highly diluted or highly viscous solutions for ultra-thin ($<$20 nm) or ultra-thick ($>$1000 nm) quantum dot films [1]. Another important limitation is material shrinkage during drying and consolidation steps, often leading to cracks in QD films. Additionally, during drying, the wet film formed during dip coating is susceptible to factors such as temperature and relative humidity.

3.2.3 Scaling up by roll-to-roll fabrication

R2R fabrication is an efficient process in which inks are continuously deposited onto flexible substrates, which are injected through the system in a constant, unbroken stream, eliminating the need for manual substrate replacement and thereby differentiating it from discrete, sheet-to-sheet methods [45]. This technique transforms the manufacturing landscape of solar cell production, by significantly lowering costs, achieved through a continuous, efficient process. The R2R method minimizes initial investment costs by eliminating the need for multiple, separate processing steps, thereby reducing equipment and maintenance expenses. Additionally, by promoting the use of low-cost, high-performance materials such as carbon inks in place of expensive evaporated precious metals, R2R directly decreases the costs associated with raw materials. Its scalability automation and reduction in material waste potential further amplify these savings by enabling rapid production expansion without a proportional increase in operational costs.

Recently, a significant development in solution-processed solar cells involving the first demonstration of entirely roll-to-roll fabricated perovskite solar cell modules under ambient room conditions with a record-high 15.5% PCE was reported by Vak and co-workers [46]. This was achieved by developing: (1) a robust and scalable deposition technique; (2) perovskite-friendly carbon inks to replace vacuum-based electrodes; and (3) an R2R-based high-throughput experimental platform, as illustrated in figure 3.10(a). This involved use of a printing-friendly sequential deposition (PFSD) technique comprising R2R-deposited ETL, light-absorbing

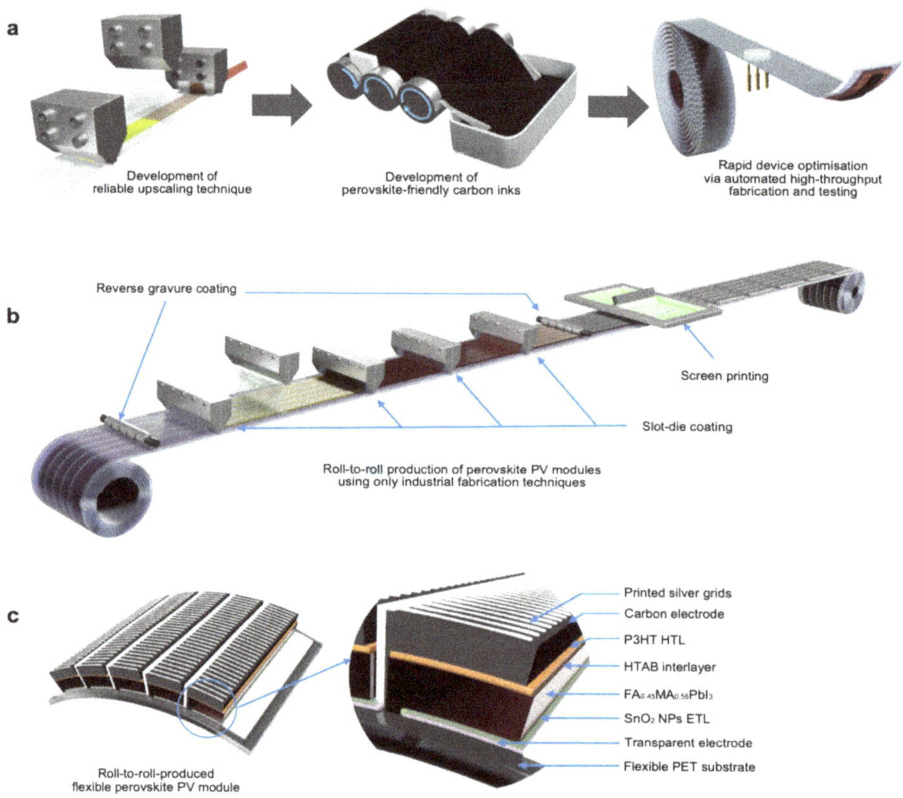

Figure 3.10. Schematic illustration of the roll-to-roll process. (a) A reliable slot die (SD) coating process and a perovskite-friendly carbon ink are developed to enable vacuum-free perovskite PV production. The carbon ink is upscaled using a three-roll mill and used to optimize device parameters by fabricating and testing numerous research cells using an automated roll-to-roll research platform. (b) Schematic illustration of roll-to-roll production of modules using SD coating, reverse gravure (RG) coating and screen printing. (c) The detailed structure of the series connected module, which is fully roll-to-roll fabricated on commercially available transparent electrodes. Reprinted from [46] CC BY 4.0.

layer, and HTL. Additionally, the optimized device fabrication parameters were used to produce large-area modules.

3.3 Conclusion and prospects

Solution-processed solar cells, like perovskite, colloidal quantum dot and organic solar cells, are fabricated by depositing thin films of absorber and transport materials onto a substrate from a solution. This general approach offers low-cost and scalable production, making them promising alternatives to traditional crystalline silicon solar cells. These processes typically involve creating a precursor solution, depositing thin films using methods like spin coating or inkjet printing, and then annealing or post-treating the film to improve its properties. However, in order for these technologies to be realized in competition with traditional devices, these methods must be amenable to scaling up. Scaling up solar cell fabrication involves transitioning from small-scale lab production to large-area, high-throughput manufacturing while maintaining high efficiency and cost-effectiveness. This requires innovative deposition techniques, material selection, and process optimization. R2R fabrication offers a promising pathway for scaling up next-generation solar cell such as perovskites, organic/polymer and colloidal quantum dots and module production, and will be achieved by focussing on continuous coating, cost reduction, and performance enhancement. Also, innovations must consider strategies for improving efficiency, stability, and environmental sustainability in order to transform the renewable energy landscape, facilitating progress towards a sustainable and energy-efficient future.

References

[1] Zhao Q *et al* 2022 Colloidal quantum dot solar cells progressive deposition techniques and future prospects on large-area fabrication *Adv. Mater.* **34** 2107888

[2] Sharma A, Masoumi S, Gedefaw D, O'Shaughnessy S, Baran D and Pakdel A 2022 Flexible solar and thermal energy conversion devices: organic photovoltaics (OPVs), organic thermoelectric generators (OTEGs) and hybrid PV-TEG systems *Appl. Mater. Today* **29** 101614

[3] Na J Y, Kang B, Sin D H, Cho K and Park Y D 2015 Understanding solidification of polythiophene thin films during spin-coating: effects of spin-coating time and processing additives *Sci. Rep.* **5** 13288

[4] Eguchi N, Fukazawa T, Kanda H, Yamamoto K, Miyake T and Murakami T N 2025 Performance optimization of perovskite solar cells with an automated spin coating system and artificial intelligence technologies *EES Solar* **1** 320–30

[5] Zuo C, Scully A D and Gao M 2021 Drop-casting method to screen Ruddlesden–popper Perovskite formulations for use in solar cells *ACS Appl. Mater. Interfaces* **13** 56217–25

[6] Troughton J, Hooper K and Watson T M 2017 Humidity resistant fabrication of $CH_3NH_3PbI_3$ perovskite solar cells and modules *Nano Energy* **39** 60–8

[7] Zuo C and Ding L 2021 Drop-casting to make efficient Perovskite solar cells under high humidity *Angew. Chem. Int. Ed.* **60** 11242–6

[8] Zuo C *et al* 2020 Crystallisation control of drop-cast quasi-2D/3D Perovskite layers for efficient solar cells *Commun. Mater.* **1** 33

[9] Jiao J, Yang C, Wang Z, Yan C and Fang C 2023 Solvent engineering for the formation of high-quality perovskite films:a review *Res. Eng.* **18** 101158

[10] Mejri H, Haidisch A, Krebsbach P, Seiberlich M, Hernandez-Sosa G and Perevedentsev A 2022 Gas-assisted blade-coating of organic semiconductors: molecular assembly, device fabrication and complex thin-film structuring *Nanoscale* **14** 17743–53

[11] Tong S *et al* 2017 Large-area and high-performance $CH_3NH_3PbI_3$ perovskite photodetectors fabricated via doctor blading in ambient condition *Org. Electron.* **49** 347–54

[12] Bishop J E, Smith J A and Lidzey D G 2020 Development of spray-coated perovskite solar cells *ACS Appl. Mater. Interfaces* **12** 48237–45

[13] Arumugam S, Li Y, Senthilarasu S, Torah R, Kanibolotsky A L, Inigo A R, Skabara P J and Beeby S P 2016 Fully spray-coated organic solar cells on woven polyester cotton fabrics for wearable energy harvesting applications *J. Mater. Chem.* A **4** 5561–8

[14] Barrows A T, Pearson A J, Kwak C K, Dunbar A D F, Buckley A R and Lidzey D G 2014 Efficient planar heterojunction mixed-halide perovskite solar cells deposited via spray-deposition *Energy Environ. Sci.* **7** 2944–50

[15] Karunakaran S K, Arumugam G M, Yang W, Ge S, Khan S N, Lin X and Yang G 2019 Recent progress in inkjet-printed solar cells *J. Mater. Chem.* A **7** 13873–902

[16] Lin X, Madhavan V E, Kavalakkatt J, Hinrichs V, Lauermann I, Lux-Steiner M C, Ennaoui A and Klenk R 2018 Inkjet-printed CZTSSe absorbers and influence of sodium on device performance *Sol. Energy Mater. Sol. Cells* **180** 373–80

[17] Wang M, Obene P, Questianx M, Harris M, Singh R and Choy K L 2025 Electrostatic inkjet printed silver grids for non-vacuum processed CIGS solar cells *Sci. Rep.* **15** 11048

[18] Saleem A, Zhang Y, Usman M, Haris M and Li P 2022 Tailored architectures of mesoporous carbon nanostructures: from synthesis to applications *Nano Today* **46** 101607

[19] Workie A B, Ningsih H S and Shih S-J 2023 An comprehensive review on the spray pyrolysis technique: historical context, operational factors, classifications, and product applications *J. Anal. Appl. Pyrolysis* **170** 105915

[20] Shahiduzzaman M *et al* 2021 Spray pyrolyzed TiO_2 embedded multi-layer front contact design for high-efficiency Perovskite solar cells *Nano-Micro Lett.* **13** 36

[21] Nguyen T H, Septina W, Fujikawa S, Jiang F, Harada T and Ikeda S 2015 Cu_2ZnSnS_4 thin film solar cells with 5.8% conversion efficiency obtained by a facile spray pyrolysis technique *RSC Adv.* **5** 77565–71

[22] Hashemi M, Saki Z, Dehghani M, Tajabadi F, Ghorashi S M B and Taghavinia N 2022 Effect of transparent substrate on properties of $CuInSe_2$ thin films prepared by chemical spray pyrolysis *Sci. Rep.* **12** 14715

[23] Di Mari G M, La Matta V, Strano V, Reitano R, Cerruti P, Filippone G, Mirabella S and Bruno E 2024 Optimized chemical bath deposition for low cost, scalable, and environmentally sustainable synthesis of star-like ZnO nanostructures *ACS Omega* **9** 38591–8

[24] Lekiket H and Aida M S 2013 Chemical bath deposition of nanocrystalline ZnS thin films: influence of pH on the reaction solution *Mater. Sci. Semicond. Process.* **16** 1753–8

[25] Najm A S *et al* 2022 An in-depth analysis of nucleation and growth mechanism of CdS thin film synthesized by chemical bath deposition (CBD) technique *Sci. Rep.* **12** 15295

[26] Huse N P, Dive A S, Mahajan S V and Sharma R 2018 Facile, one step synthesis of non-toxic kesterite Cu_2ZnSnS_4 nanoflakes thin film by chemical bath deposition for solar cell application *J. Mater. Sci., Mater. Electron.* **29** 5649–58

[27] Pathan H M and Lokhande C D 2004 Deposition of metal chalcogenide thin films by successive ionic layer adsorption and reaction (SILAR) method *Bull. Mater. Sci.* **27** 85–111

[28] Ristov M, Sinadinovski G and Grozdanov I 1985 Chemical deposition of Cu2O thin films *Thin Solid Films* **123** 63–7

[29] Sun K, Su Z, Yan C, Liu F, Cui H, Jiang L, Shen Y, Hao X and Liu Y 2014 Flexible Cu_2ZnSnS_4 solar cells based on successive ionic layer adsorption and reaction method *RSC Adv.* **4** 17703–8

[30] Al-Katrib M, Perrin L, Flandin L and Planes E 2023 Electrodeposition in perovskite solar cells: a critical review, new insights, and promising paths to future industrial applications *Adv. Mater. Technol.* **8** 2300964

[31] Bhattacharya R N, Oh M-K and Kim Y 2012 CIGS-based solar cells prepared from electrodeposited precursor films *Sol. Energy Mater. Sol. Cells* **98** 198–202

[32] Yelshibay A, Bukari S D, Baptayev B and Balanay M P 2024 Conducting polymers in solar cells: insights, innovations, and challenges *Organics* **5** 640–69

[33] Fomo G, Waryo T, Feleni U, Baker P and Iwuoha E 2019 Electrochemical polymerization *Functional Polymers* ed M A Jafar Mazumder, H Sheardown and A Al-Ahmed (Cham: Springer International Publishing) pp 105–31

[34] Maji B, Barik B and Dash P 2021 Methods for design and fabrication of nanosensors *Nanosensors for Smart Manufacturing* ed S Thomas, T A Nguyen, M Ahmadi, A Farmani and G Yasin (Amsterdam: Elsevier) ch 1 pp 3–18

[35] Gorup L F, Amorin L H, Camargo E R, Sequinel T, Cincotto F H, Biasotto G, Ramesar N and La Porta F d A 2020 Methods for design and fabrication of nanosensors: the case of ZnO-based nanosensor *Nanosensors for Smart Cities* ed B Han, V K Tomer, T A Nguyen, A Farmani and P Kumar Singh (Amsterdam: Elsevier) ch 2 pp 9–30

[36] Veerender P, Koiry S P, Saxena V, Jha P, Chauhan A K, Aswal D K and Gupta S K 2012 Electrochemical polymerization of tetra-(4-hydroxyphenyl) porphyrin for organic solar cells *AIP Conf. Proc.* **1447** 727–8

[37] Del-Oso J A, Frontana-Uribe B A, Maldonado J-L, Rivera M, Tapia-Tapia M and Roa-Morales G 2018 Electrochemical deposition of poly[ethylene-dioxythiophene] (PEDOT) films on ITO electrodes for organic photovoltaic cells: control of morphology, thickness, and electronic properties *J. Solid State Electrochem.* **22** 2025–37

[38] Ro H W *et al* 2016 Morphology changes upon scaling a high-efficiency, solution-processed solar cell *Energy Environ. Sci.* **9** 2835–46

[39] Mazzolini E, Pacalaj R A, Fu Y, Patil B R, Patidar R, Lu X, Watson T M, Durrant J R, Li Z and Gasparini N 2024 Pathways to upscaling highly efficient organic solar cells using green solvents: a study on device photophysics in the transition from Lab-to-Fab *Adv. Sci.* **11** 2402637

[40] Yan J, Savenije T J, Mazzarella L and Isabella O 2022 Progress and challenges on scaling up of perovskite solar cell technology *Sustain. Energy Fuels.* **6** 243–66

[41] Wang Y, Duan C, Lv P, Ku Z, Lu J, Huang F and Cheng Y-B 2021 Printing strategies for scaling-up perovskite solar cells *Natl. Sci. Rev.* **8** nwab075

[42] Casareto M and Rolston N 2024 Designing metal halide perovskite solar modules for thermomechanical reliability *Commun. Mater.* **5** 74

[43] Vesce L, Stefanelli M, Rossi F, Castriotta L A, Basosi R, Parisi M L, Sinicropi A and Di Carlo A 2024 Perovskite solar cell technology scaling-up: eco-efficient and industrially compatible sub-module manufacturing by fully ambient air slot-die/blade meniscus coating *Prog. Photovolt.: Res. Appl.* **32** 115–29

[44] Ma Y-F, Wang Y-M, Wen J, Li A, Li X-L, Leng M, Zhao Y-B and Lu Z-H 2023 Review of roll-to-roll fabrication techniques for colloidal quantum dot solar cells *J. Electron. Sci. Technol.* **21** 100189

[45] Parvazian E and Watson T 2024 The roll-to-roll revolution to tackle the industrial leap for perovskite solar cells *Nat. Commun.* **15** 3983

[46] Weerasinghe H C *et al* 2024 The first demonstration of entirely roll-to-roll fabricated perovskite solar cell modules under ambient room conditions *Nat. Commun.* **15** 1656

Chapter 4

Copper-based chalcogenide solar cells

Thin film solar cells composed of copper-based chalcogenide Cu(In, Ga)(S, $Se)_2$ (CIGS) absorbers have reached efficiencies of over 22.0%, nearly on a par with crystalline Si solar cells. Thin film solar cells are attractive compared to crystalline Si solar cells because of their light weight, ability to be coated over a flexible substrate, reduced material usage and amenability for solution processing. Solution processing of copper-based chalcogenide thin film solar cells has been one of the sought-after fabrication techniques in recent years as it shows promise to be cost-effective for large-scale production. Despite Cu(In, Ga)(S, $Se)_2$ being the front runner in terms of energy conversion efficiency among this class of materials, it is not free of challenges. Indium (In), the primary constituent of Cu(In, Ga)(S, $Se)_2$ absorber is scarce in the earth's crust. This makes Cu(In, Ga)(S, $Se)_2$ solar cells expensive for large-scale adoption and deployment. To address this challenge, isostructural compound Cu_2ZnSnS_4 (CZTS), composed of non-toxic and sustainable elements has been investigated. CZTS crystallizes in the kesterite structure, and it has similar optical properties to CIGS. Nevertheless, the highest power conversion efficiency from CZTS is still only at 12.7%, significantly lagging behind CIGS thin films solar cells. The fabrication of absorber thin films has been one of the detrimental factors in growing high quality thin films and thereby achieving high efficiency. Thus far CIGS and CZTS thin films have been deposited either by traditional vacuum-based approaches such as co-evaporation and sputtering from metallic elements or by solution processing using molecular precursors or nanoparticles. The focus of this chapter is predominantly on various solution processing techniques of CIGS and CZTS thin films solar cells.

4.1 Structure and properties of copper-based chalcogenide materials

$CuInS_2$ (CIS) and Cu_2ZnSnS_4 (CZTS) are derived from the zinc blende structure of ZnS. CIS crystallizes in chalcopyrite-type and CZTS in kesterite-type of tetragonal structure [1]. Ternary compound CIS, consists of I and III group cations and VI

doi:10.1088/978-0-7503-3255-2ch4

© IOP Publishing Ltd 2025. All rights, including for text and data mining (TDM), artificial intelligence (AI) training, and similar technologies, are reserved.

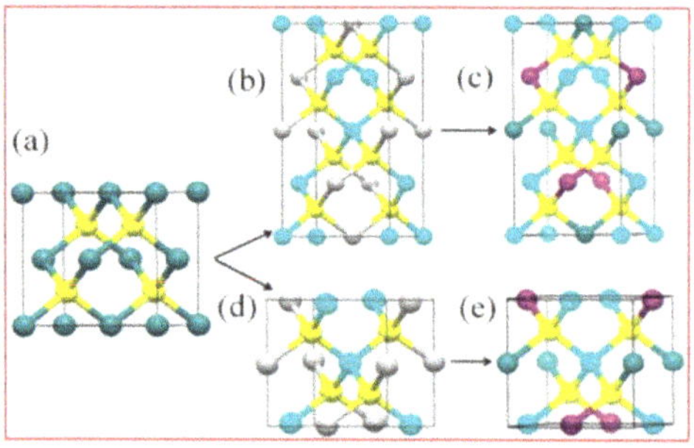

Figure 4.1. Unit cell structures of (a) cubic ZnS, (b) conventional cell for $CuInS_2$ (c) conventional cell for Cu_2ZnSnS_4 (d) cubic cell for $CuInS_2$ (e) cubic cell for Cu_2ZnSnS_4 (Zn-green, In-grey, Cu-blue, Sn-pink, S-yellow).

group anion. This is derived from cross substituting II group Zn(+2) in ZnS with I group Cu(+1) and III group In(+3) [2]. Similarly III group In(+3) in CIS is replaced with II group Zn(+2) and IV group Sn(+4) to form a quaternary compound CZTS (figure 4.1). Unlike CIS, the quaternary semiconductor CZTS can be obtained in two different crystalline forms kesterite and stannite [3]. The kesterite form has the space group I_4, whereas stannite structure is I_4 2m. One of the Cu atoms occupies the 2a position, Zn and the remaining Cu are ordered at 2d and 2c in the kesterite structure. In the stannite structure both Cu atoms occupy the 4d position and the Zn is at 2a. The position of Sn is at 2b in both the structures. As mentioned before, these two structures are viewed conventionally as a 1 × 1 × 2 tetragonal expansion of zinc blende with different ordering in the II–IV cation sublattice. Because there is only a slight difference in total energy (1.3 eV/atom) and similar lattice parameters, both phases almost exist together depending on the synthesis methods. Due to copper vacancies (V_{cu}), these type of copper compounds are *p*-type semiconductors in which holes are the majority carriers. The bandgap of CIS can be tuned between 1.0 and 1.7 eV by introducing Ga in the cation position or Se in anion position in the tetragonal crystalline lattice [4]. Similarly, the bandgap of Cu_2ZnSnS_4 can be tuned from 1.0 to 1.5 eV by substituting sulfur with selenium [5]. The combination of direct bandgap with absorption cross-section over 10^4 makes these materials attractive candidates for solar cell application.

Albeit similar in structural and electronic properties, CIS and CZTS are quite different in their intrinsic defects, thanks to diversely ordered structured compounds such as $CuIn_5S_8$, $CuIn_3S_5$ and $Cu_3In_5S_9$. Because of this, CIS can endure large variation in its stoichiometric composition, which are benign defects [6], whereas CZTS exists in a narrow phase stability window and is often intermixed with isostructural impurities ZnS, Cu_2SnS_3 or secondary impurities SnS, SnS_2 and Cu_xS [7]. The presence of these impurity phases introduces traps and mid-gap states leading to a poor device performance. In addition to V_{Cu}, due to similar ionic radii and low formation energy, Cu_{Zn} antisite defect is prevalent in CZTS. This antisite defect is found to exist at 0.12 eV above the valence band acting as an acceptor level and reducing the open-circuit voltage (V_{oc}) in CZTS-based solar cells [8]. Such a defect is not observed in CIGS solar cells.

Furthermore, grain boundary (GB) defects have a large implication in CZTS solar cells than in CIGS as they pin the Fermi level at the defect states.

4.2 Solar cell architecture

Thin films solar cells generally are a stack of materials on a substrate such as glass. The first one is a back contact metal layer and then a thin film of absorber followed by *n*-type material, transparent conducting oxide (TCO) and top contact electrodes over them complete the device. Figure 4.2 shows a schematic of thin films solar cell architecture.

4.2.1 Back contact metal layer

The back-contact layer is sandwiched between the glass and the absorber layer, and it serves as a hole-collector. Due to very good electrical conductivity (5×10^{-6} Ω cm) and corrosion resistance properties, molybdenum (Mo) has been a metal of choice for this layer [9]. The quality of the metal contact is one of the detrimental factors in solar cell performance. Alternative to Mo, other metals such as Ag, Cu, Al, Ni and metal alloys have been considered, but the device performance using those metal contacts is inferior to that using Mo [10]. The excellent performance of Mo back contacts in chalcogenide thin film solar cells is due to its ability to form a $MoS_2(Se_2)$ intermediate layer between Mo and chalcogenide absorber layer, which is found to be providing a quasi-Ohmic contact. Investigation of solution deposition of Mo back contact is limited. Since Mo sits on an insulating substrate (glass or polymers), only electroless deposition is viable. Thus far, Mo back contact coated by the magnetron sputtering approach is the cheapest considering other solution or vacuum-based deposition methods [11].

Figure 4.2. Schematic of thin film architecture.

4.2.2 *p*-type absorber layer

The next layer in the thin films solar cells stack is *p*-type absorber layer. Chalcogenide compounds CIGS and CZTS are the materials of focus in this chapter. Direct bandgap coupled with excellent absorption coefficient and tunable bandgap are the attractive properties of these materials for solar cell applications. Various solution processing routes such as using molecular precursors or nanoparticles will be discussed in the next section.

4.2.3 *n*-type buffer layer

The *n*-type buffer layer creates a *p–n* junction at the interface and thereby helps to separate photogenerated charges in the solar cells. This buffer layer is thinner (30–100 nm) than the thickness of absorber layer (~μm). A good *n*-type buffer layer should have less lattice mismatch with the absorber layer (CIGS or CZTS) and importantly should possess an optimum band alignment with the absorber layer and transparent conducting oxide (TCO) [12]. Cadmium sulfide (CdS) is generally employed as a buffer layer in CIGS and CZTS thin films solar cells. Cd-free buffers, ZnS and In_2S_3 have also been explored [13].

Chemical bath deposition (CBD) is the most used method for deposition of a CdS buffer layer [14]. CBD deposition process conventionally involves mixing of water solution of cadmium salts such as cadmium acetate or sulfate with thiourea and ammonia at 60 °C–80 °C. The device stack that needs to be coated with CdS is then dipped into the bath for uniform coating of CdS. Due to the wide bandgap and being free from Cd, ZnS is also studied as a buffer layer which resulted in the highest efficiency of 20.9% [15]. The observed improvement in the efficiency is related to low absorption of ZnS in the visible region that led to higher short circuit in the solar cells. CBD processing of ZnS shares a similar method to that for deposition of CdS. Briefly, the process involves using zinc sulfate or zinc acetate as a zinc source and thiourea or thioacetamide as a sulfur source. Another buffer material that has been studied as an alternative for CdS is In_2S_3. Spray pyrolysis method is often employed for the deposition of In_2S_3 buffer layer on the absorber layer using indium(III) chloride and thiourea in acetone. A PCE of ~13.0% is the highest recorded using In_2S_3 buffer layer and CIGS absorber layer [16].

4.2.4 Transparent conducting oxide (TCO)

Transparent conducting oxide (TCO) is generally a wide bandgap semiconductor with a good electrical conductivity and high optical transparency in the visible region to let sunlight through. A good TCO should conduct electricity easily, and this is affected by carrier concentration and mobility. However, high carrier concentration reduces optical transparency that leaves only the mobility, which is a way to get good TCO [17]. High mobility in TCO is in turn achieved by good crystallinity. In addition, TCOs should also have good mechanical properties, thermal and chemical stabilities. Some of the commonly used TCOs include, tin oxide, indium doped tin oxide (ITO), fluorine doped tin oxide (FTO), gallium doped tin oxide (IGO), cadmium tin oxide (CTO), zinc tin oxide (ZTO) and indium doped zinc oxide. Among them ZTO possesses the highest optical transparency [18]. CIGS and CZTS solar cells generally utilize sputter-coated zinc oxide or ITO TCOs in their structure.

4.3 Solution processing of Cu-based chalcogenide materials

Solution processing of solar cells has been developed in recent years to bring down manufacturing cost, energy use, material utilization and to improve the manufacturing throughput. There have been several methods investigated to deposit thin films of copper-based chalcogenides. Generally, the methods follow wet coating of precursors on a substrate and heat treatment in chalcogen/inert atmosphere to obtain the absorber layer.

4.3.1 Solution processing of CIGS absorbers

Solution processing of CIGS is broadly defined by either directly growing on the substrate or by liquid deposition of molecular solutions or nanoparticle ink. CBD and electrodeposition (ED) are the most commonly explored methods for direct growth of CIGS absorber layers. In CBD, the nucleation of absorber layer onto the substrate is induced by supersaturation of precursor solutions, which in turn is achieved by tuning precursor concentration, solution temperature, pH and growth duration. Likewise, in the electrodeposition method, metal salts are electro-reduced onto the substrates. Liquid deposition of CIGS absorbers is detailed in the following sections.

4.3.1.1 Hydrazine based precursor approach for CIGS

Motivated by the absence of non-volatile impurities such as carbon or oxygen in hydrazine, Mitzi's group at IBM first explored the use of hydrazine as a dissolution agent to form a stable precursor solution of SnS_2 and $SnSe_2$ [19]. This initial exploration has led to the creation of stable hydrazine solutions of Cu_2S and In_2Se to synthesize CISe ink, which eventually achieved over 18% efficient CIGS solar cells [20]. The strong ionic nature of hydrazine facilitates the breaking of the metal-chalcogen framework to form stable metal chalcogenide anions in the presence of excess of chalcogens. Raman spectroscopy was employed to observe the formation of $(N_2H_5)_2S$ molecules in hydrazine by locating a distinct peak at 2560 cm^{-1} corresponding to S–H bond vibration. By mixing Cu_2S powder into the above sulfur solution, an additional Raman peak at 335 cm^{-1} for Cu–S stretching was observed. Further Raman spectroscopy measurements showed that in stoichiometrically mixed indium and copper precursor solutions $[In_2Se_4]^{2-}$, ions were found to attach to $[Cu_6S_4]^{2-}$ ions via the formation of In–S bonding to form CISe precursor inks [21].

To prepare CIS precursor solution, first Cu_2S powder was mixed with excess of sulfur in hydrazine. Due to limited solubility of Cu_2Se in hydrazine, Cu_2S is typically used as a starting material. In the next step, In_2Se_3 powder was added to an appropriate amount of sulfur and selenium in hydrazine. Mixing the above solutions for several days, copper sulfide produced a pale yellow solution and indium selenide produced a clear viscous solution.

Further, the observation of both Cu–S and S–H peaks in Raman spectra helped determine the composition of soluble copper chalcogenide in the solution. For example, the intensity of the S–H peak increased linearly on increasing S/Cu_2S ratio, while the intensity of the Cu–S peak was unchanged. This suggested that some amount of elemental sulfur was consumed to form a solvated copper sulfide complex and was no longer capable of producing a S–H vibrational mode. Excess of sulfur was found to be in the form of $(N_2H_5)_2S$ and was contributing to the S–H peak. This Raman observation was helpful for determining the formation of $[Cu_6S_4]^{2-}$ and to propose the following chemical reaction in hydrazine.

$$6Cu_2S + 2S + 5N_2H_4 \rightarrow 2[Cu_6S_4]^{2-} + 4N_2H^{+}{}_5 + N_2(g)$$

To crystallize the molecule in hydrazine, Cu_2S/S hydrazine solution was evaporated at room temperature by flowing nitrogen gas. The isolated crystalline structure was found to be composed of Cu_7S_4 sheets separated by a mixture of hydrazine and hydrazinium ($N_2H_5^+$) molecules (figure 4.3) [22].

Raman analysis of In_2Se_3 precursor solutions prepared by dissolving In_2Se_3 and elemental selenium in hydrazine showed peaks around 240 and 260 cm^{-1} corresponding to symmetric Se–Se stretching modes. Further, the selenium solution lost its dark green colour upon addition of In_2Se_3 suggesting formation of indium selenide complexes, which was observed by a Raman peak at 192 cm^{-1} for the In–Se bonding vibrational mode. The dominant indium–selenium phase in hydrazine is observed to be $[In_2Se_4]^{2-}$ with a Se/In_2Se_3 ratio of 1, which is formed through the following chemical reaction [23].

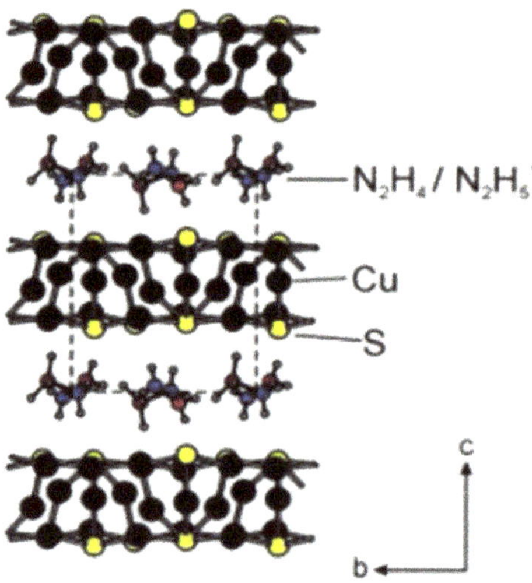

Figure 4.3. Crystal structure of $N_4H_9Cu_7S_4$ complex viewed along with (100) direction, showing the 2D Cu_7S_4 anionic slabs separated by N_2H_4/$N_2H_5^+$ spacers. Adapted with permission from [22]. Copyright 2007 American Chemical Society.

$$2In_2Se_3 + 2Se + 5N_2H_4 \rightarrow 2[In_2Se_4]^{2-} + 4N_2H_5^+ + N_2(g)$$

Evaporation of the above solution yielded an amorphous compound with proposed composition of $(N_2H_4)_2(N_2H_5)_2In_2Se_4$.

A new Raman vibrational peak around 315 cm^{-1} for In–S was observed from the solution containing $[Cu_6S_4]^{2-}$ and $[In_2Se_4]^{2+}$ suggesting the possible anion exchange between these two species. Observation of the In–S peak further emphasizes the close interaction of copper, indium, sulfur and selenium at a molecular level even before deposition. An amorphous compound was obtained with a possible composition of $(N_2H_4)0.7(N_2H_5)_3(In_2Cu_2Se_4S_3)$ on evaporation of In_2Se_3/Se and Cu_2S/S solutions. The amorphous compound containing copper, indium, selenium and sulfur was found to be responsible for the formation of the chalcopyrite phase below 200 °C.

The deposition of absorber layer starts with preparation of individual precursor solutions and mixing them together at different ratios to achieve variable stoichiometry. This enabled composition control including the Cu/In and Se/S ratios in the final solution. A multilayer solar cell stack with different gallium concentration in each layer was deposited from hydrazine solution by a layer-by-layer approach. This stack exhibited solar energy conversion efficiency of 10% with 605 mV V_{oc}, 25.7 mA short-circuit current and 66% *FF*. This was the first reported CIGS device made using hydrazine solution. Building on this success, by removing Ga in the solar stack and introducing bandgap grading in CISSe by tuning sulfur to selenium ratio, a solar cell device with a conversion efficiency of 12.2% was reported [24]. This device exhibited 73% fill factor, 550 mV open-circuit voltage, 29.8 mA short-circuit current. The uniform distribution of sulfur and selenium in the grains helped to improve the efficiency. Further, by increasing the sulfur content, the bandgap of the device was increased, that reflected in the open-circuit voltage value increasing from ~0.43 V for 5% sulfur to ~0.55 V for 15% sulfur content in the material. However, the increase in sulfur content decreased the short-circuit current value from ~37 to ~30 mA cm^{-2}, which was observed in the EQE measurements showing the shift in photon cutoff edge towards shorter wavelength. Later, to improve the efficiency by this hydrazine process, Sb was introduced into the absorber layer up to 1 mol% [25]. Sb_2S_3/S solution in hydrazine was used for doping of Sb. The introduction of Sb into the absorber layer did not change the phase or the phase purity. Powder XRD data showed no shift in the peak position, however, there was a noticeable reduction in full width at half maximum of (112) and (220)/(204) reflections by increasing the Sb content, indicating the larger grain growth. Although no changes in optical properties of the samples were observed on introducing Sb into the absorber layer, the solar cell device performance parameters had not improved significantly. Open-circuit voltage increased from 0.55 V to 0.57 V, short-circuit current value increased from 27.7 to 30.3 mA cm^{-2}, fill factor value expanded to 71.2% from 67.6% and the energy conversion efficiency improved to 12.3% upon adding Sb. Further, through optimizing Sb doping, the grain size exceeding 1 μm was obtained [26]. This increased grain size helped push the efficiency over 13% in the Sb-doped CIGS solar cells. The improved device exhibited 0.667 V

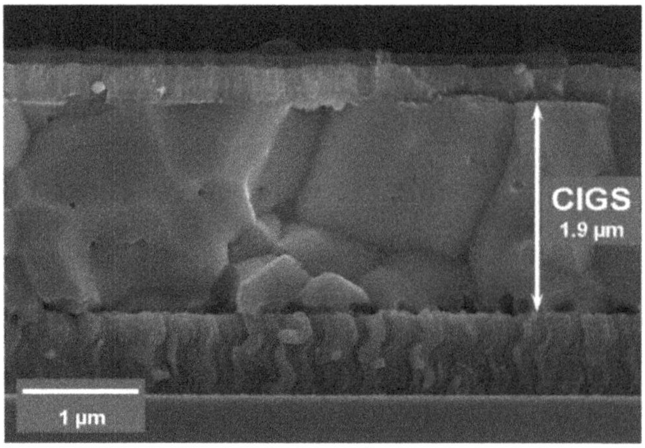

Figure 4.4. Cross-sectional scanning electron microscope image of hydrazine-processed CIGS device. Adapted with permission from [27]. John Wiley & Sons. Copyright 2013 John Wiley & Sons, Ltd.

open-circuit voltage, 27.53 mA cm^{-2} short-circuit current, 74.1% fill factor and 13.6% power conversion efficiency.

Another improvement in hydrazine-processed CIGS solar cell performance was reported with the efficiency exceeding 15% [27]. To prepare the solar cell device, In_2Se_3 with Se and Cu_2S with S were mixed in hydrazine and the mixture was spin coated at 800 rpm and annealed at 540 °C. Initially, 120 nm thick films using 0.3 M CIGS solution containing slightly Ga rich content were used and then the absorber thickness was increased up to 1.9 μm using 1.1 M CIGS solution with four layers of coating (figure 4.4). This device was doped with 0.5 at% Sb in the absorber layer. The solar cell stack was completed with 60 nm thick CBD coated CdS, 80 nm thick sputtered ZnO and 130 nm thick indium doped tin oxide and MgF_2 antireflection coating. Bandgap value measured using external quantum efficiency (EQE) and photoluminescence (PL) showed ~1.16 eV, closer to the established empirical optimum bandgap of 1.14 eV. This recorded performance of the device was found to be 0.623 V of open-circuit voltage, 32.6 mA cm^{-2} short-circuit current, 75% of fill factor and 15.2% power conversion efficiency.

A record efficiency solar cell device was reported using the hydrazine-based solution deposition method with efficiency over 18% [20]. This was achieved by V-shaped Ga composition grading profile with higher Ga content in the back and front of the absorber layer. It was explained that the higher Ga content towards the Mo back contact will drive the electrons away from the Mo back contact and suppress the recombination, and Ga in the front will enhance the separation and collection of photogenerated charge carriers. Lower Ga content in the middle layer helps obtain an optimum bandgap to obtain maximum solar absorption. Further, introducing solution-processed intrinsic ZnO layer instead of sputter-coated ZnO layer showed improved efficiency. The solution coating of ZnO layer showed a smoother interface between CIGS/CdS *p–n* junction and ITO electrode that helped to improve shunt resistance. The device fabricated using solution-processed ZnO layer showed 15.6% efficiency in comparison to 13.1% efficiency from a sputter-coated ZnO device. The biggest improvement was obtained from the device that was fabricated with Ga composition grading. An efficiency of 18.1% was reported from the compositional grated device in comparison to 15.6% from homogeneous composition device. Further, to emphasize the reproducibility of the devices, over 90 devices were fabricated. An average certified efficiency of 17.1% was recorded from these devices. Other performance parameters for the devices have been reported as 35.54 mA cm^{-2} short-circuit current, 0.66 V open-circuit voltage, 77.2% fill factor. Notably, incident photon-to-current conversion efficiency of up to 92% was obtained between wavelengths, 350 and 900 nm.

4.3.1.2 Non-hydrazine-based precursor approach: amines for CIGS

Due to the toxicity of hydrazine, there have been parallel efforts to find alternative solvents to fabricate solar cells through the solution-processing route. One of many solvents/precursors are amine-based ones. It was proposed that the ethanol solution of Cu(I) and In(III) acetates with diethanolamine or triethanolamine could be a potential precursor solution [28]. The idea behind this approach is that the strong coordinating ability of tertiary amines would reduce Cu(II) to Cu(I) in the solution and to Cu(0) during the annealing step. Subsequent sulfurization of amorphous Cu and In oxides would deposit CIS. Although there was no solar cell device reported by this approach, it paved the way for further investigation. A solar cell device with 7.7% efficiency was reported, in which the device was prepared using monoethanolamine (MEA) as a chelating ligand along with ethanol [29]. An interesting approach was taken for the solution processing of CISSe solar cells [30], where hydrazinium complexes of copper and indium were dissolved in a mixture of ethanolamine (EA) and dimethyl sulfoxide (DMSO). This method does not completely avoid the use of hydrazine, but reduces the amount significantly, yet enabling the solution-processing fabrication. Interestingly, hydrazinium complexes of Cu and In are not soluble either in EA or in DMSO as an individual solvent, but produce a processable solution in their mixture. In addition to the EA/DMSO mixture, EA and thiourea, acetylacetone and DMSO mixtures also could dissolve hydrazinium salts of Cu and In. The devices fabricated by this route showed solar energy conversion efficiency between 2.9 and 3.8%. Formate complexes of copper, indium and gallium were dissolved in tetramethylguanidine and methanol and 4-amino-1,2,4-triazole as an additive in a glove box at 110 °C for 2 min. The process yielded a viscous solution which was used for doctor-blading on Mo-coated soda lime glass [31]. The device was fabricated following the conventional solar stack structure. The cells exhibited 11% and 10.7% solar energy conversion efficiencies from 2- and 6-minutes annealed samples. Longer annealed sample, due to deeper incorporation of Ga, showed a higher open-circuit voltage (533 mV), compared to the two-minutes annealed sample (483 mV). However, this reduced the short-circuit current for the longer annealed sample (29.0 mA cm^{-2}). In the follow-up work, the group introduced sodium into the absorber structure by mixing NaCl or NaSCN or $NaHCO_2$ into the tetramethylguanidine/methanol inks [32]. The solar cell device fabricated using NaCl along with longer duration of selenization showed 13.3% efficiency, 532 mV open-circuit voltage, 365 mA cm^{-2} short-circuit current with 68.5% fill factor. However, the device fabricated using sodium formate showed poor performance due to loss of gallium in the structure. This was due to oxidation of gallium by formate ions during the processing. One of the features of solution-processed solar cells using molecular ink is deposition of residual carbon. This has been believed to be affecting the device performance negatively.

A recent investigation showed that the carbon residual layer is electrically benign to the device performance [33]. However, the carbon residual layer was found to be detrimental for elemental distribution of the final film, in which carbon selectively hinders the diffusion of Cu during selenization, resulting in a copper deficient top layer while improving the film morphology (figure 4.5). Devices were deposited using varying levels of residual carbon, namely low, intermediate and high carbon. Current–voltage (*I–V*) curves of the device with intermediate level of carbon showed reasonable rectifying behaviour, whereas devices with low and high residual carbon exhibited large leakage currents and complete short-circuit characteristics, respectively. The high carbon devices, due to lack of copper diffusion to the top layer showed linear *I–V* characteristics with no photovoltaic behaviour, suggesting only In–Se composition and lack of *p–n* junction. However, in the case of low-carbon content devices, very rough and porous films with imperfect coverage of CdS and/or i-ZnO on the CISe film yielded significant shunting. Low-carbon devices exhibited

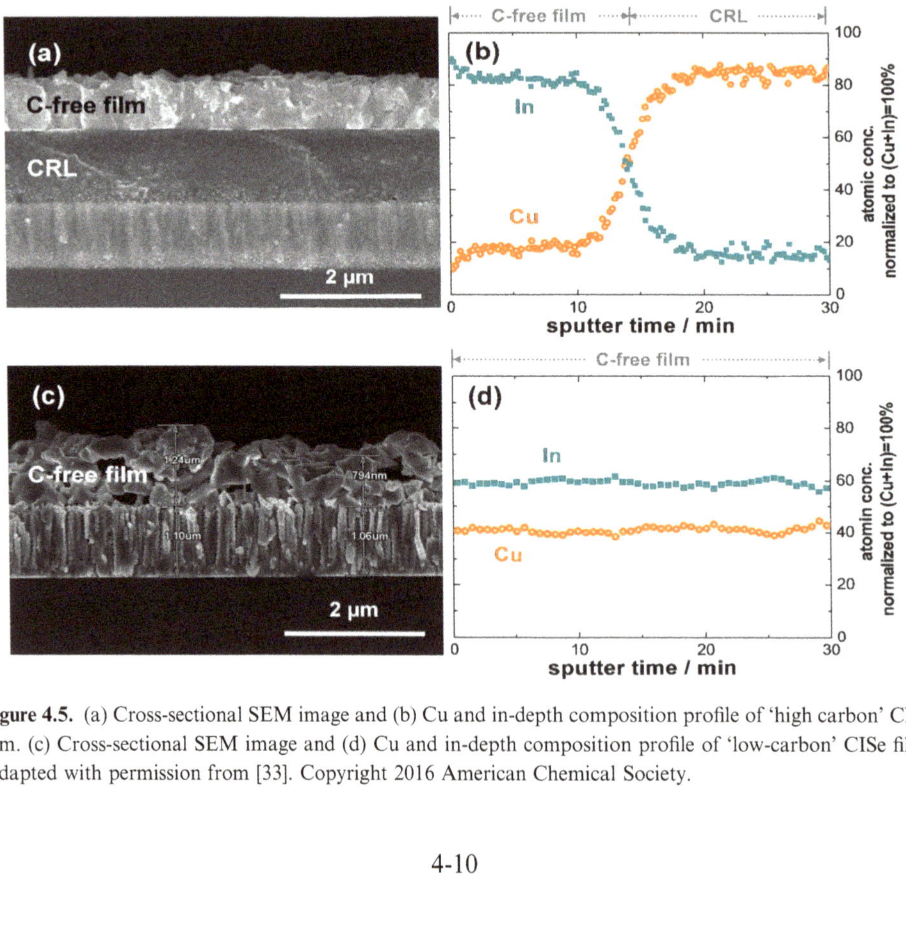

Figure 4.5. (a) Cross-sectional SEM image and (b) Cu and in-depth composition profile of 'high carbon' CISe film. (c) Cross-sectional SEM image and (d) Cu and in-depth composition profile of 'low-carbon' CISe film. Adapted with permission from [33]. Copyright 2016 American Chemical Society.

an average efficiency of 1.5% ± 0.4, but intermediate-carbon content devices showed solar energy conversion efficiency up to 9.15%.

Another approach was taken for solution state growth of CIGS [34], in this case using dithiocarbamate solutions of copper, indium and gallium. To prepare the precursor mixture, 1-butylamine was reacted with carbon disulfide in ethanol and then, $In(OH)_3$ and $Ga(acac)_3$ were added to the mixture and heated to 60 °C to dissolve all solid. In the next step, Cu_2O was added to the above solution. The resultant homogenous solution was evaporated under vacuum at 80 °C to remove ethanol from the mixture. The viscous mixture was further diluted using ethanol to obtain 0.4 M solution. For fabrication of solar cells, the solution was spin coated on the Mo cover slip and sintered at 350 °C. A 1.2 μm thin film was obtained by repeating the spin coating process seven times and annealing at 400 °C. Further, selenization was carried out at 540 °C for 1 h. The complete solar cell stack was fabricated by CBD growth of CdS, RF sputtering of ZnO and DC sputtering of indium tin oxide. The photovoltaic performance of the device was found to be 8.92% efficiency, with 0.609 V open-circuit voltage, 20.86 mA cm^{-2} short-circuit current and 68.9% fill factor. The external quantum efficiency of the device was measured to be ~80% in the visible region. Further, expanding this approach and by exploring other short carbon chain amine compounds, authors have managed to obtain up to 10.1% efficiency, 0.56 V open-circuit voltage, 27.64 mA cm^{-2} short-circuit current and a fill factor of 65% [35].

Continuing the progress using amine-based compounds for dissolution of metal precursors for solution processing of solar cells, another versatile approach was explored in which a mixture of amine and thiols was used [36]. This amine–thiol combination is unique because it enables a wide range of volatilities suitable for different chalcogenide materials, by tuning the ratio of amine to thiol, and control over the solubility of chalcogenides can be achieved. Importantly, a variety of sources such as metals, metal oxide, metal salts, metal chalcogenides can be dissolved. Nevertheless, the second highest efficiency device with energy conversion efficiency up to 16.4% was reported by this approach [37]. The amine–thiol dissolution method is motivated by the original work that was reported in 1963, in which dissolution of sulfur using amine–thiol mixture was demonstrated [38]. The proposed mechanism was that the amine deprotonates the thiol and then the deprotonated thiolates react with elemental sulfur to form a polysulfide. However, the same dissolution method was not investigated for dissolving selenium or tellurium, until recently. For selenium, amine was found to be forming polyselenide in the presence of thiols. This was demonstrated in a mixture containing oleylamine and dodecanethiol [39]. A mixture of 1,2-ethylenediamine and 1,2-ethanedithiol dissolved up to 38 wt% selenium and 9.3 wt% tellurium [14]. This dissolution approach was further extended to dissolve Cu_2S/Se, $In_2/Se/S_3$, SnSe, CdS or ZnSe in a mixture of 1,2-ethylenediamine and 1,2-ethanedithiol [40, 41].

The first solar cell device fabricated with this amine–thiol approach was using thioacetic acid and ammonia [42]. Ammonia deprotonates thioacetic acid and deprotonated thioacetate coordinates with metal salts to form a molecular ink. For preparation of CIS thin films, copper(I) oxide and indium(III) hydroxide were added to a mixture of isopropanol, ethylene glycol, propylene glycol, then thioacetic acid was introduced to the mixture, after which the colour mixture turned black. Further, on addition of ammonia the solution turned back to the orange-red clear precursor solution. This solution was spin coated onto a fluorine doped tin oxide substrate and then annealed at 300 °C to evaporate the organics. Further, to form CISSe thin films, sulfurization and selenization were carried out at 500 °C using H_2S and Se powders. Three different devices with varying Cu/In ratio were fabricated. Device with Cu/In ratio of 0.8 showed energy conversion efficiency of 6.75%, 0.44 V open-circuit voltage, 30.39 mA cm^{-2} short-circuit current and 50.23% fill factor.

One of the highest efficiency solar cell devices was fabricated by dissolving elements of Cu, In, Ga and Se in a mixture of 1,2-ethylenediamine and 1,2-ethanedithiol at 60 °C [43]. Notably, the ink prepared by this approach stayed clear and homogeneous under ambient conditions, which enabled spin coating at room temperature. The molecular ink was spin coated onto the Mo substrate and annealed at 350 °C between each layer of spin coating nine times to form 1 μm thick films. As-spin coated devices appeared with shallow cracks and voids on the surface. The cracks were effectively reduced by selenization at 550 °C, while maintaining the elemental composition. The power conversion efficiency of the best performing device was measured to be 9.5% at 1.5 AM illumination condition. The same device showed 26.64 mA cm^{-2} short-circuit current and 528 mV open-circuit voltage with 67.5% fill factor. Continuing this 1,2-ethylenediamine and 1,2-ethanedithiol dissolution approach, Cu_2S, In_2S_3, elemental Ga and Se were dissolved in the solvent mixture and the resultant molecular ink was spray coated, annealed and selenized to fabricate a device. The best device performed at 8.0% power conversion efficiency [44]. It was emphasized that the spray coating of molecular inks was better than spin coating as it avoids redissolution of absorber layer and thereby retains the elemental composition, particularly, Cu.

Replacing bidentate 1,2-ethylenediamine with monodentate hexylamine or 1,2-ethanedithiol with propanethiol in the amine–thiol mixture, metal salts such as $Cu(OAc)_2$, $In(OAc)_3$ or $Cu(acac)_2$, $In(acac)_3$, or CuCl, $InCl_3$, $GaCl_3$, and GaI_3 were readily dissolved [45]. However, dissolution of binary chalcogenides, Cu_2Se or In_2Se_3 needed a mixture of 1,2-ethanedithiol and hexylamine. An ultrathin layer (~560 nm) of CIGS absorber was obtained from eight rounds of spin coating the molecular ink composed of Cu_2Se, $In(OAc)_3$, $Ga(acac)_3$ and Se in hexylamine and 1,2-ethanedithiol. A solar cell device fabricated with this absorber layer exhibited 10.3% PCE with 59% *FF* at 0.54 V V_{oc} and 32.7 mA cm^{-2} I_{sc}. Because the amine–thiol mixture provides very good solubility, spin coating the precursor mixture usually leaves copper deficient thinner films. To address this, doctor-blading of the solution was attempted to grow a thicker film. The doctor-bladed layer was found to be 1.1 μm thick with ~400 nm of a fine-grain layer. On optimizing the chemical composition and grain growth, a better performing device was fabricated that exhibited 12.2% efficiency with an improved fill factor of 65.4%.

Chemical composition is one of the key factors determining the efficiency of the devices. Double gradient concentration of Ga in the absorber layer was fabricated using the amine–thiol solution [46]. Four different types of devices were constructed by controlling the composition in the molecular solution. The efficiency of the devices varied from 11.4% for homogeneous composition to 13.12% for a double Ga graded device. Although the presence of $MoSe_2$ at the bottom contact was found to be important for improving the efficiency of the device, a thick $MoSe_2$ affected the adhesion of absorber layer to the Mo back contact [47]. Introducing Mo–N in the middle of the Mo layer helped to improve solution-processed device efficiency. The presence of Mo–N layer in the Mo back contact acted as a diffusion barrier for selenium and thereby restricted the conversion of conductive Mo to insulating $MoSe_2$. However, the Mo–N layer did not affect the beneficial sodium migration from soda-lime glass to the absorber layer. Because the selenium diffusion is restricted, there was sufficient selenium to react with Cu to promote grain growth and achieve higher external quantum efficiency at longer wavelength. The Mo–N film was grown by the magnetron sputtering method introducing nitrogen with argon. To demonstrate the effect of $MoSe_2$ thickness effect, four different devices were fabricated with and without the Mo–N diffusion layer. Duration of selenization was varied for growth of $MoSe_2$ thickness. A device selenized for 90 min without Mo–N barrier layer showed a poor PCE of 4.23%, whereas the device with Mo–N barrier layer showed a best performance of 12.05% efficiency, supporting the argument that the thicker $MoSe_2$ affects cell performance negatively. Notably, the best performing device had a low series resistance (0.967 Ω cm^{-2}) and high shunt resistance (331 Ω cm^{-2}) compared to the poor performance device.

The amine–thiol mixture is effective for dissolving many metals, metal salts and metal chalcogenides. Although it has been exploited for fabrication of solar cells by dissolving precursors, it is equally corrosive for processing apparatus. To reduce the excess use of this mixture and to understand the molecular chemistry, Cu nanoparticles, In and Se were dissolved in a mixture of hexylamine and 1,2-ethanedithiol separately [48]. After obtaining complete dissolution, the solvent was evaporated in vacuum at room temperature first and then at raised temperature. The resultant residue was a golden yellow paste from copper, white powders from indium and a dark red viscous gel from selenium containing solutions. The residues were analyzed using mass spectroscopy, nuclear magnetic resonance and x-ray absorption spectroscopy and they revealed the presence of thiolate complexes with ammonium counter ions. Interestingly, the residues were soluble in dimethyl sulfoxide, dimethylformamide, 2-methoxyethanol, acetonitrile, etc. Utilizing the solubility, a solar cell device was fabricated by spin coating the molecular mixtures in dimethyl sulfoxide. The CISSe device exhibited 9.7% efficiency for 0.44 cm^2 active area. Extending this approach, introducing Ga, replacing hexylamine with propylamine and using acetonitrile instead of dimethyl sulfoxide, a better performing device with 12.9% efficiency was reported [49]. This improvement was reported to be due to the reduced carbon content in the absorber layer.

Another interesting device structure was investigated by the amine–thiol dissolution method [50]. Conventionally, in a *n*-type semiconductor, CdS layer was deposited on top of a selenized CIGS layer to complete the *p–n* junction. In the modified structure, a layer of thioacetamide was deposited to form a sulfide on top of selenide. The introduction of the sulfide layer passivated the interfacial defect between CIGSe/CdS layers, which improved the solar conversion efficiency significantly. The device performance without thioacetamide layer showed 13.14% efficiency, but 15.15% was obtained with a layer of thioacetamide. Another technique for suppressing the interfacial defect was to introduce ordered vacancy sites (OVCs) in the absorber structure [37]. OVCs were introduced into the structure in the absorber fabrication step. The Cu/(In + Ga) ratio was systematically tuned through the precursor ratio in the ink solution. The first eight layers were coated by using Cu/(In + Ga) in a ratio of 0.9 and then two or three layers were coated with inks containing Cu(In + Ga) ratio from 0.8 to 0.6. These films were selenized in the rapid thermal processing furnace at 520 °C for 60 s. The device with OVC content $(A_{OVC}/A_{CIGS}) = 0.18$ showed the highest energy conversion efficiency of 16.39% with 73.83% fill factor, 0.65 V open-circuit voltage and 33.94 mA cm^{-2} short-circuit current. Performance of the devices with lower or higher OVC content was progressively reduced. Formation of hole barrier at CIGS/CdS interface due to the downward valence band maximum suppressed the electron–hole recombination at the interface and this helped achieve the highest efficiency in a non-hydrazine solution-processed device.

In the hydrazine-processed solar cells, it was shown that the introduction of doping elements improved the crystal growth and that benefitted improvement in cell efficiency. This was again tested in the non-hydrazine process by mixing antimony acetate (2 mol%) in the ethylenediamine/1,2-ethanedithiol mixture. The CIGS device with antimony doping exhibited fine grains in the bottom part of the device with reasonably bigger crystals on the top. It performed at 10.3% efficiency [51]. Similarly, a CISSe device with antimony doping showed 1.2% improvement from the undoped CISSe device [52]. Sodium from the soda-lime glass was found to be migrating to the absorber layer promoting larger grain growth. In an attempt to introduce sodium from the top, a piece of soda-lime cover glass was used on top of a CIGSe layer during the post-deposition treatment step [53]. Sodium migration from both bottom and top side of the CIGSe layer assisted to get larger crystals than the absorbers with a quartz glass on the bottom and top sides of the layer. In addition to grain growth, sodium occupies the copper site by forming shallow defect Na_{Cu}, hence increasing the hole carrier concentration. The sodium-doped device exhibited 11.53% efficiency with 69.8% fill factor, 28.36 mA cm^{-2} short-circuit current and 0.58 V open-circuit voltage.

4.3.1.3 Non-hydrazine-based precursor approach: thiourea for CIGS

Thiourea binds to copper strongly through the sulfur, making it readily soluble in dimethyl sulfoxide (DMSO). Interestingly, when indium chloride is mixed in the solution containing thiourea, copper chloride and DMSO, indium binds more to DMSO than thiourea. X-ray single-crystal structure determination identified that the copper–thiourea complex containing three molecules of thiourea is attached to copper and with a chloride atom to complete the coordination [54]. Likewise, indium is coordinated to three molecules of DMSO and three chloride atoms. These coordination complexes provided a clear solution while maintaining +1 and +3 oxidation states for copper and indium. Spin coating of the solution followed by annealing deposited CIS and then selenization at 550 °C converted CIS to a CISSe absorber layer without any observable secondary binary phases. The CISSe devices deposited using thiourea solution in DMSO performed at 13.0% efficiency with 68.3% fill factor, 37.4 mA cm^{-2} short-circuit current and notable open-circuit voltage of 507 mV. Introducing gallium into the device structure, the efficiency improved to 14.7% and fill factor to 71.5%, and open-circuit voltage to 660 mV, however, short-circuit current was reduced to 31.2 mA cm^{-2}. The bandgap of CISSe and CIGSSe absorber layers is measured to be 1.0 and 1.15 eV [53]. Continuing with the DMSO solution approach, but using rather undisclosed metal chalcogenide complexes, 2 μm thick absorber with larger grains were deposited by the spin coating process [55]. The device with this layer showed one of the highest efficiencies deposited by the solution state approach with an efficiency of 17% for undoped and 18.68% for doped (doping element is not provided) with remarkable open-circuit voltage of 633 and 660 mV for undoped and doped devices. It was reported that replacing DMSO with dimethylformamide (DMF)-isopropanol (IPA) mixture the solubility of CuCl, $InCl_3$ and thiourea concentration could increase to 1.6 M from 0.45 M in DMSO [56]. This allowed for a micron thick film using only four rounds of spin coating instead of eight layers. In addition, the presence of isopropanol reduced the surface tension which helped to spread the precursor solution over 5 × 5 cm^2. The CISSe device exhibited 1.36 eV bandgap with 3.4% power conversion efficiency. However, a solar cell device with an improved efficiency up to 10.25% was reported by using only DMF and selenizing at higher pressure [57]. The bandgaps of the absorber layer selenized at 0.04 and 0.06 MPa measured were 1.01 and 0.98 eV indicating selenium-rich composition at higher pressure. Building on these results, high pressure selenization was further continued on the CISSe devices fabricated with varying [Cu]/[In] ratio from 0.85 to 1.20 [58]. By dissolving thiourea, CuCl and $InCl_3$ in DMF the precursor solution was prepared. The solution was spin coated on Mo-soda lime glass at 4000 rpm for 60 s and then immediately annealed at 340 °C for 1 min. This spin coating and annealing steps were repeated 10 times to reach a desired thickness. Selenization was carried out at 1.0 and 1.6 atm argon pressure at 580 °C. Photoconversion efficiency of the devices increased from 9.79% for a 0.89 [Cu]/[In] ratio sample to 11.25% for a 1.02 [Cu]/[In] ratio sample that were selenized at 1.0 atm pressure. On further increasing the [Cu]/[In] ratio, the efficiency dropped significantly. The same trend was observed for the devices fabricated at 1.6 atm selenization pressure with highest efficiency of 13.29%. However, the champion performance was obtained from a device that was fabricated by substituting Ga for In. The device fabricated with a 1.05 ratio of [Cu]/[In] and 108 nm of ARC coating exhibited 15.2% efficiency, 71.5% fill factor, 0.604 V open-circuit voltage and 35.2 mA cm^{-2} short-circuit current.

4.3.1.4 Non-hydrazine-based precursor approach: water and alcohol for CIGS

Water and alcohol-based solvents are the best solvents in terms of toxicity, ease of handling, environmental friendliness and cost. Of course, water was used for dissolving metals salts of copper and indium for spray deposition of $CuInS_2$ films using thiourea as the sulfur source [59]. Films were deposited at different temperatures from 200 °C to 600 °C at every 50 °C. The films deposited from 250 °C to 450 °C were of good crystallinity, but films deposited over 450 °C were poor with very little deposit. Later, it was identified that methyl-substituted thiourea was more soluble in water along with copper chloride and indium chloride than nascent thiourea. This approach was extended for depositing $CuInSe_2$, $CuGaS_2$, $CuGaSe_2$ and their alloys by using dimethyl thiourea and dimethyl selenourea as sulfur and selenium sources. A solar cell device fabricated by spray coating a water solution of copper, indium and thiourea solution onto ITO substrate exhibited a low conversion efficiency of 2.0%, which was mainly due to high resistivity and less compatible work function of ITO [60]. When replacing ITO with the commonly investigated Mo, oxidation of Mo was observed at high temperature. Nevertheless, this issue was addressed by gradual heating and spraying from low temperatures [61]. Further, excess of sulfur in the precursor solution promoted the growth of MoS_2 instead of MoO_2. The precursor solution was prepared by dissolving copper chloride, indium chloride and thiourea in deionized (DI) water. Films were further selenized using selenium vapour, produced from heating of selenium pellets at 480 °C–500 °C. The device fabricated by this method showed 5.9% efficiency, 55% fill factor, 418 mV open-circuit voltage and 25.48 mA cm^{-2} short-circuit current. Introducing gallium into the water solution by dissolving gallium chloride formed CIGSSe devices with the improved efficiency of 10.5% [62]. Sodium doping was shown to improve grain growth [63]. In this approach, sodium was introduced to the structure by using a sodium nitrate water solution. Use of a nitrogen carrier gas helped avoid MoO_2 formation. Excess of thiourea prevented the immediate precipitation of the Cu thiourea complex. However, it did not provide a stable solution beyond 5 min. This solution stability was further increased to 30 min by replacing thiourea with 1-methylthiourea [64]. It was proposed that 1-methylthiourea avoided metal nitrate thiourea complex precipitation. Because 1-methylthiourea provided a stable solution, it prevented the use of excess sulfur and that helped reduce the carbon impurities in the films. The solar cell device fabricated with this metal precursor solution yielded an 8.7% efficient device with 54% fill factor, 31.0 mA cm^{-2} short-circuit current and 520 mV open-circuit voltage. Hydrolysis of $InCl_3$ is found to be another issue that occurs in the water-based fabrication approach. This was addressed by preforming a complex of $Cu(thiourea)_3Cl$ and $In(thiourea)_3Cl_3$ and then dissolving them in water [65]. The water solution was spin coated onto Mo-coated substrate and then annealed at 270 °C to form a precursors film and then selenized at 595 °C for 20 min. Powder x-ray diffraction, Raman spectroscopy and SEM characterization of the films showed highly crystalline, smooth films with densely packed grains. Device fabricated using these films showed 12.3% PCE, 70.5% fill factor, 501 mV open-circuit voltage and 1.0 eV bandgap. Further, CIGSSe cell was fabricated using water solution of metal chloride, nitrate salts with thiourea for the sulfur source [66]. The ratio of sulfur to selenium was varied during the selenization and sulfurization processes that was carried out in nitrogen atmosphere at 550 °C. The sulfo-selenium alloyed device exhibited 10.9% efficiency with 62.0% fill factor, 27.4 mA cm^{-2} and 640 mV open-circuit voltage.

Like water, alcohol is also a protic polar solvent known for dissolving metals salts. Use of alcohol solution of metal salts for solution preparation of CIS, CIGS, CIGSSe cells has been investigated in recent years due to its low-toxicity, low-cost and ease of processing. Devices up to 14.4% efficiency have been fabricated using this alcohol-based solution-processing approach. Despite the promising solubility offered by alcohol, it is not free of challenges. Since the alcohol evaporates at low temperatures, it most often needs a polymer binder material to obtain good rheology. Because of that, the deposited films were mostly contaminated with carbon and oxygen impurities inhibiting a good grain growth and causing poor adhesion. For instance, an inkjet-printed device using the alcohol solution containing copper acetate, indium acetate, gallium chloride, ethylene glycol and ethanolamine showed the presence of carbon impurities [67]. Nevertheless, the device exhibited 5.0% efficiency with 29.78 mA cm^{-2} short-circuit current, 386 mV open-circuit voltage and only 44% fill factor. Another report was published showing device efficiencies with up to 8.0% [68]. Salts of copper acetate, indium acetate and gallium acetylacetonate were added to the solvent mixture containing propylene glycol and ethanol. This alcohol-containing metal salt solution was spin coated and annealed at 350 °C in air repeatedly to get the desired thickness. Selenization was carried out to convert metal films to metal chalcogenides using selenium vapour with 20% hydrogen and 80% nitrogen at 500 °C–530 °C. It was noted that the formation of CuSe and $Cu_{2-x}Se$ occurred at low temperature and conversion to $CuInSe_2$ and $CuInGaSe_2$ at high temperature selenization. The device fabricated with low selenization pressure showed 6.1% efficiency and high selenization pressure device performed at 8.01% efficiency. Notably, the high selenium pressure helped improve the fill factor from 49% to 59% with a slight improvement of open-circuit voltage from 473 to 525 mV. To reduce the carbon impurities from the solvent-binder decomposition, multistep annealing was explored [69]. Films coated from the nitrate salts solution in methanol with polyvinyl alcohol were air annealed at 300 °C and then sulfurized using 1% of H_2S in nitrogen at 500 °C followed by selenization using Se in argon at 420 °C. This process limited the carbon impurities to ~3%, analyzed by atom probe microanalyzer. CIGS cells with 1.57 eV bandgap exhibited 8.28% efficiency and CIGSeS device with 1.12 eV showed 8.81% efficiency. Further optimizing the three-step heat treatment process by annealing in nitrogen atmosphere, and then selenization followed by sulfurization improved the device efficiency to 12.7% with a CdS *n*-type layer and to 14.4% with a (Zn,Cd)S layer. The sulfurization step on top of selenized metals yielded sulfur rich films at the CIGS/CdS (or) (Zn,Cd)S interface and Ga rich film towards the bottom of the stack [70]. A schematic of a three-step heat treatment process is given in figure 4.6. This graded structure was found to be beneficial in aligning the band structure for better charge collection. Further, to understand the reason for improved efficiency in CIGS/(Zn,Cd)S structure compared to CIGS/CdS structure, external quantum efficiency (EQE) measurement was carried out. The measurement showed higher EQE between 300 and 550 nm, responsible for 0.79 mA cm^{-2} short-circuit current improvement. Further, significant improvements in EQE in the wavelength region of 820–1170 nm were also noted, which helped harness another 1.6 mA cm^{-2} short-circuit current. Other notable performance parameters of these devices are 70.9%, 71.0% fill factor, 32.61 and 34.73 mA cm^{-2} current densities and 0.549, 0.584 V open-circuit voltages for CIGS/CdS and CIGS/(Zn,Cd)S devices, respectively. Nitrate salts of copper, indium and gallium were dissolved in isopropanol and

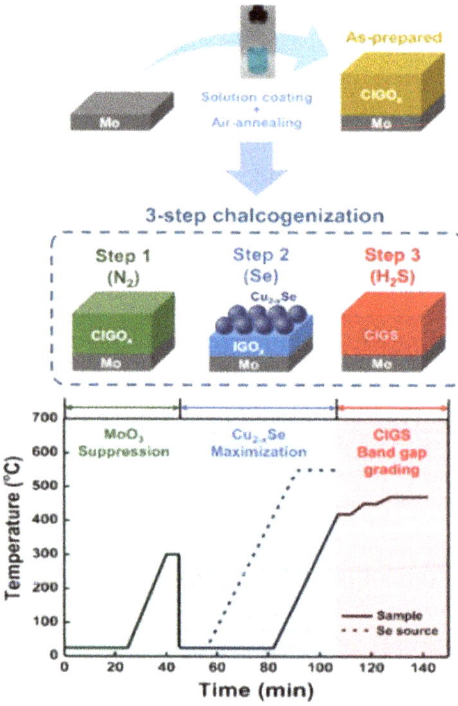

Figure 4.6. Schematic of three-step heat treatment process in different atmosphere. Adapted from [70]. Copyright 2018 American Chemical Society.

ethylene glycol to form a molecular ink [71]. The ink was jet printed onto the Mo-coated substrates prior to annealing and selenization and sulfurization to form Cu(In,Ga)(S,Se)$_2$. The cell was found to contain large grains at the surface and fine grains towards the bottom and was gallium rich near the back contact. The cell exhibited 11.3% efficiency with 67% fill factor, 31.1 mA cm^{-2} and 0.541 V open-circuit voltage.

Sodium doping is known for increasing the grain growth in CIGS cells. This was explored in the alcohol-based approach as well. Sodium acetate was mixed with copper acetate, indium acetate and gallium nitrate in 2-methoxyethanol and propylene glycol at 80 °C [72]. Further, varying amounts of sodium citrate from 0.002 to 0.02 mmol were added to the precursor solution to incorporate sodium into the CIGS absorber. The solution was spin coated onto Mo-coated glass and then dried at 300 °C, and repeated 12 times to obtain film thickness up to 1.0 μm. The device structure was completed with *n*-type CdS, i-ZnO, Al:ZnO and Al after selenization at 570 °C. A sodium-doped device showed ~8.2% efficiency with an improvement of 10.2% from the undoped CIGS device with a notable increase in open-circuit voltage and fill factor on increasing sodium doping. Continuing the doping approach, the effect of potassium doping was investigated using potassium fluoride in ethanol solution containing copper nitrate, indium nitrate and gallium nitrate salts [73]. It was observed that the incorporation of potassium up to 1.0 mol%

reduced the presence of $Cu_{2-x}Se$ compound while improving the grain size and smoothness. The cell with 1.0 mol% potassium doping showed a remarkable performance with efficiency of 11%, whereas the device without potassium showed only 4.70% efficiency. All other performances, such as open-circuit voltage, short-circuit current and fill factor values were significantly higher for the potassium-doped device. Further, increasing the potassium content from 1.0 to 5.0 mol% reduced the grain size and induced carrier recombination, leading to poor performance.

4.3.2 Nanocrystals ink-based approach for CIGS

Nanoparticles of absorber materials with particle size less than 10 nm are amenable for solution processing. These nanoparticles are either composed of binary constituents such as Cu_2S, Cu_2Se, In_2S_3 or In_2Se_3 or fully formed ternaries and quaternaries. These can be synthesized with well controlled size and shape by colloidal hot-injection, heat-up, solvothermal, hydrothermal, melt atomization, and spray pyrolysis methods. It has been shown that the commodity chemicals of metal salts such as metal-halides, oxides, nitrates, acetates, acetylacetonates, hydroxides and preformed metal chalcogenide complexes can be used for the synthesis. Typical synthesis of nanoparticles of absorber materials involves either heating up metal salts and chalcogen sources to over 250 °C or keeping metal salts and chalcogen sources separately until the target temperature is reached and swiftly injecting the chalcogen source to induce the nucleation. Because of the un-terminated dangling bonds, nanoparticles have high surface energy, which makes them agglomerated. To prevent the agglomeration, these are capped with non-polar, long-chain capping agents such as oleylamine, oleic acid, and 1-dodecanthiol ligands. These ligands have a polar functional head on one side to coordinate to the metal/chalcogen sites and non-polar long carbon chains on the other to keep coordinated particles apart. These ligands are always introduced during the synthesis and enable size, shape and growth control, and provide for the colloidal stability. Moreover, these ligands provide solubility in non-polar solvents such as toluene, hexane, chloroform, etc. For fabrication of solar cells, these nanoparticle solutions can be spray coated, drop-casted, inkjet-printed or doctor-bladed. Due to their high solubility, the concentration of nanoparticles in the solution can be significantly higher than the previously mentioned molecular ink solutions. This high solubility reduces the number of coating cycles required for achieving the desired absorber thickness for solar energy conversion. In addition, because of their reduced size, these nanoparticles show high absorption cross-section than their bulk. Although nanoparticles offer many benefits for solution processing of solar cells, they come with challenges. One notorious challenge with nanoparticles is the air-borne toxicity for skin and inhalation. Other challenges include difficulties in scale-up while maintaining size and shape control. The synthesis requires high temperature using often high hazard precursors and solvents. Purification of nanoparticles from their excess capping agents is often necessary. At the lab scale, dispersion and reprecipitation is the go-to approach for removing excess of capping agents and unreacted precursors from the nanoparticle solution. However, this purification method uses centrifuges which limits scalability and provides poor purification control at large scale. Because of these challenges, while processing nanoparticles for solar cells, impurities are left behind in the films. These impurities hinder photogenerated charge carriers reaching the collection sites. Efforts to remove capped ligands during the post-deposition steps by annealing at high temperatures leave voids and pinholes in films due to the loss of long-chain molecules and carbon impurities. These cracks and pinholes act as trap sites for charge carriers.

There have been several efforts to address the shortcomings of nanoparticles approach for solar cells including ligand exchange, thermal treatment, and chemical treatments. To improve the film's density, lattice expansion was tried by selenization of sulfide nanoparticles, where smaller sulfur atoms are replaced by larger selenide atoms. Increasing the flow viscosity of solutions of monodentate ligands involves replacement with chelating organic ligands. Ligand exchange has been investigated to replace the bulky long-chain, non-polar ligands with short-chain smaller ligands or metal chalcogenide complexes. Further, a hybrid approach where portions of nanoparticles are replaced with traditional molecular inks has also been investigated. Nevertheless, 17.1% efficient solar cell has been reported using a nanoparticles solution-processing approach [74]. Various methods of synthesis, processing, ligand exchange methods are covered in detail in different reports [75–78]. This chapter will cover fabrication and performance of devices using nanoparticles in detail.

One of the earliest reports of using $CuInSe_2$ nanocrystals for solar cells involved making them by the colloidal synthesis method [79]. The device exhibited 3.2% efficiency, and it was fabricated using hexagonal nanoring $CuInSe_2$. The nanorings were synthesized by heating up CuCl, $InCl_3$ and Se in oleylamine at 265 °C for 1 h. This process yielded chalcopyrite structure $CuInSe_2$ nanorings. By slightly modifying the process and injecting selenium dissolved in oleylamine at 285 °C, sphalerite $CuInSe_2$ nanocrystals were synthesized. Interestingly, replacing selenium in oleylamine with tri-octylphosphine (TOP) selenium and by partly replacing oleylamine with TOP, hexagonal nanorings were obtained (figure 4.7). For device fabrication, nanoparticles were drop casted on top of Mo substrates and then annealed in Se atmosphere at 500 °C to remove organic ligands and to sinter the films. There was no noticeable change in structure, morphology or composition. The remaining layers of the solar cell stack were completed by the conventional approach. Following this work, utilizing densification due to lattice expansion when sulfur was replaced with selenium in $CuInS_2$ nanocrystals, a slightly improved device was fabricated (4.76% efficiency) [80]. $CuInS_2$ nanocrystals were synthesized using a similar approach detailed for $CuInSe_2$ nanocrystals [78], but using sulfur in oleylamine. This approach was further extended for synthesizing $CuGaS_2$ and $CuIn_xGa_{1-x}S_2$ nanocrystals. A significant improvement in device efficiency was reported when the bulky oleylamine ligand was replaced with a short-chain hexanethiol with efficiency of up to 7.7%. Further, by doping sodium into the absorber layer, the efficiency jumped to 12.0% [81]. In this method, $CuInGaS_2$ nanocrystals were synthesized using acetylacetonate

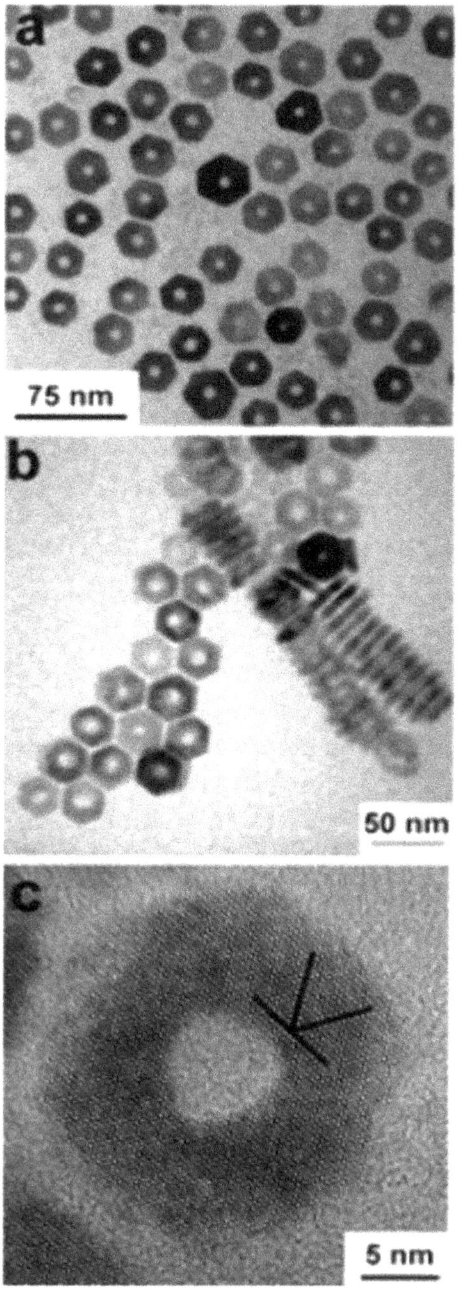

Figure 4.7. Transmission electron microscope images of CIS nanorings. Adapted with permission from [79]. Copyright 2008 American Chemical Society.

complexes of copper, indium and gallium. The nanocrystals were nucleated by injecting sulfur in oleylamine at 285 °C and kept at that temperature for 30 min. Isolated particles were purified using hexane and isopropanol, and the dried sample was redispersed in hexanethiol for knife coating onto Mo substrate to form thin films. Prior to selenization, the film was soaked in NaCl solution for sodium incorporation and then annealed in selenium vapour at 500 °C for 20 min. The sodium-doped device exhibited 12.0% efficiency with 65.7% fill factor, 28.8 mA cm^{-2} short-circuit current and 630 mV open-circuit voltage, whereas the undoped device showed 7.7% efficiency, 48.9% fill factor, 26.8 mA cm^{-2} short-circuit current and 590 mV open-circuit voltage. The observed improvement in NaCl treated device was due to dense packing of large grains that helped charge transport.

Expanding this ligand exchange approach, a CISe device was constructed onto a gold-coated substrate using CISe nanocrystals capped with metal chalcogenide complexes (MCC) [82]. As synthesized, oleylamine capped CISe nanocrystals were ligand exchanged with a hydrazinium complex of $Cu_2In_2S_3Se_4$. The MCC capped nanocrystals were soluble in hydrazine, and the solution was spin coated to form a thin film. The film was not further annealed or selenized to improve the grain size or densification. Although the ligand exchange with MCC ligands for solar cells was shown to be effective for solubility, these rather displayed a poor power conversion efficiency. However, a decent efficiency of 7.9% was reported for a device that was fabricated using $CuInS_2$ nanocrystals ligand exchanged with ethanolamine and selenized to improve the film density [83]. Another effort was reported using MCC ligand exchanged $CuInSe_2$ nanocrystals with a very good performance of up to 10.85% efficiency [84]. Nanoparticles were synthesized by a low-temperature method, in which CuI and InI_3 in pyridine were reacted with Na_2S in methanol at 0 °C for a minute in nitrogen atmosphere. Nanoparticles with and without MCC were investigated using 2-methoxyethanol and monoethanolamine solvents. For preparation of MCC ligands, Cu(II) acetate and In(III) acetate were dissolved in 2-methoxyethanol and monoethanolamine. The nanoparticle solution was spin coated, heat treated and selenized for the completion of solar cells. The device without the MCC ligand showed 7.75% efficiency, whereas with the MCC ligand the efficiency improved to 10.85% with a significant increase in the short-circuit current as a direct result of better film quality using the MCC ligands. Exchanging a nanoparticle's native ligand with short-chain ligands often leaves poor surface termination which affects their solubility, limiting their solution processability. A hybrid ligand exchange method was explored in which portions of native oleylamine was exchanged with pyridine on $CuInGaS_2$ nanoparticles [85]. The exchanged nanoparticles were dissolved in dimethyl sulfoxide for blade coating onto Mo-coated substrates and the film was selenized at 500 °C–550 °C before completing other layers. Hybrid ligand exchanged nanoparticles device showed 12.0% efficiency with 29.7 mA cm^{-2} short-circuit current compared to 9.2% efficiency with 26.0 mA cm^{-2} from the device fabricated using unchanged nanoparticles. One of the highest efficiency devices was reported using CIGS nanoparticles [86]. Although the synthesis and processing methods were not changed much, care over controlling their exposure to oxygen and moisture and composition variation helped achieve 15.0% solar energy conversion efficiency. The nanoparticles were dissolved in hexanethiol for doctor-blade coating and then annealed at 350 °C before selenization at 550 °C and MgF_2 was coated on top of the complete stack. The device showed impressive 15.0% efficiency with a remarkable fill factor value of 73.4%, 32.1 mA cm^{-2} short-circuit current and 630 mV open-circuit voltage.

4.3.3 Solution processing of CZTS absorbers

Solution processing of CZTS involves wet coating of molecular precursors onto a substrate and thermal treatment to form a thin absorber layer. Success in CIGS coating by solution processing has been extended for CZTS considering the structure and composition commonality. The hydrazine dissolution approach has been one of the most successful solution-processing approaches in terms of obtaining high-efficiency devices. Nevertheless, there have been other solvent systems investigated to address toxicity and flammability issues with hydrazine. Further, like CIGS, nanoparticles of CZTS have also been explored for coating of high-performing solar cells. Approaches to solution processing of CZTS devices using various dissolution methods are detailed in the following sections.

4.3.3.1 Hydrazine approach for CZTS

Hydrazine is found to be a good solvent for dissolving many different metal salts and metal chalcogenides including CZTS constituents such as $SnSe_{2-x}S_x$, Cu_2S [87, 88]. Combination of volatility and being free of carbon, oxygen and halide in hydrazine is attractive for solar cells coating. Thus far, the highest efficiency CZTS solar cells fabricated using hydrazine molecular precursors stands at 12.6%. Dissolution of tin chalcogenide involves mixing of tin sulfide or tin selenide in hydrazine along with excess of sulfur. It has been found that the dissolution is occurring mainly through the following reaction:

$$5N_2H_4 + 2X + 2SnX_2 \rightarrow N_2 + 4N_2H_5^+ + Sn_2X_6^{4-} \ (X = S, Se)$$

The single-crystal structure of hydrazine complex of tin chalcogenide obtained by evaporating the solution under nitrogen flow was found to be $(N_2H_5)_4Sn_2S_6$ or $(N_2H_4)_3(N_2H_5)_4Sn_2Se_6$—two analogous structures comprising dimers of edge sharing SnX_4 tetrahedra ($Sn_2X_6^{4-}$) alternating with hydrazinium cations [87]. Cu_2S dissolution pathway in the hydrazine is outlined in the CIGS section. A CZTS device using hydrazine molecular precursors was reported in 2010 by spin coating a hydrazine solution of copper, tin and zinc chalcogenides [89]. The solution was prepared by individually dissolving Cu_2S–S and SnSe–Se in hydrazine. $ZnSe(N_2H_4)$ was obtained *in situ* by reacting zinc powder with a tin chalcogenide hydrazine solution. The ratio of the Sn/Zn and Cu solution was adjusted to close to nominal composition $Cu_{2-x}Zn_{1-y}Sn(S,Se)_4$. The solution was spin coated five times at 800 rpm and annealed at 540 °C. To adjust the sulfur content, the film was heat treated in sulfur vapour. Powder x-ray diffraction analysis of the coated film showed the presence of phase pure kesterite without secondary impurities such as Cu_2S or

Sn_2S_3. However, elemental analysis showed copper poor and zinc rich composition composed of large grains with some isolated voids. The solar cell device exhibited a remarkable 9.66% efficiency from the mixed sulfo-seleno device with 516 mV open-circuit voltage, and 28.6 mA cm^{-2} short-circuit current and 65% fill factor. Likewise, the device made with only sulfur composition exhibited 9.3% efficiency.

Continuing this initial work, a slightly improved device with efficiency of over 10% was reported [90]. The molecular precursors were prepared by the hydrazine dissolution approach that was reported previously. The device performed at 10.1% efficiency with 63.7% fill factor, 517 mV open-circuit voltage, 30.8 mA cm^{-2} short-circuit current. Also, the bandgap value of sulfo-seleno layer was measured to be 1.15 eV close to the value measured for the CIGS layer (1.14 eV). Although CZTS and CIGS devices exhibited very close bandgap values, other performance parameters such as open-circuit voltage, short-circuit current and fill factor values were significantly lower for the CZTS device. To understand the reason for lower performance, time resolved photoluminescence (TR-PL) measurement was carried out on the device, which showed a short minority carrier lifetime of 3.1 ns. Further, EQE measurement revealed low collection efficiency in the long wavelength range. Also, high series resistance was observed suggesting blocking of back contact as the reason for lower fill factor. One of the advantages with the ternary and quaternary chalcogenide materials is their compositional flexibility. In CIGS solar cells, bandgap tuning by adjusting the composition of sulfur to selenium or indium to gallium has been demonstrated. However, in CZTS solar cells, the presence of the fourth element provides additional flexibility. Tuning the bandgap of CZTS by substituting germanium for tin was explored [91]. The Ge-substituted device was fabricated by following the previous reports [89, 90], but using $GeSe_2$ as a Ge source. EQE measurement of the device showed 1.15 eV for a 40% Ge-substituted device and 1.06 eV for pure CZTS. The performance parameters of the corresponding devices exhibited a marginal improvement in efficiency for a Ge-substituted device (9.14%) in comparison to the unsubstituted CZTS device (9.07 eV) with a significant reduction in short-circuit current (31.8 mA cm^{-2} for Ge Vs 35.4 mA cm^{-2} for CZTS). Although the bandgap of the Ge-substituted device was higher than the unsubstituted CZTS device, open-circuit voltage was found to be $\sim$50 mV higher; still it was considerably lower than the expected open-circuit voltage of 657 mV for this composition.

Further optimizing the device structure and by growing larger grains, CZTSSe devices with efficiencies over 10% have been fabricated [92]. The devices showed a significant improvement in fill factor and short-circuit current relative to previous generation devices, but without notable improvement in open-circuit voltage. Despite hydrazine dissolving all major metal chalcogenides for solar cells, dissolution of zinc sulfide or zinc selenide in hydrazine was difficult. This was addressed by converting hydrazine into hydrazinocarboxylic acid ($NH_2NHCOOH$) and directly dissolving zinc powder [93]. Hydrazinocarboxylic acid was prepared by reacting hydrazine and carbon dioxide. This exothermic reaction was highly corrosive which helped to dissolve zinc powder by stirring at room temperature. This zinc solution enabled control of the stoichiometry of

Zn/Sn, Cu/(Zn + Sn), and S/Se independently. A solar cell device fabricated by spin coating the hydrazine solution of Cu, Zn, Sn onto Mo-coated substrate showed 8.08% efficiency with 61% fill factor, 409 mV open-circuit voltage and 32.25 mA cm^{-2} short-circuit current.

Although, there has been significant improvement in the performance of hydrazine-processed CZTS devices, they still lag in theoretical efficiency limit. In particular, open-circuit voltage and short-circuit current values are only 56% and 79% of the Shockley–Queisser (SQ) limit of 820 mV and 43.4 mA cm^{-2}. Short-circuit current showed improvement when the CdS buffer layer and transparent conducting oxide (TCO) layer's thickness reduced to 50 nm. This thickness reduction made photon penetration to CZTS layer more effective, thus obtaining 12% efficiency and 34.8 mA cm^{-2} short-circuit current [94]. Nevertheless, the open-circuit voltage value has not seen much improvement. By adopting a zinc-hydrazine solution method reported earlier, which helped get better uniformity and film structure, in addition to optimizing the TCO and CdS layers thickness to maximize light transmission, efficiency of the device was improved to 12.6% with an open-circuit voltage of 513.4 mV (figure 4.8) [95]. Addressing the interfacial defect was one of the primary reasons for improving the open-circuit voltage. Continuing the effort to improve the open-circuit voltage, a few more studies have been reported such as, sulfurization to improve surface composition [96] and separately, using In_2S_3/CdS buffer layer instead of simple CdS. This double buffer layer approach pushed to achieve the highest short-circuit current (37.1 mA cm^{-2}) along with an improvement in fill factor (70.3%) value, but open-circuit value was still stagnant below 500 mV [97]. This hydrazine solution approach has been successful in terms of growing impurity free large grains of absorber layer of CIGS nearly well as CZTS. However, the toxicity and flammability of hydrazine are a hindrance for scaling. Thus, various solution state approaches devoid of hydrazine have been investigated for CZTS solar cells and those will be detailed in the following sections.

Figure 4.8. Top and cross-sectional SEM images of CZTS films on Mo-coated substrate. Adapted with permission from [95]. John Wiley & Sons. Copyright 2014 WILEY-VCH Verlag GmbH & Co. KGaA, Weinheim.

4.3.3.2 Non-hydrazine-based precursor approach: amine–thiols for CZTS

A mixture of 1,2-ethanedithiol (EDT) and 1,2-ethylenediamine (EN) was reported as a good solvent system for dissolving metal chalcogenides, metal chlorides and metal oxides. The effectiveness of the EDT-EN mixture was demonstrated by dissolving Cu_2Se, In_2S_3, In_2Se_3, CdS, and SnSe and also using the ink to form Cu_2Se and Cu_2S films [36, 41]. The mixture is relatively less toxic and composed of low boiling-point liquids. Several CIGS devices with highest efficiency of up to 16.4% have been reported [37]. It is a natural progression to utilize this solvent mixture for preparation of CZTS molecular inks and for fabrication of solar cell devices. First, CZTS devices utilizing an amine–thiol solvent mixture for dissolution of molecular constituents was reported with device efficiency up to 7.86% [98]. However, the device was fabricated by dissolving CuCl, $ZnCl_2$, $SnCl_2$ into a mixture of hexylamine and propanethiol, and separately dissolving sulfur and selenium in hexylamine and propanethiol. These solutions were mixed together before spin coating onto Mo-coated substrate and selenization at 500 °C. Four different devices were fabricated by varying Cu-to-Sn ratio from 1.45:1 to 1.74:1. Devices with 1.45:1 and 1.53:1 ratio of Cu and Sn showed better open-circuit voltage and short-circuit current along with higher fill factor and efficiency values (7.24% and 7.86%, respectively). Interestingly, the device fabricated using a slightly higher Cu-to-Sn ratio (1.74:), performed poorly with efficiency of only 2.47%.

This amine–thiol dissolution approach was further tested for dissolving metal oxides for fabrication of a CZTS device, but by introducing an additional solvent, 2-methoxyethanol [99]. A mixture of 1,2-ethanedithiol, ethanolamine and 2-methoxyethanol was mixed with Cu_2O, ZnO, and SnO at room temperature. The oxide solution was spin coated and selenized in selenium vapour at 540 °C for 10 min using rapid thermal processing. Notably, the thiol solvent used for dissolution, acted as a sulfur source prior to the selenization step. The CZTSSe device exhibited 7.34% power conversion efficiency, 0.436 V open-circuit voltage and 34.03 mA cm^{-2} short-circuit current with significantly lower fill factor of 49.5%. Another dissolution solvent system was explored in which a mixture of ethanolamine and cysteamine was used for dissolving metals of Cu, Zn, and Sn [100]. Mass spectroscopy (MS) and infrared multiphoton dissociation (IRMPD) spectroscopy were employed for identifying the complex formation during dissolution. It was found that copper was forming monomeric Cu(II) $(cysteamine)_2$ initially with *m/z* corresponding to 214, but *m/z* values of 481 and 622 were slowly rising with the continuing dissolution process, indicating formation of a polymeric species. Likewise, a tin complex formation through sulfur atom mirroring Cu was observed. Zinc was initially forming a complex with a mixture of cysteamine and ethanolamine, but eventually converted to a homoleptic cysteamine complex. The solution containing all three metals was spin coated and air annealed and selenized. Electron microscopy images of films before and after selenization showed the occurrence of grain growth during the selenization process with noticeable voids at the back contact. The device performed at 8.1% efficiency with 30.8 mA cm^{-2} short-circuit current, 62% fill factor and still lower open-circuit voltage of 0.42 V. Addition of dopant generally improves grain growth in solution-processed solar cells; and this was explored by adding a small amount of Ag^+ into the CZTSSe lattice using the amine–thiol dissolution method [101]. Precursor solution was prepared by dissolving Cu, Sn, Zn, S and Se in 1,2-ethylenediamine and 1,2-ethanedithiol at 60 °C for 12 h. Varying levels of Ag ranging from 0, 1, 3 and 5% were added to explore the effect of silver doping in the solar cells performance. The bandgap of silver-incorporated films narrowed with increased silver content, from 1.125 eV for the undoped film to 1.037 eV for 5% Ag doped film. This decrease in bandgap value also reflected in the device performance. A systematic increase in efficiency from 7.39 to 10.36% and open-circuit voltage from 398 to 448 mV and 30.40 to 35.19 mA cm^{-2} short-circuit current was obtained by increasing the silver content from 0% to 3%. However, further increasing the silver content to 5% resulted in a decrease in the device parameters.

As discussed earlier, CZTS offers compositional tunability, thereby offering better band matching with the other layers of the solar cell stack. A dual gradient absorber structure was constructed by substituting Ag for Cu and Ge for Sn [102]. These substitutions down-shifted the valence band maximum at the front contact and upshifted the conduction band minimum at the back contact. The solar cell structure was formed by first coating Ge-substituted CZTSSe on the Mo-coated substrate and then unsubstituted CZTSSe layer and finally an Ag substituted layer on top. This three layer absorber stack was selenized together to form a dual gradient structure. The precursor preparation for solution deposition was quite involved. First, to prepare the Ge solution, ethanolamine, 2-methoxyethanol and mercaptoacetic acid were mixed and then different amounts of GeO_2 were added to the solution and stirred at 70 °C for 24 h. In parallel, Cu, Zn, Sn salts, and S and Se powders were added with ethylenediamine and 1,2-ethanedithiol and heated to 70 °C for 1.5 h until a clear solution was obtained. Different amounts of Ge solution were mixed with the previous solution to get Ge/(Ge + Sn) ratios of 0%, 20%, 40% and 60%. Dissolution of Ag_2O was difficult using ethylenediamine and 1,2-ethanedithiol mixture, therefore, 1-butylamine and carbon disulfide were used. A precursor solution for Ag_2ZnSnS_4 was prepared by dissolving Ag_2O, ZnO and SnO in 1-butylamine, carbon disulfide, mercaptoacetic acid. To optimize the performance, a solar cell stack was fabricated initially only with Ge-substituted CZTS at different concentration of Ge. A cell stack with 40% Ge substituted exhibited highest power conversion efficiency (10.52%) with 35.37 mA cm^{-2} short-circuit current, 440 mV open-circuit voltage and 67.56% fill factor. Further, to enhance the efficiency, the 40% Ge substituted device was coated with Ag substituted films at different levels of Ag content. A device with 40% Ge in the bottom layer and ¼ of Ag in the top layer showed the highest efficiency of 12.26%, 67.47% fill factor, 489.2 V open-circuit voltage and 37.14 mA cm^{-2} short-circuit current, considerably higher than Ge alone or unsubstituted CZTS device.

Another tri-layer absorber structure was investigated, but by varying the copper content in the layers [103]. The tri-layer was composed of fine grains sandwiched between large grains with higher copper content near the back contact and low copper content near the front contact. This type of architecture drives carrier diffusion towards the front contact and decreases the valence band edge offset in

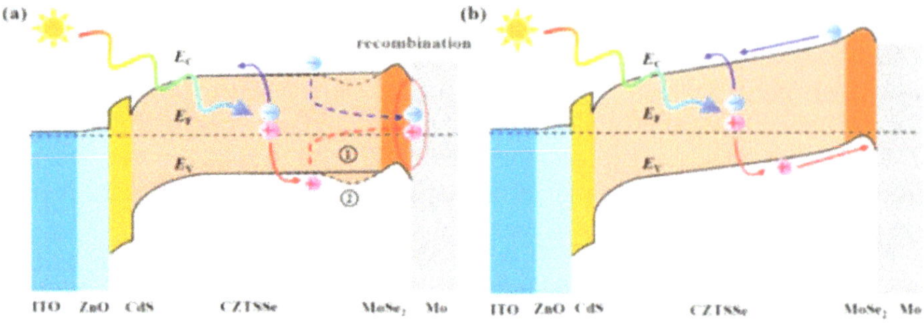

Figure 4.9. Schematic of specific recombination routes between electron and holes in (a) Cu-poor, (b) Cu-rich CZTSSe solar cells. Adapted with permission from [103]. Copyright 2021 American Chemical Society.

the rear of the device to aid in hole extraction (figure 4.9). The precursor solution was prepared by dissolving metal powders such as Cu, Zn, Sn, Se and S in ethylenediamine and ethylenedithiol at 65 °C for 90 min and then stabilized using thioglycolic acid and glycol methyl ether at 70 °C for 4 h. There was a systematic increase in solar cell performance as increasing Cu/(Zn + Sn) ratio from 0.72 to 1.15 with an efficiency increase from 10.48% to 12.54%. However, the efficiency dropped to 9.45% along with other parameters when the ratio further increased to 1.25. In a similar study, the effect of Cd substitution on the performance was investigated by adding up to 15% of Cd into the amine–thiol solution [104]. Maximum efficiency of 6.11% was obtained for the cell that was fabricated with 11% Cd, whereas the undoped device showed 4.63% efficiency. This doping approach was further continued with a few other elements including revisiting Ge incorporation to reduce Sn_{Zn} defect concentration and interestingly, exploring Li doping [105, 106]. Ge doping increased the efficiency from 9.15% to 11.48% with 41 mV improvement in open-circuit voltage. Notably, the density of Sn_{zn} deep traps decreased to 9.42×10^{12} cm^{-3} from 8.97×10^{13} cm^{-3}. Similarly, it was reported that Li doping was passivating non-radiative recombination centers and reducing band tailings, contributing to defect density reduction. There was a significant improvement in power conversion efficiency from 6.99% to 9.68% by Li doping. Defect passivation by doping was further extended by introducing indium instead of tin [107]. It was found that the indium incorporation improved the crystallinity and carrier concentration of the absorber layer, in addition to passivating the deep defects and reducing interface recombination and band tailing. The best performance was observed for indium doping (X = In/(Sn + In) = 9%) with 7.19% efficiency and 62 mV open-circuit voltage and the impact of Ga doping on solar cells performance was investigated [108]. First principle calculations show that Ga was occupying Zn and Sn sites and has a benign effect on suppressing the formation of the Sn_{Zn} deep donor defects. Ga doping enhanced open-circuit voltage from 473 to 515 mV with considerable improvement in power conversion efficiency from 10.5% for the undoped CZTSSe device to 12.3% for

5.0% Ga doped device. Other performance parameters such as short-circuit current, open-circuit voltage and fill factor values followed a similar trend.

4.3.3.3 Non-hydrazine-based precursor approach: thiourea for CZTS

The use of thiourea for fabrication of CIGS solar cells has been demonstrated successfully. This approach was extended for CZTS using $Cu(OAc)_2 \cdot H_2O$, $SnCl_2$, $ZnCl_2$ and thiourea as a coordinating ligand and sulfur source in dimethyl sulfoxide (DMSO) solvent [109]. DMSO was chosen because of its polarity, solubility property, thermal stability and moderate toxicity. Similar to other solution processing routes, the precursor solution was spin coated onto the Mo-coated soda-lime substrate and annealed at 580 °C. A solar cell device was completed by selenizing the film and chemical bath coating of CdS.

The device exhibited power conversion efficiency of 4.1%, with 400 mV open-circuit voltage, 24.9 mA cm^{-2} short-circuit current and 41.2% fill factor. EQE of the device was measured to be 85% in the visible wavelength range, but lower at the longer wavelength region suggesting a short minority carrier lifetime. However, the efficiency was improved to 8.32% by adding precursors stepwise [110]. In the improved process, a DMSO solution of $Cu(OAc)_2$ turned to light green solution from deep blue upon addition of $SnCl_2$ and to even lighter green when $ZnCl_2$ was added (figure 4.10). It was suggested that Cu^{2+} and Sn^{2+} underwent a disproportionation reaction to form Cu^{+} and Sn^{4+} in this stepwise addition process in comparison to a single-step mixing process where the presence of Cu^{2+} was observed. This process improved the short-circuit current density to 31.2 mA cm^{-2} and notably, fill factor value to 60.2%, which helped obtain 8.32% efficiency. Following the stepwise precursor addition, a three-step annealing process was explored for improving the crystallinity [111]. It was observed that the thiourea/DMSO process often yielded films with high level of porosity, inhomogeneity and with many grain boundaries. A three-step annealing was proposed as a solution for addressing these shortcomings. This was demonstrated by constructing four

Figure 4.10. Photographs of DMSO solution of $Cu(OAc)_2$ (deep blue) upon addition of $SnCl_2 \cdot 2H_2O$, $ZnCl_2$, and thiourea. Adapted with permission from [110]. John Wiley & Sons. © 2014 WILEY-VCH Verlag GmbH & Co. KGaA, Weinheim.

different devices, in which two of them were annealed at 300 °C and 500 °C and the other two were annealed at 300 °C, 500 °C and 550 °C during the selenization step. The improved grain growth was apparent from the electron microscope images that showed closely packed large grains from three-stage annealed samples. This improvement was reflected in their solar cell performance with 10.1% and 8.7% efficiencies from the three-stage annealed samples, whereas the two-stage annealed samples exhibited 7.1% and 6.6% with notable improvement in short-circuit current density (up to 35 mA cm^{-2}) and fill factor (61.2%) values, but open-circuit voltage was not improved. The V_{oc} deficit is an issue in kesterite that was dragging its maturity. One of many explorations to improve the V_{oc} deficit involved fabricating CZTS using Sn^{4+} [112]. Two separate CZTS devices were fabricated using $SnCl_2$ and $SnCl_4$. Powder XRD analysis of the films prepared using Sn^{2+} and Sn^{4+} salts showed the presence of SnS from Sn^{2+}, but only CZTS was observed from the film that was prepared using Sn^{4+}, suggesting CZTS was formed through the multiphase secondary fusion grain growth when Sn^{2+} was used. The devices exhibited performances with a significant difference in open-circuit voltage (0.507 V from Sn^{4+} and 0.452 V from Sn^{2+}) fill factor (71.2% from Sn^{4+} 44.1% from Sn^{2+}) and power conversion efficiencies (12.3% from Sn^{4+} and 6.6% from Sn^{2+}). Another attempt was made towards improving the V_{oc} by introducing GeO_2 on top of Mo substrate and before coating CZTS [113]. The idea behind the GeO_2 coating was to incorporate Ge into the CZTS layer, believing that Ge would help reduce defect density, band tailing, and facilitate a quasi-Fermi level split with relatively high hole concentration. A CZTS device with GeO_2 layer showed 13.14% efficiency with a relatively higher open-circuit voltage of 547 mV. Doping CZTS with other elements was explored in a few investigations to improve the performance.

Lithium was introduced in the form of lithium chloride in the thiourea and DMSO solution along with other molecular precursors. Scanning probe microscopy analysis showed that lithium doping of CZTSeS changes the polarity of the electric field at the grain boundaries in a way the minority carrier electrons are repelled from the grain boundaries. It was proposed that due to the isoelectronic nature of lithium and copper, lithium was competing for the copper vacancies, thereby decreasing Zn_{cu} donor concentration, which was also competing for zinc vacancies. The solar cell device fabricated using lithium exhibited 11.8% and 11.6% efficiencies in two separate studies [114, 115]. Although lithium doping in CZTS improved the efficiency, lithium loss was observed in the solar cell structure [116]. To retain the lithium, it was introduced during the selenization step along with Se using LiF. This showed a decent efficiency of 11.63% with an open-circuit voltage of 0.583 V. Introducing silver into the CZTS device showed a bigger impact on improving the efficiency [117]. A device prepared using AgCl along with other precursors in 2-methoxyethanol solution showed 12.87% efficiency with 39.3 mA cm^{-2} short-- circuit current, 505.5 mV open-circuit voltage and 64.8% fill factor.

Due to the similarities between DMSO and dimethylformamide (DMF) in terms of polarity and their ability to dissolve metal salts and chalcogen sources, there have been a few investigations using DMF as a solvent but using thiourea as a sulfur source. Notable efficiencies of 11.76% and 11.5% have been reported with a fill factor value of 70.6% [118, 119]. The solar cell device was constructed by using CuCl, $Zn(OAc)_2$, $SnCl_4$

and thiourea in DMF. Interestingly, it was found that $SnCl_4$ was reacting with DMF and formed a $Sn(DMF)_2Cl_4$ complex, whereas CuCl and $Zn(OAc)_2$ formed complexes with thiourea. One of the observations reported in the solar cells that are fabricated using DMSO or DMF solvents was the bi- or tri-grain layers. The solvent, 2-methoxyethanol was used instead to dissolve chloride salts of Cu, Zn, Sn, and thiourea which helped obtain a fine-grain layer-free structure [120]. The device exhibited 5.0% efficiency before selenization and that doubled to 10.1% after selenization. As described earlier, open-circuit voltage deficit is one of the issues that affect the CZTS material's progress. Low open-circuit voltage has been identified as an issue of antisite defect and unfavourable heterojunction interface. Post-annealing treatment was carried out to promote diffusion of elements among the layers to address the defects [121]. Devices were fabricated using a 2-methoxyethanol solution of copper acetate, zinc acetate, cadmium acetate, tin chloride and thiourea and spin coating the solution onto Mo substrate. The performance of devices that were annealed after the CdS layer and after the full stack at 300 °C for 8 min was compared. The full stack annealed device exhibited 12.6% efficiency and the device that was annealed after CdS showed 10.9% efficiency, significantly higher performance than the unannealed device efficiency of 6.7%. Remarkably, the open-circuit voltage was improved nearly by 100 mV.

4.3.3.4 Non-hydrazine: thioglycolic acid and ammonia approach for CZTS

The use of thioglycolic acid for dissolution of metal precursors for CZTS has been investigated in a few different studies due to its benign nature, non-toxicity, water solubility and ability to coordinate metals from metal salts. In one of the studies, ammonium thioglycolate was used for dissolving tin oxide first and then copper oxide and zinc oxides were added to solution [122]. The solution was spin coated onto a Mo-coated substrate for 40 s and annealed at 360 °C for 2 min. This coating and annealing process was repeated seven times to obtain the desired thickness of 1.5 μm. The effect of silver doping on the performance was investigated by introducing silver oxide at different concentrations in the solution. The solar cell devices exhibited a decent performance with 5.86% power conversion efficiency from the neat CZTS and up to 7.38% from a 2% silver doped CZTS device (figure 4.11).

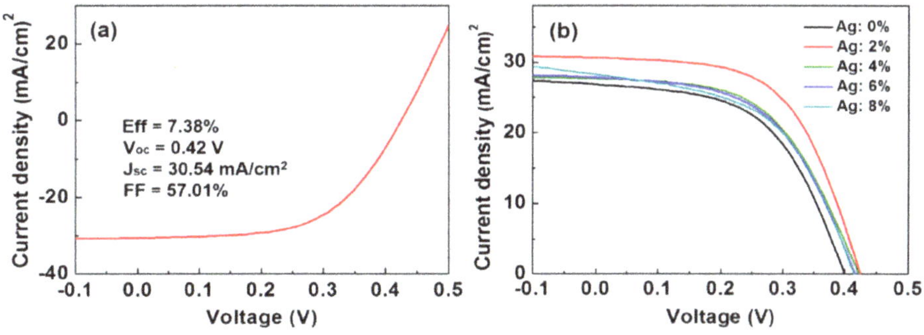

Figure 4.11. (a) *J–V* curves of the $(Cu_{1-x}Ag_x)_2ZnSn(S,Se)_4$ ($x = 2$) solar cells. (b) *J–V* curves of devices with different doping levels of silver. Adapted with permission from [122] John Wiley & Sons. Copyright 2018 WILEY-VCH Verlag GmbH & Co. KGaA, Weinheim.

Later, the device efficiency was improved over 12% by air annealing at 280 °C [123]. Recently, this was further improved to achieve one of the highest-efficiency CZTS devices (12.8%) on a par with DMSO-processed devices and higher than hydrazine-processed devices [124, 125].

4.3.4 Nanocrystals ink-based approach for CZTS

The use of nanoparticles for the fabrication of CZTS provides many advantages including avoiding secondary phases and eliminating defects. One of the earlier reports involving the fabrication of solar cells from nanoparticles utilized the hot-injection colloidal synthesis process. In brief, acetylacetonate complexes of copper (II), zinc(II) and tin(IV) were heated in oleylamine to 225 °C and then sulfur dissolved in oleylamine was injected into the metal complexes solution to initiate the nucleation and was kept at that temperature for 30 min for growth [126]. The nanoparticle ink was drop casted onto a Mo-coated substrate and annealed at 350 ° C for 1 h to remove organic ligands. Nanoparticle film after removing the ligands was selenized between 400 °C and 500 °C for fabrication of CZTS cells. Although, the initial device exhibited only 0.74% efficiency, it introduced the CZTS nano-particles concept for fabrication of solar cells.

The same group of researchers reported an improved device using the CZTS nanoparticles in the following year [127]. To improve the efficiency, the nano-particles' capping ligand was removed from the surface by a repeated solvent washing procedure using hexane and isopropanol solvent mixture. These residual dry nanoparticles were redispersed in hexanethiol to form a stable ink. The ink was knife-coated onto the substrate for selenization and for completion of remaining layers. This modification helped achieve PCE of 7.23%. Bandgap tunability through composition variability has been explored for tuning CZTS nanoparticles' bandgap, but with introduction of Ge [128]. The device fabricated using Ge-substituted nanocrystals did not show any improvement in efficiency from pure CZTS-based cells, but the bandgap was tuned from 1.5 eV for pure CZTS to 1.94 eV for CZGS. Controlling the stoichiometry of the elements in CZTS nanoparticles has been difficult. One of the approaches explored for stoichiometry control was to fabricate CZTS using binary and ternary metal sulfide nanocrystals (figure 4.12) [129]. Nanocrystals of copper tin sulfide (CTS), ZnS, SnS, CuS and Cu_7S_4 were synthesized by reacting metal salts with sulfur in oleylamine and tri-octylphosphine oxide. A CZTSSe device was prepared mixing the appropriate ratio of binary and ternary sulfide nanocrystals and annealing in selenium atmosphere. The device exhibited a respectable efficiency of 8.5% with 29 mA cm^{-2} short-circuit current, 451 mV open-circuit voltage and 64.9% fill factor values. A few more investigations have been reported by modifying the nanocrystals' surface using polar ligands (figure 4.13) and synthesizing CZTS nanocrystals with alternate metals salts. The modifications yielded 7.7% and 9.0% efficiency solar cell devices (figure 4.13) [130, 131].

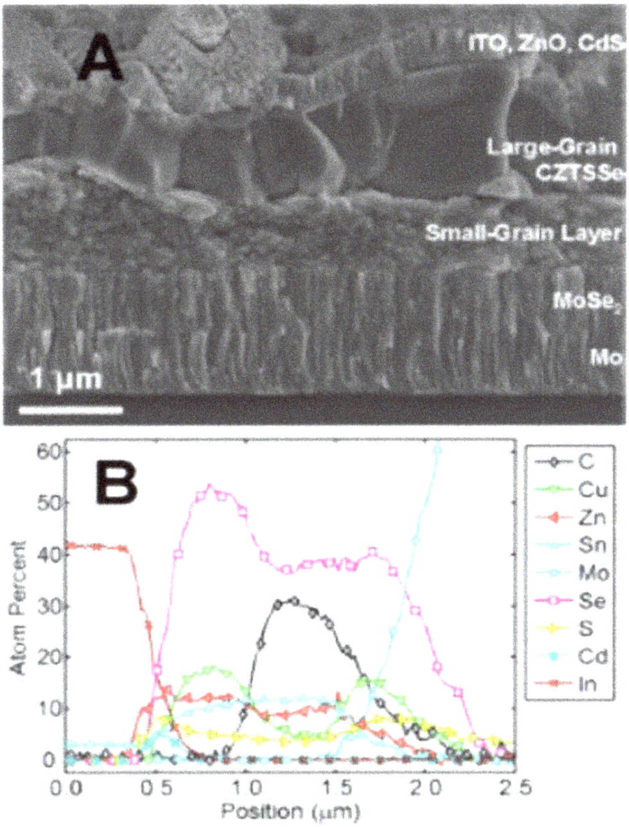

Figure 4.12. (a) Cross-sectional SEM image and (b) Auger depth profile of CZTSSe solar cells fabricated using CZTS nanocrystals. Adapted with permission from [129]. Copyright 2012 American Chemical Society.

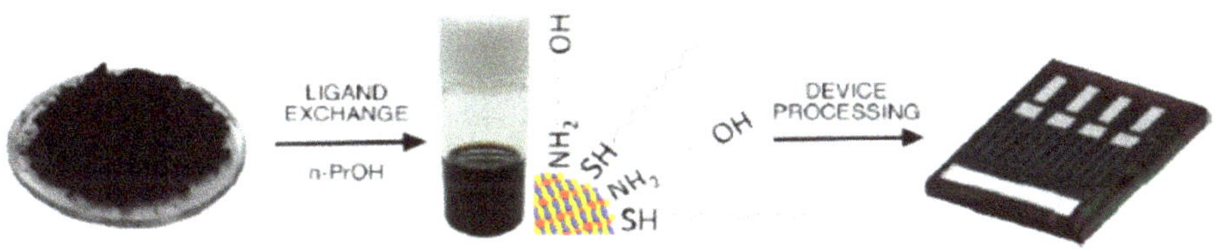

Figure 4.13. Schematic of CZTS device fabrication process using polar ligand exchanged nanoparticles. Adapted with permission from [130]. Copyright 2014 American Chemical Society.

4.4 Cell degradation and failure diagnostics

Thin film solar cell technologies are advancing rapidly with power conversion efficiencies reaching close to matured single-crystal silicon cells. The reliability of cells is equally important to ensure that the performance is retained in their working lifetime. There have been a few investigations directed towards understanding of the degradation behaviour and mechanism in thin films solar cells. However, very limited studies exist on thin film solar cells from solution fabrication methods.

4.4.1 CIGS cell degradation and failure diagnostics

The reliability of CIGS solar has been very encouraging relative to other competing technologies. For these, there has been yearly power degradation rate of 0.5% observed from some cells and a majority of the cells studied lost between 0% and 1% per year. Surprisingly, some cells showed performance improvement [132]. CIGS cells can be degraded due to humidity, biases and partial shading. There have been three different degradation pathways identified in CIGS cells, namely: water oxidation of molybdenum back contact and TCO conductivity loss; alkali elements migration due to internal and external biases; and partial shading-induced worm-like defect formation. Reliability studies of CIGS cells under damp heat (85% RH/ 85°) conditions showed loss of open-circuit voltage and fill factor values, whereas they are stable in dry heat conditions suggesting water's role in degradation [133]. In addition, increased resistivity of ZnO:Al TCO layer was observed from the unencapsulated devices indicating atmospheric impurities such as water and CO_2 impacting the carrier mobility. Interestingly, this was reversible when the cells were annealed under vacuum at elevated temperature. Indium doped tin oxide TCO was found to be more stable than ZnO:Al under high humid and high heat conditions. However, recrystallization and local concentration segregation of Sn and In was observed [134]. Further, oxidation of molybdenum at the back contact is found to be occurring in the presence of humidity that causes loss of contact.

Alkali elements such sodium and potassium are readily available in CIGS solar cells due to the use of soda lime glass substrates. These elements are known for impacting the performance positively. However, under damp heat conditions, migration and accumulation of sodium at the *p–n* junction and ZnO:Al was observed to be affecting shunt resistance negatively [135]. External voltage stress is also found to be inducing alkali migration causing reduction in charge carrier concentration, built-in voltage, TCO corrosion, leading to open-circuit voltage and fill factor drop [136]. Monolithically interconnected devices due to reverse bias exposure partial shading causes irreversible worm-like defects. The presence of worm-like defects leads to the formation of localized shunts and negatively affects the output. These are some of the commonly observed degradations in CIGS cells. Specific degradation studies of solution-processed cells are yet to be conducted in detail.

4.4.2 CZTS cell degradation and failure diagnostics

CZTS and CIGS share most of the components in their architecture, therefore one could expect them to share degradation mechanisms related to non-absorber layers such as Mo oxidation, alkali migration and partial shading. Specific to CZTS absorber degradation is very limited in the literature. It has been found to have a fast non-Ohmic barrier at the backside at the back contact and a slow long-term degradation at the front side and both caused by copper migration [137]. Copper migration to the CdS layer leads to a high dark resistance which hinders the carrier extraction and loss of photocurrent in the CdS close to the *p–n* junction.

4.5 Conclusion, challenges and future prospects

Solution-processed copper-based chalcogenide thin film solar cells are emerging as a promising alternative to traditional silicon-based solar cells, particularly for flexible and lightweight applications, and have shown impressive advancement in terms of cost-effective manufacturing, material utilization and high throughput. These cells utilize various copper-based chalcogenide materials, including $CuInGaSe_2$ (CIGS) and its substitutes like Cu_2ZnSnS_4 (CZTS), for their light-absorbing layer. They offer advantages like high charge carrier mobility, excellent absorption coefficient, and tunable bandgaps, leading to good efficiency and the potential for flexible devices. However, the performance of these solution-processed solar cells is still lagging behind the traditional vacuum-based deposition methods. The best performance of CIGS and CZTS cells deposited utilizing various molecular inks is listed in table 4.1. It can be noted that the best efficiency from solution-processed CIGS is only 18.7% compared to 23.35% from vacuum-based deposition methods. Moreover, the efficiency is even lower for solution-processed CZTS with the highest recorded efficiency being only ~13.0%. There are challenges associated with solution processing of solar cells affecting the performance including the presence of carbon impurities, voids and surface roughness, deep defects, creating difficulties in incorporating alkali elements effectively. For instance, the presence of carbon impurities has been found to be affecting the series resistance of cells by formation of the charge blocking layer towards the back contact. It was shown that the carbon impurities can be reduced to vacuum deposition level by low temperature and nitrogen annealing using thiourea and DMF solution. Likewise, alkali metals such as sodium and potassium incorporation has improved the cell efficiency in vacuum-

Table 4.1. Best performance of CIGSSe and CZTSSe devices fabricated using different solution-processing approaches.

Material	Dissolution agent	PCE (%)	J_{sc} (mA cm^{-2})	V_{oc} (mV)	FF (%)	References
CIGSSe	Hydrazine	18.1	35.5	660	77.2	[20]
CIGSSe	1,2-Ethylenediamine and 1,2-ethanedithiol	16.4	33.9	650	73.8	[36]
CIGSSe	Dimethyl sulfoxide	18.7	33.6	694	76.0	[54]
CIGSSe	Ethanol	14.4	35.0	591	69.0	[69]
CIGSSe	Nanoparticles	15.0	32.1	630	73.4	[85]
CZTSSe	Hydrazine	12.6	35.2	513	69.8	[94]
CZTSSe	1,2-Ethylenediamine and 1,2-ethanedithiol	12.5	37.8	480	69.0	[102]
CZTSSe	Dimethyl sulfoxide	13.1	34.3	547	70.0	[112]
CZTSSe	Thioglycolic acid and ammonia	12.7	36.2	494	67.5	[124]
CZTSSe	Nanoparticles	9.0	35.1	404	63.7	[130]

based deposition methods. However, adopting this in the solution-processing route needs more development. A potential pathway for incorporating alkali metals into the cell structure is to anneal under alkali vapour in an oxygen-free environment. Further, the presence of micro voids and surface roughness affects the carrier collection, thereby reducing the efficiency. A strategy would be to optimize heat treatment during the selenization or post-processing annealing steps. Bandgap engineering is another strategy that can be optimized for better collection of charges at the front and back contacts. Nevertheless, solution processing of CIGS and CZTS solar cells is attractive, and there has been significant development in the past decade. We can expect the performance of solution-processed cells on a par or surpassing the vacuum-deposited cells in a few years making them commercially lucrative for scale-up.

References

[1] Katagiri H, Jimbo K, Maw W S, Oishi K, Yamazaki M, Araki H and Takeuchi A 2009 Development of CZTS-based thin film solar cells *Thin Solid Films* **517** 2455–60

[2] Nakazawa K I 1988 Electrical and optical properties of stannite-type quaternary semiconductor thin films *Jpn. J. Appl. Phys.* **27** 2094

[3] Persson C 2010 Electronic and optical properties of Cu_2ZnSnS_4 and $Cu_2ZnSnSe_4$ *J. Appl. Phys.* **107** 053710

[4] Neisser A, Hengel I, Klenk R, Matthes T W, Alvarez-Garcia J, Perez-Rodriguez A, Romano-Rodriguez A and Lux-Steiner M-C 2001 Effect of Ga incorporation in sequentially prepared $CuInS_2$ thin film absorbers *Sol. Energy Mater. Sol. Cells* **67** 97–104

[5] Ananthoju B, Mohapatra J, Jangid M K, Bahadur D, Medhekar N V and Aslam M 2016 Cation/anion substitution in Cu_2ZnSnS_4 for improved photovoltaic performance *Sci. Rep.* **6** 35369

[6] Zhang S B, Wei S-H, Zunger A and Katayama-Yoshida H 1998 Defect physics of the $CuInSe_2$ chalcopyrite semiconductor *Phys. Rev.* B **57** 9642

[7] Kumar M, Dubey A, Adhikari N, Venkatesan S and Qiao Q 2015 Strategic review of secondary phases, defects and defect-complexes in kesterite CZTS-Se solar cells *Energy Environ. Sci.* **8** 3134–59

[8] Nagoya A, Asahi R, Wahl R and Kresse G 2010 Defect formation and phase stability of Cu_2ZnSnS_4 photovoltaic material *Phys. Rev.* B **81** 113202

[9] Wada T, Kohara N, Negami T and Nishitani M 1996 Chemical and structural characterization of $Cu(In,Ga)Se_2$/Mo interface in $Cu(In, Ga)Se_2$ solar cells *Jpn. J. Appl. Phys.* **35** L1253–6

[10] Orgassa K, Schock H W and Werner J H 2003 Alternative back contact materials for thin-film $Cu(In, Ga)Se_2$ solar cells *Thin Solid Films* **431–2** 387–91

[11] Aydin S S C 2021 Effects of different back contacts on photovoltaic performances of CdTe/CdS thin film solar cells *J. Optoelectron. Adv. Mater.* **23** 157–60

[12] McCandless B E and Hegedus S S 1991 Influence of CdS window layer on thin film CdS/CdTe solar cell performance *Proc. 22nd IEEE Photovoltaic Specialists Conf.* pp 967–72

[13] Chu T L, Chu S S, Chen G, Britt J, Ferekides C and Wu C Q 1992 Zinc selenide films and heterojunctions *J. Appl. Phys.* **71** 3865–9

[14] Dona J M and Herrero J 1997 Chemical bath deposition of CdS thin films: an approach to the chemical mechanism through study of the film microstructure *J. Electrochem. Soc.* **144** 4081
[15] Kobayashi T, Kao Z J L, Kato T, Sugimoto H and Nakada T 2016 A comparative study of Cd- and Zn- compound buffer layers on $Cu(In_{1-x},Ga_x)(S_y, Se_{1-y})_2$ thin film solar cells *Prog. Photovolt.* **24** 389–96
[16] Spiering S, Eicke A, Hariskos D, Powalla M, Naghavi N and Lincot D 2004 Large area Cd-free CIGS solar modules with In_2S_3 buffer layer deposited by ALCVD *Thin Solid Films* **451–52** 562–6
[17] Chopra K L, Major S and Pandya D K 1983 Transparent conductors—a status review *Thin Solid Films* **102** 1–46
[18] Major S, Banerjee A and Chopra K L 1983 Highly transparent and conducting indium-doped zinc oxide films by spray pyrolysis *Thin Solid Films* **108** 333–40
[19] Mitzi D B, Kosbar L L, Murray C E, Copel M and Afzali A 2004 High-mobility ultrathin semiconducting films prepared by spin coating *Nature* **428** 299–303
[20] Zhang T, Tang Y, Liu D, Tse S C, Cao W, Feng Z, Chen S and Qian L 2016 High efficiency solution-processed thin-film $Cu(In,Ga)(Se,S)_2$ solar cells *Energy Environ. Sci.* **9** 3674–81
[21] Chung C-H, Li S-H, Li B, Yang W, Hou W W, Bob B and Yang Y 2011 Identification of the molecular precursors for hydrazine solution processed $CuIn(Se,S)_2$ films and their interactions *Chem. Mater.* **23** 964–9
[22] Mitzi D B 2007 $N_4H_9Cu_7S_4$: a hydrazinium-based salt with a layered $Cu_7S_4^-$ framework *Inorg. Chem.* **46** 926–31
[23] Mitzi D B, Copel M and Chey S J 2005 Low-voltage transistor employing a high-mobility spin-coated chalcogenide semiconductor *Adv. Mater.* **17** 1285–9
[24] Liu W, Mitzi D, Yuan M, Kellock A J, Chey S J and Gunawan O 2010 12% Efficiency $CuIn(SeS)_2$ photovoltaic device prepared using a hydrazine solution process *Chem. Mater.* **22** 1010–4
[25] Yuan M, Mitzi D B, Liu W, Kellock A J, Chey S J and Deline V R 2010 Optimization of CIGS based PV device through antimony doping *Chem. Mater.* **22** 285–7
[26] Mitzi D B *et al* 2010 *Proceedings of the 35th Photovoltaic Specialists Conference (Waikoloa, HI)* (Piscataway, NJ: IEEE) pp 000640–5
[27] Todoro T K, Gunawan O, Gokmen T and Mitzi D B 2013 Solution processed $Cu(In,Ga)(S, Se)_2$ absorber yielding a 15.2% efficient solar cell *Prog. Photovolt.: Res. Appl.* **21** 82–7
[28] Todorov T, Cordoncillo E, Snachez-Royo J F, Carda J and Escribano P 2006 $CuInS_2$ films for photovoltaic applications deposited by a low cost-method *Chem. Mater.* **18** 3145–50
[29] Ahn S, Son T H, Cho A, Gwak J, Yun J H, Shin K, Ahn S K, Park S H and Yoon K H 2012 $CuInSe_2$ thin film solar cells with 7.72% efficiency prepared via direct coating of a metal salts/alcohol-based precursor solution *Chem Sus Chem.* **5** 1773–7
[30] Zhou H, Hsu C-J, Hsu W-C, Duan H-S, Chung C-H, Yang W and Yang Y 2013 Non-hydrazine solutions in processing $CuIn(S,Se)_2$ photovoltaic devices from hydrazinium precursors *Adv. Eng. Mater.* **3** 328–36
[31] Berner U and Widenmeyer M 2015 Solution-based processing of $Cu(In, Ga)Se_2$ absorber layers for 11% efficiency solar cells via a metallic intermediate *Prog. Photovolt.: Res. Appl.* **23** 1260–6
[32] Berner U, Colombara D, de Wild J, Robert E V C, Schutze M, Hergert F, Valle N, Widenmeyer M and Dale P J 2016 13.3% Efficient solution deposited $Cu(In, Ga)Se_2$ solar cells processed with different sodium salt sources *Prog. Photovolt.:* **24** 749–59

[33] Rehan S, Kim K Y, Han J, Eo Y-J, Gwak J, Ahn S K, Yun J H, Yoon K, Cho A and Ahn S 2016 Carbon-impurity affected depth elemental distribution in solution processed inorganic thin films for solar cell application *ACS Appl. Mater. Interfaces* **8** 5261–72

[34] Wang G, Wang S, Cui Y and Pan D 2012 A novel and versatile strategy to prepare metal–organic molecular precursor solutions and its application in Cu(In,Ga)(S,Se)$_2$ solar cells *Chem. Mater.* **24** 3939–97

[35] Zhao W, Cui Y and Pan D 2013 Air-stable, low-toxicity precursors for CuIn(SeS)$_2$ solar cells with 10.1% efficiency *Energy Technol.* **1** 131–4

[36] Webber D H and Brutchey R L 2013 Alkahest for V$_2$VI$_3$ chalcogenides: dissolution of nine bulk semiconductors in a diamine-dithiol solvent mixture *J. Am. Chem. Soc.* **135** 15722–5

[37] Zhao Y, Yuan S, Chang Q, Zhou Z, Kou D, Zhou W, Qi Y and Wu S 2021 Controllable formation of ordered vacancy compound for high efficiency solution processed Cu(In,Ga)Se$_2$ solar cells *Adv. Funct. Mater.* **2007928** 31

[38] Vineyard B D 1967 Versatility and the mechanism of the n-butyl-amine-catalyzed reaction of thiols with sulfur *J. Org. Chem.* **32** 3833–6

[39] Liu Y, Yao D, Shen L, Zhang H, Zhang X and Yang B 2012 Alkylthiol-enabled Se powder dissolution in oleylamine at room temperature for the phosphine-free synthesis of copper-based quaternary selenide nanocrystals *J. Am. Chem. Soc.* **134** 7207–10

[40] Arnou P, Cooper C S, Malkov A V, Bowers J W and Walls J M 2014 Solution-processed CuIn(S,Se)$_2$ absorber layers for application in thin film solar cells *Thin Solid Films* **582** 31–4

[41] Lin Z, He Q, Yin A, XU Y, Wang C, Ding M, Cheng H-C, Papandrea B, Huang Y and Duan X 2015 Cosolvent approach for solution-processable electronic thin films *ACS Nano* **9** 4398–405

[42] Xie Y, Liu Y, Wang Y, Zhu X, Li A, Zhang L, Qin M, Lu X and Huang F 2014 CuIn(S,Se)$_2$ thin films prepared from a novel thioacetic acid-based solution and their photovoltaic application *Phys. Chem. Chem. Phys.* **16** 7548–54

[43] Zhao D, Tian Q, Zhou Z, Wang G, Meng Y, Kou D, Zhou W, Pan D and Wu S 2015 Solution-deposited pure selenide CIGSe solar cells from elemental Cu, In, Ga and Se *J. Mater. Chem.* A **3** 19263–7

[44] Arnou P, van Hest M F A M, Cooper C S, Malkov A V, Walls J M and Bowers J W 2016 Hydrazine-free solution deposited CuIn(S,Se)$_2$ solar cells by spray deposition of metal chalcogenides *ACS Appl. Mater. Interfaces* **8** 11893–7

[45] Zhao X, Lu M, Koeper M J and Agrawal R 2016 Solution-processed sulfur depleted Cu(In,Ga)Se$_2$ solar cells synthesized from a monoamine-dithiol solvent mixture *J. Mater. Chem.* A **4** 7390–7

[46] Fan Q, Tian Q, Wang H, Zhao F, Kong J and Wu S 2018 Regulating the starting location of front-gradient enabled highly efficient Cu(In,Ga)Se$_2$ solar cells via a facile thiol-amine solution approach *J. Mater. Chem.* A **6** 4095–101

[47] Ulicna S, Arnou P, Abbas A, Togay M, Welch L M, Bliss M, Malkov A V, Walls J M and Bowers J W 2019 Deposition and application of a Mo–N back contact diffusion barrier yielding a 12.0% efficiency solution-processed CIGS solar cell using an amine-thiol solvent system *J. Mater. Chem.* A **7** 7042–52

[48] Zhao X, Deshmukh S D, Rokke D J, Zhang G, Wu Z, Miller J T and Agrawal R 2019 Investigating chemistry of metal dissolution in amine–thiol mixtures and exploiting it towards benign ink formulation for metal chalcogenide thin films *Chem. Mater.* **31** 5674–84

[49] Rokke D, Deshmukh S D and Agrawal R 2019 A novel approach to amine-thiol molecular precursors for fabrication of high efficiency thin film CISSe/CIGSSe devices *Proceedings of the 46th Photovoltaic Specialists Conference (Chicago, IL)* (Piscataway, NJ: IEEE) pp 1813–5

[50] Yuan S, Wang X, Zhao Y, Chang Q, Xu Z, Kong J and Wu S 2020 Solution processed Cu (In, Ga)$(S,Se)_2$ solar cells with 15.25% efficiency by surface sulfurization *ACS, Appl. Energy Mater.* **3** 6785–92

[51] Togay M, Ulicna S, Bukhari S, Lisco F, Bliss M, Eeles A, Walls J M and Bowers J W 2018 Exploring metastable defect behavior in solution-processed antimony doped CIGS thin film solar cells *Proceedings of the 45th Photovoltaic Specialist Conference (Waikoloa, HI)* (Piscataway, NJ: IEEE)

[52] Mazalan E, Chaudhary K T, Nayan N and Ali J 2019 Influence of antimony dopant on $CuIn(S,Se)_2$ solar thin absorber layer deposited via solution-processed route *J. Alloys Compd.* **772** 710

[53] Wang X-S, Fan Q-M, Tian Q-W, Zhou Z-J, Kou D-X, Zhou W-H, Meng Q-B, Zheng Z and Wu S-X 2019 $CuInGaSe_2$ thin-films solar cells with 11.5% efficiency: an effective and low-cost way of Na-incorporation for grain growth *Sol. Energy* **185** 34–40

[54] Uhl A R, Katahara J K and Hillhouse H W 2016 Molecular-ink route to 13.0% efficiecnt low-bandgap $CuIn(S,Se)_2$ and 14.7% efficient Cu(In, Ga)$(S,Se)_2$ solar cells *Energy Environ. Sci.* **9** 130–4

[55] Aramoto T, Kawaguchi Y, Liao Y-C, Kikuchi Y, Ohhashi T, Lida H and Nakamura A 2016 High efficiency solution coated Cu(In,Ga)$(Se,S)_2$ thin film solar cells *32nd European Photovotaic Solar Energy Conf. and Exhibition (Munich)* pp 1108–11

[56] Tiwari D, Koehler T, Lin X, Sarua A, Harnimab R, Wang L, Klenk R and Fermin D J 2017 Single Molecular precursor solution for $CuIn(S,Se)_2$ thin films photovoltaic cells: structure and device characteristics *ACS Appl. Mater. Interfaces* **9** 2301–8

[57] Jiang J, Yu S, Gong Y, Yan W, Zhang R, Liu S, Huang W and Xin H 2018 10.3% efficient $CuIn(S,Se)_2$ solar cells from DMF molecular solution with the absorber selenized under high argon pressure *Sol. RRL* **2** 1800044

[58] Jiang J *et al* 2020 Highly efficient copper-rich chalcopyrite solar cells from DMF molecular solution *Nano Energy* **69** 104438

[59] Pamplin B and Feigelson R S 1979 Spray pyrolysis of $CuInSe_2$ and related ternary semiconducting compounds *Thin Solid Films* **60** 141–6

[60] Tomar M S and Garcia F J 1982 A ZnO/p-$CuInSe_2$ thin film solar cell prepared entirely by spray pyrolysis *Thin Solid Films* **90** 419–23

[61] Ho J C W, Zhang T, Lee K K, Batabyal S K, Tok A I Y and Wong L H 2014 Spray pyrolysis of $CuIn(S,Se)_2$ solar cells with 5.9% efficiency: a method to prevent Mo oxidation in ambient atmosphere *ACS Appl. Mater. Interfaces* **6** 6638–43

[62] Hossain M A, Tianliang Z, Keat L K, Xianglin L, Prabhakar R R, Batabyal S K, Mhaislkar S G and Wong L H 2015 Synthesis of Cu(In, Ga)(S, $Se)_2$ thin films using an aqueous spray pyrolysis approach, and their solar cell efficiency of 10.5% *J. Mater. Chem. A* **3** 414–4154

[63] Septina W, Kurihara M, Ikeda S, Nakajima Y, Hirano T, Kawasaki Y, Harada T and Matsumura M 2015 Cu(In,Ga)$(S,Se)_2$ thin film solar cell with 10.7% conversion efficiency obtained by selenization of the Na-doped spray pyrolyzed sulfide precursor film *ACS Appl. Mater. Interfaces* **7** 6472–9

[64] Kurihara M, Septina W, Hirano T, Nakajima Y, Harada T and Ikeda S 2015 Fabrication of $Cu(In,Ga)(S,Se)_2$ thin film solar cells via spray pyrolysis of thiourea and 1-methylthiourea-based aqueous precursor solution *Jpn. J. Appl. Phys.* **54** 091203
[65] Wu S, Jiang J, Yu S, Gong Y, Yan W, Xin H and Huang W 2019 Over 12% efficient low-bandgap $CuIn(S,Se)_2$ solar cells with the absorber processed from aqueous metal complexes solution in air *Nano Energy* **62** 818–22
[66] Kim S Y, Mina M S, Lee J and Kim J H 2019 Sulfur-alloying effects on $CU(In,Ga)(S,Se)_2$ solar cell fabricated using aqueous spray pyrolysis *ACS Appl. Mater. Interfaces* **11** 45702–8
[67] Wang W, Su Y-W and Chang C-H 2011 Inkjet printed chalcopyrite $CuIn_xGa_{1-x}Se_2$ thin film solar cells *Sol. Energy Mater. Sol. Cells* **95** 2616–20
[68] Wang W, Han S-Y, Song S-J, Kim D-H and Chang C-H 2012 8.01% $CuInGaSe_2$ solar cells fabricated by air-stable low-cost inks *Phys. Chem. Chem. Phys.* **14** 11154–9
[69] Park S J, Cho Y, Moon S H, Kim J E, Lee D-K, Gwak J, Kim J, Kim D-W and Min B K 2014 A comparative study of solution processed low- and high bandgap chalcopyrite thin film solar cells *J. Phys. D: Appl. Phys.* **47** 135105
[70] Park G S, Chu V B, Kim B W, Kim D-W, Oh H-S, Hwang Y J and Min B K 2018 Achieving 14.4% alcohol based solution processed $Cu(In,Ga)(S,Se)_2$ thin film solar cell through interface engineering *ACS Appl. Mater. Interfaces* **10** 9894–9
[71] Lin X, Klenk R, Wang L, Kohler T, Albert J, Fiechter S, Ennaoui A and Lux-Steiner M C 2016 11.3% efficiency $Cu(In,Ga)(S,Se)_2$ thin film solar cells via drop-on-demand inkjet printing *Energy Environ. Sci.* **9** 2037–43
[72] Choi I J, Jang J W, Mohanty B C, Lee S M and Cho Y S 2016 Improved photovoltaic characteristics and grain boundary potentials of $CuIn_{0.7}Ga_{0.3}Se_2$ thin films spin-coated by Na-dissolved nontoxic precursor solution *ACS Appl. Mater. Interfaces* **8** 17011–5
[73] Sung J-C and Lu C-H 2017 Potassium-ion doped $Cu(In, Ga)Se_2$ thin films solar cells: phase formation, microstructures and photovoltaic characteristics *Appl. Surf. Sci.* **409** 270–6
[74] Brown G, Stone P, Woodruff J, Cardozo B and Jackrel D 2012 Device characteristics of a 17.1% efficient solar cell deposited by a non-vacuum printing method on flexible foil *Proc. of 38th IEEE Photovoltaic Specialist Conf. (Austin, TX)*
[75] Walker B C and Agrawal R 2014 Contamination-free solutions of selenium in amines for nanoparticle synthesis *Chem. Commum.* **50** 8331–4
[76] Hillhouse H W and Beard M C 2009 Solar cells from colloidal nanocrystals: fundamentals, materials, devices and economics *Curr. Opin. Colloid Interface Sci.* **14** 245–59
[77] Thanh N T K, Maclean N and Mahiddine S 2014 Mechanisms of nucleation and growth of nanoparticles in solution *Chem. Rev.* **114** 7610–30
[78] Jungemann A M *et al* 2019 The role of ligands in the chemical synthesis and applications of inorganic nanoparticles *Chem. Rev.* **119** 4819–80
[79] Guo Q, Kim S J, Kar M, Shafarman W N, Birkmire R W, Stach E A and Agrawal R 2008 Development of $CuInSe_2$ nanocrystal and nanoring inks for low-cost solar cells *Nano Lett.* **8** 2982–7
[80] Guo Q, Ford G M, Hillhouse H W and Agrawal R 2009 Sulfide nanocrystal inks for dense $Cu(In_{1-x}Ga_x)(S_{1-y}Sey)_2$ absorber films and their photovoltaic performance *Nano Lett.* **9** 3060–5
[81] Guo Q, Ford G M, Agrawal R and Hillhouse H W 2013 Ink formulation and low-temperature incorporation of sodium to yield 12% efficient $Cu(In,Ga)(S,Se)_2$ solar cells from sulfide nanocrystals inks *Prog. Photovolt.: Res. Appl.* **212** 64–71

[82] Stolle C J, Panthani M G, Harvey T B, Akhavan V A and Korgel B A 2012 Comparsion of the photovoltaic response of oleyamine and inorganic ligand-capped $CuInSe_2$ nanocrystals *ACS Appl. Mater. Interfaces* **4** 2757–61

[83] Ahn S, Choi Y J, Kim K, Eo Y-J, Cho A, Gwak J, Yun J H, Shin K, Ahn S K and Yoon K 2013 Amorphous Cu–In–S nanoparticles as precursors for $CuInSe_2$ thin-film solar cells with a high efficiency *Chem. SuS. Chem.* **6** 1282–7

[84] Ahn S J, Rehan S, Moon D G, Eo Y-J, Ahn S, Yun J H, Cho A and Gwak J 2017 An amorphous Cu–In–S nanoparticle based precursor ink with improved atom economy for $CuInSe_2$ solar cells with 10.85% efficiency *Green Chem.* **19** 1268–77

[85] Ellis R G, Turnley J W, Rokkem D J, Fields J P, Alruqobah E H, Deshmuck S D, Kisslinger K and Agrawal R 2020 Hybrid ligand exchange of Cu(In, Ga)S_2 nanoparticles for carbon impurity removel in solution-processed solar cells *Chem. Mater.* **32** 5091–103

[86] McLeod S M, Hages C J, Carter N J and Agrawal R 2015 Synthesis and characterization of 15% efficient CIGSSe solar cells from nanoparticle inks *Prog. Photovolt.: Res. Appl.* **23** 1550–6

[87] Yuan M and Mitzi D B 2009 Solvent properties of hydrazine in the preparation of metal chalcogenide bulk materials and films *Dalton Trans.* 6078–88

[88] Mitzi D B 2009 Solution processing of chalcogenide semiconductors via dimensional reduction *Adv. Mater.* **21** 3141–58

[89] Todorov T K, Reuter K B and Mitzi D B 2010 High-efficiency solar cell with earth-abundant liquid processed absorber *Adv. Mater.* **22** E156–9

[90] Barkhouse D A R, Gunawan O, Gokmen T, Todorov T K and Mitzi D B 2012 Device characteristics of a 10.1% hydrazine-processed $Cu_2ZnSn(Se,S)_4$ solar cell *Prog. Photovolt.: Res. Appl.* **20** 6–11

[91] Bag S, Gunawan O, Gokmen T, Zhu Y and Mitzi D B 2012 Hydrazine-processed Ge-substituted CZTSe solar cells *Chem. Mater.* **24** 4588–93

[92] Todorov T K, Tang J, Bag S, Gunawan O, Gokmen T, Zhu Y and Mitzi D B 2013 Beyond 11% efficiency: characteristics of state-of-the art $Cu_2ZnSn(S,Se)_4$ solar cells *Adv. Energy Mater.* **3** 34–8

[93] Yang W, Duan H-S, Bob B, Zhou H, Lei B, Chung C-H, Li S-H, Hou W W and Yang Y 2012 Novel solution processing of high-efficiency earth abundant $Cu_2ZnSn(S,Se)_4$ solar cells *Adv. Mater.* **24** 6323–9

[94] Winkler M T, Wang W, Gunawan O, Hovel H J, Todorov T K and Mitzi D B 2014 Optical designs that improve efficiency of $Cu_2ZnSn(S,Se)_4$ solar cells *Energy Environ. Sci.* **7** 1029–36

[95] Wang W, Winkler M T, Gunawan O, Gokmen T, Todorov T K, Zhu Y and Mitzi D B 2014 Device characteristics of CZTSSe thin-film solar cells with 12.6% efficiency *Adv. Energy Mater.* **4** 1301465

[96] Zhong J *et al* 2014 Sulfurization induced surface constitution and its correlation to the performance of solution-processed $Cu_2ZnSn(S,Se)_4$ solar cells *Sci. Rep.* **4** 6288

[97] Kim J *et al* 2014 High efficiency $Cu_2ZnSn(S,Se)_4$ solar cells by applying a double In_2S_3/CdS emitter *Adv. Mater.* **26** 7427–31

[98] Zhang R, Szczepaniak S M, Carter N J, Handwerker C A and Agrawal R 2015 A versatile solution route to efficient $Cu_2ZnSn(S,Se)_4$ thin film solar cells *Chem. Mater.* **27** 2114–20

[99] Tian Q, Cui Y, Wang G and Pan D 2015 A robust and low-cost strategy to prepare Cu_2ZnSnS_4 precursor solution and its application in $Cu_2ZnSn(S,Se)_4$ solar cells *RSC Adv.* **5** 4184–90

[100] Lowe J C, Wright L D, Eremin D B, Burykina J V, Martens J, Plasser F, Anaikov V P, Bowers J W and Malkov A V 2020 Solution processed CZTS solar cells using amine-thiol systems: understanding the dissolution process and device fabrication *J. Mater. Chem.* C **8** 10309–18

[101] Qi Y, Tian Q, Meng Y, Kou D, Zhou Z, Zhou W and Wu S 2017 Elemental precursor solution processed $(Cu_{1-x}Ag_x)_2ZnSn(S,Se)_4$ photovoltaic devices with over 10% efficiency *ACS Appl. Mater. Interfaces* **9** 21243–50

[102] Fu J, Kou D, Zhou W, Zhou Z, Yuan S, Qi Y and Wu S 2020 Ag, Ge dual-gradient substitution for low-energy loss and high-efficiency kesterite solar cells *J. Mater. Chem.* A **8** 22292–301

[103] Zhao Y, Zhao X, Kou D, Zhou W, Zhou Z, Yuan S, Qi Y, Zheng Z and Wu S 2021 Local Cu component engineering to achieve continuous carrier transport for enhanced kesterite solar cells *ACS Appl. Mater. Interfaces* **13** 795–805

[104] Yan Q, Cheng S, Yu X, Jia H, Fu J, Zhang C, Zheng Q and Wu S 2020 Mechanism of current shunting in flexible $Cu_2Zn_{1-x}Cd_xSn(S,Se)_4$ solar cells *Sol. RRL* **4** 1900410

[105] Deng Y, Zhou Z, Zhang X, Cao L, Zhou W, Kou D, QI Y, Yuan S, Zheng Z and Wu S 2021 Adjusting the Sn_{zn} defects in $Cu_2ZnSn(S,Se)_4$ absorber layer via Ge^{4+} implanting for efficient kesterite solar cells *J. Energy Chem.* **61** 1–7

[106] Yan Q, Sun Q, Deng H, Xie W, Zhang C, Wu J, Zheng Q and Cheng S 2022 Enhancing carrier transport in flexible CZTSSe solar cells via doping Li strategy *J. Energy Chem.* **75** 8–15

[107] Yu X, Cheng S, Yan Q, Fu J, Jia H, Sun Q, Yang Z and Wu S 2020 Efficient flexible Mo foil-based $Cu_2ZnSn(S,Se)_4$ solar cells from In doping technique *Sol. Energy Mater. Sol. Cells* **209** 110434

[108] Do Y, Wang S, Tian Q, Zhao Y, Chang X, Xiao H, Deng Y, Chen S, Wu S and Liu S 2021 Defect engineering in earth abundant $Cu_2ZnSn(S,Se)_4$ photovoltaic materials via Ga^{3+} doping for over 12% efficient solar cells *Adv. Funct. Mater.* **31** 2010325

[109] Ki W and Hillhouse H W 2011 Earth-abundant element photovoltaics directly from soluble precursors with high yield using a non-toxic solvent *Adv. Energy Mater.* **1** 732–5

[110] Xin H, Katahara J K, Braly I L and Hillhouse H W 2014 8% efficient $Cu_2ZnSn(S,Se)_4$ solar cells from redox equilibrated simple precursors in DMSO *Adv. Energy Mater.* **4** 1301823

[111] Haass S G, Diethelm M, Werner M, Bissig B, Romantuk Y E and Tiwari A N 2015 11.2% efficient solution processed kesterite solar cell with a low voltage deficit *Adv. Energy Mater.* **5** 1500712

[112] Gong Y, Zhang Y, Zhu Q, Zhou Y, Qiu R, Niu C, Yan W, Huang W and Xin H 2021 Identifying the origin of the Voc dfecit of kesterite solar cell from the two grain growth mechanisms induced by Sn^{2+} and Sn^{4+} precursors in DMSO solution *Energy Environ. Sci.* **14** 2369–80

[113] Wang J *et al* 2022 Ge bidirectional diffusion to simultaneously engineer back interface and bulk defects in the absorber for efficient CZTSSe solar cells *Adv. Mater.* **34** 2202858

[114] Xin H, Vorpahl S M, Collord A D, Braly I L, Uhl A R, Krueger B W, Giner D S and Hillhouse H W 2015 Lithium-doping inverts the nanoscale electric field at the grain

boundaries in $Cu_2ZnSn(S,Se)_4$ and increases photovoltaic efficiency *Phys. Chem. Chem. Phys.* **17** 23859–66
[115] Cabas-Vidani A *et al* 2018 High-efficiency $(Li_xCu_{1-x})_2ZnSn(S,Se)_4$ kesterite solar cells with lithium alloying *Adv. Energy Mater.* **34** 1801191
[116] Guo H *et al* 2020 An efficient Li^+-doping strategy to optimize the band alignment of a $Cu_2ZnSn(S,Se)_4$/CdS interface by a Se & LiF co-selenization process *J. Mater. Chem.* A **8** 22065–74
[117] Geng H *et al* 2023 Two-step cooling strategy for synergistic control of Cu_{zn} and Sn_{zn+} defects enabling 12.87% efficiency $(Ag, Cu)_2ZnSn(S,Se)_4$ solar cells *Adv. Funct. Mater.* **33** 2210551
[118] Niu C, Gong Y, Qiu R, Zhu Q, Zhou Y, Hao S, Yan W, Huang W and Xin H 2021 11.5% efficient $Cu_2ZnSn(S,Se)_4$ solar cell fabricated from DMF molecular solution *J. Mater. Chem.* A **9** 12981–7
[119] Cui Y, Wang M, Dong P, Zhang S, Fu J, Fan L, Zhao C, Wu S and Zheng Z 2022 DMF-based large-grain spanning $Cu_2ZnSn(S_xSe_{1-x})_4$ device with a PCE of 11.76% *Adv. Sci.* **9** 2201241
[120] Wu S-H, Chang C-W, Chen H-J, Shih C-F, Wang Y-Y, Li C-C and Chan S-W 2017 High-efficiency $Cu_2ZnSn(S,Se)_4$ solar cells fabricated through a low-cost solution process and a two-step heat treatment *Prog. Photovolt.* **25** 58–66
[121] Su Z *et al* 2020 Device postannealing enabling over 12% efficient solution-processed Cu_2ZnSnS_4 solar cells with Cd^{2+} substitution *Adv. Mater.* **32** 2000121
[122] Tian Q, Lu H, Du Y, Fu J, Zhao X, Wu S and Liu S 2018 Green atmospheric aqueous solution deposition for high performance $Cu_2ZnSn(S,Se)_4$ thin film solar cells *Sol. RRL* **2** 1800233
[123] Zhao X, Pan Y, Zuo C, Zhang F, Huang Z, Jiang L, Lai Y, Ding L and Liu F 2021 Ambient air-processed $Cu_2ZnSn(S,Se)_4$ solar cells with over 12% efficiency *Sci. Bull.* **66** 880–3
[124] Xu X, Guo L, Zhou J, Duan B, Li D, Shi J, Wu H, Luo Y and Meng Q 2021 Efficient and composition-tolerant kesterite $Cu_2ZnSn(S,Se)_4$ solar cells derived from an *in situ* formed multifunctional carbon framework *Adv. Energy Mater.* **11** 2102298
[125] Yin K, Xu X, Wang M, Zhou J, Duan B, Shi J, Li D, Wu H, Luo Y and Meng Q 2022 A high-efficiency (12.5%) kesterite solar cell realized by crystallization growth kinetics control over aqueous solution based $C_2ZnSn(S,Se)_4$ *J. Mater. Chem.* A **10** 779–88
[126] Guo Q, Hillhouse H W and Ahrawal R 2009 Synthesis of Cu_2ZnSnS_4 nanocrystal ink and its use for solar cells *J. Am. Chem. Soc.* **131** 11672–3
[127] Guo Q, Ford G M, Yang W-C, Walker B C, Stach E A, Hillhouse H W and Agrawal R 2009 Fabrication of 7.2% efficient CZTSSe solar cells using CZTS nanocrystals *J. Am. Chem. Soc.* **131** 12054–5
[128] Ford G M, Guo Q, Agrawal R and Hillhouse H W 2011 Earth abundant element $Cu_2Zn(S_{1-x}Ge_x)S_4$ nanocrystals for tunable bandgap solar cells: 6.8% efficient device fabrication *Chem. Mater.* **23** 2626–9
[129] Cao Y *et al* 2012 High-efficiency solution-processed $Cu_2ZnSn(S,Se)_4$ thin-film solar cells prepared from binary and ternary nanoparticles *J. Am. Chem. Soc.* **134** 15644–7
[130] van Embden J, Chesman A S R, Gaspera E D, Duffy N W, Watkins S E and Jasieniak J J 2014 $Cu_2ZnSnS_{4x}Se_{4(1-x)}$ solar cells from polar nanocrystal inks *J. Am. Chem. Soc.* **136** 5237–40

[131] Miskin C K, Yang W-C, Hages C J, Carter N J, Joglekar C S, Stach E A and Agrawal R 2015 9.0% Efficient $Cu_2ZnSn(S,Se)_4$ solar cells from selenized nanoparticle inks *Prog. Photovolt. Res. Appl.* **23** 654–9

[132] Jordan D C, Kurtz S R, VanSant K and Newmiller J 2016 Compendium of photovoltaic degradation rates *Prog. Photovoltaics Res. Appl.* **245** 978–89

[133] Theelen M and Daume F 2016 Stability of Cu(In, Ga)Se_2 solar cells: a literature review *Sol. Energy* **133** 586–627

[134] Mei-Zhen G, Ke X and Fahmer W R 2009 Study of the morphological change of amorphous ITO films after temperature-humidity treatment *J. Non-Cryst. Solids* **355** 2682–7

[135] Theelen M, Hans V, Barreau N, Steijvers H, Vroon Z and Zeman M 2015 The impact of alkali elements on the degradation of CIGS solar cells *Prog. Photovolt. Res. Appl.* **23** 537–45

[136] Fjallstrom V, Salome P M P and Hultqvist A 2013 Potential-induced degradation of $CuIn_{1-x}Ga_xSe_2$ *IEEE J. Photovolt.* **3** 1090–4

[137] Neubauer C, Samieipour A, Oueslati S, Danilsaon M and Meissner D 2019 Ageing of kesterite solar cells 1: degradation processed and their influence on solar cell parameters *Thin Solid Films* **669** 595–9

Chapter 5

Colloidal quantum dot solar cells

Colloidal quantum dot solar cells (CQD-SCs) leverage the unique properties of QDs as semiconductor nanocrystals for solar conversion. Of all the solution-processed emerging photovoltaic (PV) solutions, CQD-SCs have seen a drastic increase in power conversion efficiencies (PCEs) upwards of 20% in conventional configurations and 30% for perovskite/CQD tandem solar cells, and with improvements in device engineering are expected to have even more competitive efficiencies in the coming years. While there are different device architectures, the key principle involves absorption of light photons which excite electrons from the valence band (VB) to the conduction band (CB), creating electron–hole pairs within the QD, and then separating and extracting these charge carriers to generate a current. This separation can be achieved by using a heterojunction, where a *p*-type CQD layer is combined with an *n*-type material like zinc oxide, where a depletion region forms at the interface, driving electrons to one electrode and holes to the other, effectively creating a potential difference and driving current flow. Alternatively, a Schottky contact can be used for charge extraction. Figure 5.1 illustrates some essential features of CQD-SCs including, solution-processing methods, absorption features and quantum-confined nanosized dimensions [1]. A salient feature of CQD-SCs is that CQDs are synthesized in, and processed from, the solution phase, and their processing into thin films is typically carried out at or near room temperature [2]. Fundamentally, CQDs offer strong optical absorption due to their direct bandgap as well as the light weight and low materials and fabrication costs. The most essential feature of why CQDs are applicable for solar cells is that their bandgap is readily tuned by controlling their size and shape during synthesis. This feature facilitates the use of a single-materials system to fabricate multiple-junction solar cells with potential efficiencies upwards of 49%.

doi:10.1088/978-0-7503-3255-2ch5

© IOP Publishing Ltd 2025. All rights, including for text and data mining (TDM), artificial intelligence (AI) training, and similar technologies, are reserved.

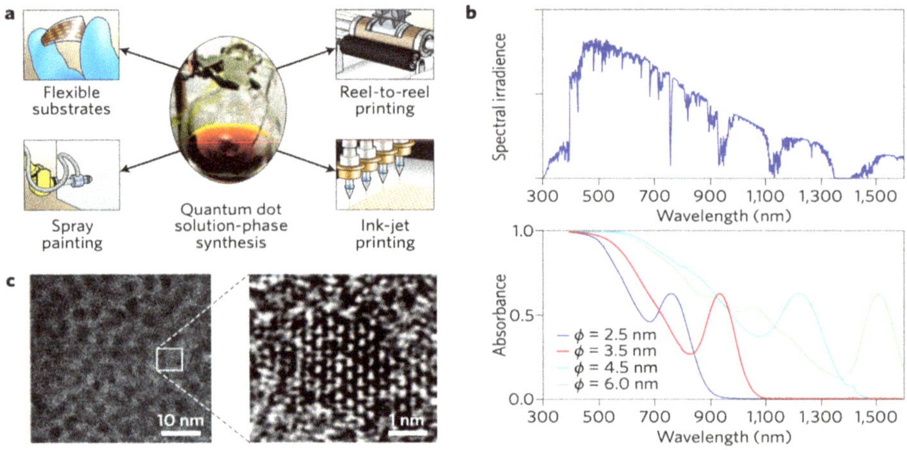

Figure 5.1. (a) After synthesis, the colloidal QDs are deposited onto a flexible substrate using different processing techniques. (b) Absorption spectrum of a colloidal QD film (bottom) matched to the power spectrum of the Sun reaching the Earth (top). (c) Transmission electron microscopy showing monodispersed (left) and crystalline (right) nature of colloidal QDs. Reprinted from [1] with permission from Springer Nature.

5.1 Structure and chemical properties of CQDs

Semiconductor QDs were initially reported in the early 1980s by Ekimov [3] and Brus [4] as nanoscale-sized crystals in the glassy matrix and colloidal solutions, respectively. After more than 30 years since their discovery, the growing progress in QD materials has led to their use in a vast array of optoelectronic applications. The so-called colloidal quantum dots (CQDs), inorganic semiconductor nanocrystals (typically <10 nm) typically capped with surfactant organic molecules and in some cases inorganic capping agents, dispersed in solution have provided a powerful platform for the development of several classes of cost-effective, solution-processed optoelectronic devices over the past decade, including PV cells, photodetectors and light-emission devices [5, 6].

A unique advantage of CQDs is the characteristic quantum confinement effect—size- and shape-dependent optoelectronic properties which are particularly useful for photo-absorption, especially in respect of optimized absorption and spectral adaptation for solar cells. Importantly, their ease of solution processing makes them attractive and advantageous candidates for third generation solar PV. Additionally, CQDs have large intrinsic dipole moments ideal for rapid charge carrier separation and extraction in solar cells and they exhibit unique multiple exciton generation (MEG), by emitting up to three electrons per photon and theoretically could increase PV efficiency from 20% to 65% [7].

5.1.1 Classification of CQDs

CQDs are classified by their composition, structure, and application. Common types include core-type (single material), core–shell (semiconducting core with a shell), and alloyed or heterostructured QDs. Other classifications include perovskite, chalcogenide, and carbon-based QDs. The most commonly used QDs for solar cell applications are the metal chalcogenide QDs which comprise a range of elements, such as I and VI (e.g., Ag_2X, X: S, Se, and Te), II and VI (ZnX and CdX, X: S, Se, and Te), III and V (InAs, InP, InSb, GaAs, and GaSb) IV and VI (PdX, X: S, Se, and Te), and I, III and VI_2 groups ($AgInX_2$, $CuInX_2$, X: S, Se, and Te). Other types of group IV QDs, such as carbon, silicon, and germanium QDs with size- and composition-dependent optoelectronic properties, have also been realized via wet-chemical approaches.

5.1.1.1 Single-core CQDs

Single-core homo-structural CQDs of uniform composition and phase comprising an inorganic crystalline core, surface atoms and surface capping/passivating organic ligands are the simplest and most widely studied. Their electronic and charge transport properties are controlled by their size, shape, stoichiometry and surface chemistry, and they are useful for thin film solid-state electro-optical devices such as solar cells. A critical characteristic is that the surface atoms do not have complete valence states arising from incomplete bonding (unsaturated or *dangling bonds*) and due to the high surface-to-volume ratio manifest as surface defects allowing for donor or acceptor electronic trapping states within the bandgap [5]. These trapping states facilitate non-radiative charge carrier recombination which increase luminescence decay lifetimes, thereby decreasing quantum yield and efficiency. The surface-active capping ligands passivate the dangling bonds, reducing the density of states associated with these defects improving emission properties with longer radiative recombination lifetime recombination. Even though the most effective passivating ligands that produce good quality, stable CQDs with relatively minimal surface defects are long insulating ligands such as dodecanethiol, oleylamine and oleic acid, they can be detrimental to efficient electronic transport in QD thin films because of larger interdot spacing and low film density [5, 6]. To minimize this issue, post-synthetic ligand exchange with short, conductive ones or inorganic anions such as halides to produce denser, more compact thin films with smaller interdot spacing is a commonly employed strategy to improve electrical properties.

5.1.1.2 Core–shell CQDs

The distinctive optoelectronic properties of CQDs by virtue of their composition and structure can be modulated by the relative conduction and valence band-edge alignment of the core semiconductor and shell semiconductor in core–shell heterostructures, as shown in figure 5.2 [5, 8, 9]. The inorganic shell enables complete and long-lasting isolation of the core atoms from the surrounding environment. Importantly, the optoelectronic properties can be designed and tuned by tailoring the shell thickness. The choice of the core and shell materials depends on two parameters: band offset alignment and their lattice mismatch [5]. Typically, core–shell CQDs display improved photoconductivity efficiency, decreased charge carrier response time, enhanced photoluminescence and increased photostability. Based on the alignment of the CB and VB in core–shell heterostructures, QDs have been classified into different types: type-I, inverse type-I, type-II, and quasi-type-II

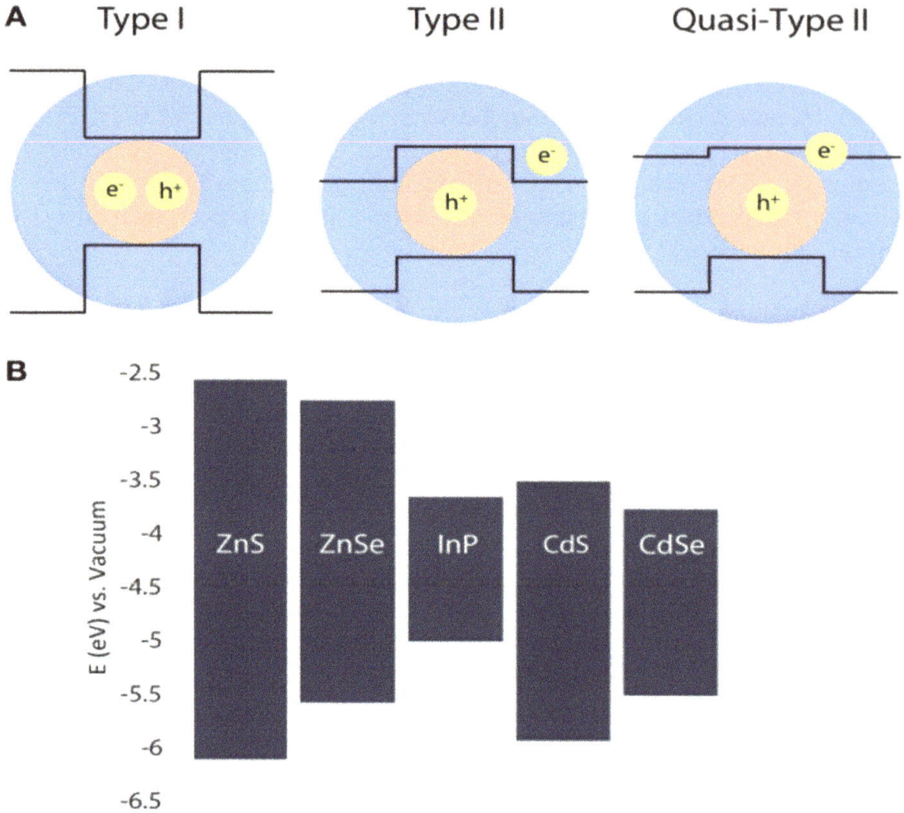

Figure 5.2. (A) Schematic representation of the relative positions of conduction and valence bands in different core–shell heterostructures. (B) Relative positions of the conduction and valance bands of the semiconductors used in this study. Reprinted from [14] CC BY 4.0.

heterostructures [5, 10]. As shown in figure 5.2, type-I heterostructures have a smaller bandgap core with a wider bandgap shell. As a result, both charge carriers are completely localized in the core and the bandgap emission is almost preserved or undergoes a small red shift [8]. In these, the shell more effectively passivates the core than the capping ligand, thereby improving its optical properties and increasing its stability against photo-bleaching [5]. For example, the wide-band-gap semiconductors, ZnS and ZnSe have become extensively used as an effective shell to improve the optical properties of a range of metal chalcogenide QD cores [11]. Contrastingly, the inverse type-I QDs have the bandgap of the shell localized in the bandgap of the core, and the electron–hole pair is confined entirely or partially in the shell [8]. For these, the charge carrier recombination occurs across the core–shell interface, and consequently, the photoluminescence (PL) emission wavelength is smaller than that of the bandgap comprising materials due to spatially indirect transitions. Also, type-II QDs exhibit longer exciton decay lifetimes than type-I and can allow access to wavelengths that would otherwise be unavailable for a single material, making them more suitable for PV applications [11]. Interestingly, there can be a change from type-I to type-II or quasi-type-II structure upon increasing the shell thickness. Importantly, it is noted that the degree of lattice mismatch in core–shell heterostructures influences the stress at the interface through dislocations that act as non-radiative recombinations which affect the quantum yield. For example, Ivanov and co-workers reported highly luminescent colloidal CdS-core/ZnSe-shell heterostructured QDs that exhibited type-II carrier localization regime by spatially separating electrons and holes between the core and the shell, respectively [12]. Their emission wavelengths were controlled by varying both core radius and shell width, from 500 to 650 nm. Their photoluminescent quantum yield (PLQY) increased by 15% and further tripled by alloying the interfacial layer of ZnSe with a small amount of CdSe. The quasi-type-II core–shell heterostructures are such that the electrons and holes are not fully confined to either the core or the shell. In essence, the band structure alignment of the core and shell materials is neither a pure type-I nor a pure type-II. For example, $CuInS_2$/ZnS QDs have type-I band alignment where both the lowest energy electron and hole wavefunctions are confined in the core [13]. The CB and VB offsets between $CuInS_2$ and CdS are ~0.05 and 0.95 eV, respectively and therefore, the offsets allow the formation of quasi-type II core–shell QDs where holes are well confined in $CuInS_2$ and electrons are delocalized among $CuInS_2$ and CdS.

5.1.2 Categories of CQDs

5.1.2.1 Transition metal chalcogenide CQDs

Parts of this subsection have been reproduced from [15] with permission from the Royal Society of Chemistry.

Metal chalcogenides are a diverse cohort of semiconductor materials that possess a pronounced covalent character because of the low electronegativity differences between metals and chalcogen elements like sulfur, selenium, or tellurium, which leads to a strong mixing of the *s* and *p* atomic orbitals of the chalcogen with the outer *s* and *p* atomic orbitals of the metal. This mixing results in broad VBs and CBs and narrower energy gaps, enabling visible light absorption [15]. Metal chalcogenide QDs are indispensable semiconductors because of their unique and functional properties, associated with both intrinsic (quantum confinement) and extrinsic (high surface area) effects, as dictated by their size, shape, and surface characteristics. Thus, they show considerable promise for diverse applications, including PV. Their diversity in structure and properties is dependent on their composition and can be classified as binary (II–VI), ternary (I–III–VI and I–IV–VI) and quaternary (I–II–IV–VI) compounds. In particular, copper-based multinary chalcogenides, which are II–VI-derived semiconductors, are composed of environmentally friendly elemental components and possess suitable bandgaps for efficient absorption of broad solar irradiation as well as high absorption coefficients [15]. Therefore, they represent one of the most promising candidates for solar energy conversion. Copper-based chalcogenides exhibit significant diversity in terms of stoichiometries and crystal structures, encompassing numerous non-stoichiometric phases and a wide array of solid solutions [16]. The reasons for this phenomenon include: (1) the small size and low electronegativity differences between chalcogens and metals; (2) the fact that the formal oxidation state of chalcogens is not always –2; (3) the ability of chalcogens to form bonds with other chalcogens; (4) the variable valences exhibited by transition metals; and (5) the higher propensity of metal atoms to form bonds with other metal atoms [15].

What makes metal chalcogenide materials attractive is that chemical synthesis methods are suitable for obtaining large quantities of size-controlled semiconductor QDs and are particularly well developed for II–VI-type semiconductor QDs [17]. This has allowed for extensive investigation of their quantum-confined optical properties, for example of CdS and CdSe QDs. However, most of the conventional binary compound semiconductors contain toxic elements such as Cd, Pb, Hg, and Se. Therefore, semiconductor QDs with non-toxic constituents such as InP, InAs, and GaAs QDs and impurity-doped ZnS QDs have been studied as alternatives [18]. Importantly, ternary compound QDs that are type I–III–VI_2 (chalcopyrite) such as $CuInS_2$, $CuInSe_2$, $AgInS_2$, $AgInSe_2$ and $AgGaS_2$ have been investigated as more environmentally friendly nanomaterials compared to the toxic II–VI Cd-based QDs. Chalcopyrite QDs have attracted much interest as strong candidates for solar cell materials because they have direct bandgaps well matched to the solar spectrum, high absorption coefficients, and high environmental stability and have been successfully synthesized via solution-phase routes.

Binary II–VI, III–V and III–VI nanocrystals

Group II–VI semiconductor compounds (II = Zn, Cd; VI = S, Se, Te) have two common crystal structures: cubic zinc blende (ZB) and hexagonal wurtzite (WZ), as illustrated in figure 5.3(a) and (b) [15]. The WZ and ZB phases show many similarities in terms of atomic arrangements where each atom has four neighbouring bonds but only the third nearest neighbour geometry can distinguish the crystal structures. The main differences between the ZB and WZ structures are determined by their dihedral conformations or the handedness of the fourth interatomic bond. Most group II–VI materials are direct bandgap semiconductors with high optical absorption and emission coefficients [19]. Group III–VI semiconductor materials show great diversity with many different stoichiometries. Some of these include a defect wurtzite structure for Ga_2E_3, a defect spinel for In_2E_3 and layered structures for compounds of stoichiometries, ME (E = S, Se) and are typically direct wide bandgap semiconductors.

Of the II–VI class, cadmium telluride (CdTe) is a leading candidate for solar cell applications because of its bandgap of 1.45 eV and high optical absorption coefficient. Like CdTe, lead sulphide, PbS is one of the widely studied chalcogenides for solar cell applications. It has a narrow bandgap of 0.41 eV and is appealing because it exhibits strong quantum confinement effects due to the large Bohr radii of both electrons and holes [20]. There have been many approaches to PbS nanoparticles with the chemical routes being favoured over physical methods due to size and shape control that can be achieved during the synthesis procedure.

The iron sulfide class exists in various and variable stoichiometries including FeS_2, Fe_2S_3, Fe_3S_4, Fe_7S_8, $Fe_{1-x}S$, FeS, and $Fe_{1+x}S$ [20]. All these sulfides have

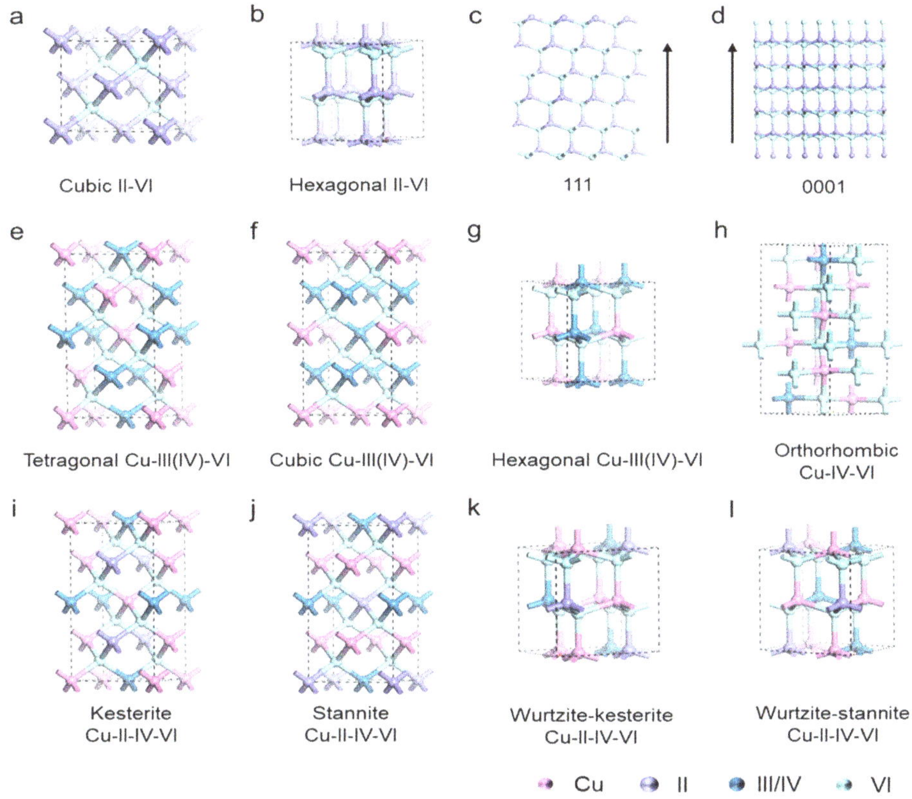

Figure 5.3. Crystal structures of copper chalcogenides: (a) cubic II–VI, (b) hexagonal II–VI, (c) and (d) atomic arrangement along [111] and [0001] directions for cubic and hexagonal II–VI, respectively, (e) tetragonal Cu–III (IV)–VI, (f) Cu–III(IV)–VI, (g) hexagonal Cu–III(IV)–VI, (h) orthorhombic Cu-IV–VI, (i) kesterite Cu–II–IV–VI, (j) stannite Cu–II–IV–VI, (k) wurtzite–kesterite Cu–II–IV–VI, and (l) wurtzite–stannite Cu–II–IV–VI. Adapted from [15] with permission from Royal Society of Chemistry.

either the NiAs–$Cd(OH)_2$ or pyrite structures with FeS_2 also adopting the marcasite structure. Pyrite FeS_2 is a nonmagnetic semiconductor with an optical bandgap of 0.9 eV and is a cheap and less toxic material for solar cells. Similarly, copper sulfide (CuS) is a less expensive material because of its earth abundant copper and is less toxic compared to Cd and Pb counterparts [20]. The class of copper sulfides are *p*-type semiconductors with a direct bandgap between 1.2 and 2.0 eV dependent on their phases and stoichiometry. Of the fourteen different identifiable phases the more common ones are chalcocite (Cu_2S), djurleite ($Cu_{1.94}S$), digenite ($Cu_{1.8}S$), roxbyite ($Cu_{1.75}S$), covellite (CuS), and villamaninite (CuS_2).

III–V semiconductors are compounds formed by combining elements from group III (B, Al, Ga, In) and group V (N, P, As, Sb) of the periodic table. Among them, the most common ones are GaAs, InAs, GaN, InN, InP, and their alloys and are known for their direct bandgaps, higher carrier mobility, and smaller electron effective mass, crucial for optoelectronic applications. The materials have either zinc blende or wurtzite crystal structures, most of which have a direct bandgap. In particular, the wurtzite nitrides have wide bandgap values ranging from 0.64 to 6.2 eV, while the zinc blende arsenides vary from 0.35 to 2.24 eV, thus are highly attractive for optoelectronic devices [21]. In addition to their direct bandgaps, III–V semiconductors are particularly attractive for applications if CQD-SCs owing to their large excitonic Bohr radius thereby exhibiting size-tunable properties as QDs [22].

Ternary I–III–VI nanocrystals

Among the different classes of transition metal chalcogenides, the synthesis and application of ternary I–III–VI metal chalcogenide QDs have been widely investigated with tremendous interest in those composed of copper and zinc which are inexpensive, abundant and more environmentally friendly. In particular, the I–III–VI copper-based congeners, copper indium disulfide and diselenide, $CuInS_2$ (CIS) and $CuInSe_2$ (CISe) QDs exhibit high absorption coefficient and comparatively narrow bandgaps that are spectrally requisite for solar energy absorption [23]. The most reported I–III–VI compound is CIS, which exhibits a stable tetragonal chalcopyrite (CH) structure at room temperature. The cations in the CH structure are ordered in the unit cell as shown in figure 5.3(e), and the anions are tetrahedrally coordinated to two In^{3+} ions and two Cu^+ ions [15]. The tetragonal structure of Cu–III–VI semiconductors is similar to that of zinc blende II–VI binary compounds, with Cu^+ and In^{3+} cations occupying the cation sublattice in an ordered alignment. As the more thermodynamically stable phase, the CH structure of CIS has been mostly studied because of its tolerance for a large range of anion and cation off-stoichiometry which allows doping defects of either p-type or n-type conductivity, tunable bandgap and PL properties [24]. These tunable properties are influenced by the presence of intrinsic defects such as S vacancies and In interstitials as donors and copper vacancies as acceptors; types implicated in the off-stoichiometric phases. In addition to the CH phase, the phase diagram of the Cu_2S–In_2S_3 ternary system suggests that $CuInS_2$ exists in the ZB and WZ phases with all three, stable in nanocrystals at room temperature. The CH structure undergoes a transformation to the ZB structure, accompanied by disordering of the cation sublattice [15]. The ZB structure exhibits similarities to the CH structure in terms of lattice parameters, except for the key difference in the cation sublattice.

Additionally, $AgInS_2$ has an orthorhombic structure or chalcopyrite-like structure, and is a promising light-absorbing material. In contrast to the usual chalcopyrite tetragonal and orthorhombic phases of $AgInS_2$ NCs, the synthesis of metastable cubic $AgInS_2$ has been reported using a solvothermal approach [25]. Also, it is well established that one phase of the ternary Ag–III–VI_2 semiconductor is an In-rich compound of the general formula, $AgIn_5S_8$, having a cubic spinel structure. $AgIn_5S_8$ has direct bandgaps of 1.80 eV at 300 K and 1.90 eV at 96 K. Copper indium gallium (di)selenide (CIGS) is a I–III–VI_2 semiconductor material of the chemical formula of $CuIn_{1-x}Ga_xSe_2$, where the value of x can vary from 0 (pure copper indium selenide) to 1 (pure copper gallium selenide). CIGS is a tetrahedrally bonded semiconductor, with the chalcopyrite crystal structure, and a bandgap varying continuously with x from about 1.0 eV to about 1.7 eV. CIGS-based thin film solar cells have been one of the prominent emerging types.

Though these ternary I–III–VI_2 semiconductors, such as $CuInS_2$ and $AgInS_2$ are useful as light-harvesting materials, the presence of sub-bandgap states from surface defects and/or donor–acceptor pairs (DAPs) introduces complexity upon photo-excitation [26]. It has been established that when photo-irradiated, the photo-generated charge carriers in the QDs undergo rapid relaxation to populate intrinsic DAP states while competing with charge carrier recombination. Interestingly, these defect-related DAP states can be activated through sub-bandgap excitation and, thus, extend the absorption range to the near-infrared region and give rise to spectrally broad emission. In cases where surface defects are prominent, they mediate emission through non-radiative decay or fast recombination. The structural diversity characteristic of ternary materials is considered to be the origin of various donor–acceptor levels [18]. As such, efforts to generate intense, defect-free semi-conductor band-edge PL from I–III–VI semiconductor nanomaterials have been particularly challenging to accomplish. It has also been established that emission originating from several sets of DAPs is inevitably generated in I–III–VI-based semiconductor nanomaterials, giving rise to PL spectra with a full width at half maximum (FWHM) larger than 100 nm, or 300 meV.

Quaternary I–II–III–VI nanocrystals

Quaternary chalcogenides are a class of materials composed of four different elements, one of which is a chalcogen, and three other elements, typically metals. They can be considered as being derived from successive substitution of the III atoms in I–III–VI ternary chalcogenide compounds with II and IV atoms with tetrahedral bonding of the formula, I–II_2–III–VI_4 (I = Cu, Ag; II = Zn, Cd; III = Al, Ga, In; VI = S, Se, Te) [15, 27]. Certainly, the increased number of elements leads to more complicated structural and electronic properties compared to binary and ternary chalcogenide compounds. Of the quaternary series of compounds, the most commonly studied ones are Cu_2ZnSnS_4 (CZTS), Cu_2CdSnS_4 (CCTS), Cu_2ZnGeS_4 (CZGS), and Cu_2CdGeS_4 (CCGS). Among these, CZTS and CZTSe have emerged as the most promising quaternary chalcogenides for QD solar cells because they consist of earth abundant and non-toxic elements, and their bandgaps are tunable in the 1.0–1.5 eV range. Another attractive attribute of the I–II_2–III–VI_4 class is that because of the variety of chemical compositions, doping on different sites is also possible [27]. These features allow for unique electronic and optical tunability, particularly relevant for spectrally adaptable solar cells.

In particular, CZTS has four different structures: (1) kesterite; (2) stannite; (3) wurtzite–stannite; (4) and wurtzite–kesterite. As shown in figure 5.3(i) and (j), the kesterite and stannite structures can serve as ZB-derived structures with differences in cation occupation [15]. The CZTS compound exhibits an ordered kesterite structure at room temperature, characterized by a $1 \times 1 \times 2$ cubic ZB anion lattice with distinct cation layers: one composed of Cu and Zn, and another composed of Cu and Sn along the *c*-axis direction. The kesterite structure of CZTS is reported to undergo a transition to the stannite structure at 533 K, which is similar to the kesterite but exhibits different cation occupation, with a layer of Cu ordered alongside a layer of Zn and Sn (figure 5.3(j)). Heating of the cubic CZTS results in a gradual and irreversible transformation into stannite. Importantly, different phases of these CQDs can be accessed under different synthetic conditions.

5.1.2.2 Perovskite QDs

The colloidal synthesis of low-dimensional metal halide perovskite QDs (PQDs) has endowed the emerging semiconductors with new peculiarities in optical and electrical properties, such as quantum confinement-induced size tunability, separated perovskite crystallization from film deposition, layer-by-layer processing, improved structural stability, etc, making them promising candidates for next-generation PV [28]. PQDs are a class of semiconductors with a broad range of outstanding optoelectronic properties of the general chemical formula, ABX_3, where A is a monovalent alkali such as Cs^+ and Rb^+ and methylammonium (MA) cation, $CH_3NH_3{}^+$ and formamidinium (FA) cation, $CH(NH)_2{}^+$; B is a bivalent cation such as Ge^{2+}, Pb^{2+}, Cu^{2+} and Sn^{2+}; and X is a monovalent anion such as Cl^-, Br^-, and I^- [29, 30]. Their unique optoelectronic properties feature high fluorescence efficiency/quantum yield, narrow emission spectrum, quantum confinement electronic effects, direct bandgaps, strong light–matter interactions, broadband absorption, reduced PL blinking, high electron–hole mobility, and long carrier diffusion length, and they are spectrally tunable from ultraviolet to near-infrared (NIR) dependent on their composition, such as by replacing the A-site monovalent cation or X-site halogen anion [8]. Another important feature is that their structures are highly tolerant to dopant impurities which facilitates optimized and tunable electronic properties. For example, Yuan and co-workers [30] reported that ytterbium, Yb^{3+} ions doped into $CsPbI_3$ QDs tended to occupy the Pb^{2+} site, as well as remove Cs^+ and I^- vacancies. As a result, the number of defects and trap states caused by surface and lattice vacancies were found to be reduced, improving QD PLQY, crystallinity, transport and thermal stability. The other advantage is that they are not difficult to prepare and are compatible with room temperature deposition processes [8]. Bandgap tunability for the lead halide perovskites (ABX_3) is often attained by changing the cation [31]. In particular, the gradual increase in the cationic radius of the A site will lead to a gradual increase in the B–X bond length and this eventually leads to the enlargement of the perovskite lattice and the decrease in bandgap. Also, an increase in the anion causes a widening of the bandgap. Furthermore, the ratio of the anions enables the tuning of the emission from visible to near-infrared regions with different fluorescence lifetimes.

Despite the attractive optoelectronic features, the potential of perovskite materials including PQDs is impeded by several issues related to their stability. In particular, lead halide perovskites are highly sensitive to polar solvents due to their inherent ionic nature [32]. They usually lose optical properties, surface ligands (for perovskite nanomaterials), and even structural integrity in polar organic solvents or water and also decompose in moist conditions. Their low formation energy makes them susceptible to environmental stress like light, oxygen and heat. A critical feature of colloidal PQDs is that the surface ligands are not strongly bound due to their ionic nature. This typically involves a dynamic exchange of surface ligands and anions which ultimately affects their optical property and colloidal stability. For example, Roo and co-workers [33] showed that as-synthesized $CsPbBr_3$ PQDs were stabilized with oleylammonium bromide or oleylammonium oleate, but in fast exchange between a free and bound state. Although the ligand density was found sufficiently high to fully passivate the surface in nonpolar media, the nanocrystals became colloidally unstable and structural integrity and PL emission were diminished when polar solvents were added in excess, presumably due to rapid desorption of the ligand. The colloidal stability, PL emission and high quantum yields were achieved with combinations of excess oleic acid and oleylamine ligands which strongly bind to the surface of the nanocrystals, as shown in figure 5.4.

5.1.3 Optical and electronic properties

5.1.3.1 Size- and shape-tunable electronic properties

The fundamental characteristic of CQDs is that their electro-optical properties display size-dependence as a consequence of quantum confinement. This phenomenon is premised on the Heisenberg uncertainty principle in which electrons and holes are confined to a specific region of space, comparable to the size of the dot, the Bohr exciton radius. This so-called confinement acts in all three spatial dimensions, making QDs essentially zero-dimensional objects with discrete electronic states, as shown in figure 5.5(a). In an intuitive approximation for the bandgap of the QD, E_g^{QD} as a function of the particle size is determined using equation [5.1], [4]

$$E_g^{QD} = E_g^{\mathrm{bulk}} + \frac{h^2\pi^2}{2r^2}\left(\frac{1}{m_e^*} + \frac{1}{m_h^*}\right) - \frac{1.8e^2}{4\pi\varepsilon_0\varepsilon_r r} \tag{5.1}$$

Here, E_g^{bulk} is the bulk semiconductor bandgap, r is the radius of the QD, and the confinement term includes the effective mass of both the hole and the electron (m_e^*; m_h^*). The excitonic term includes the permittivity in vacuum (ε_0) and in the QD material (ε_{r}). As a consequence, quantum-confined structures are generally classified in terms of their dimensionality (size and shape). Different from the bulk materials, the bandgap absorption and emission energy of quantum-confined nanostructures can be tuned over a wide range by changing the size of the quantum-confined dimensions [34]. Perhaps the best prototype is CdSe QDs, whose size-tunable bandgap ranges from 1.8 to 3.0 eV [35]. But a wide range of others including CdS, CdTe, ZnS, ZnSe, PbS/Se, PbS, InP, $CuInS_2$, and $AgInS_2$, are very attractive because their absorption and emission span a wide range, from the UV up to the NIR, determined by their size and shape. Figure 5.5(a) shows the effect of size reduction on the first exciton energy of PbSe QDs [36]. Also, a consequence of their small size is that these CQDs have a high surface area and a much larger absorption coefficient (typically in the range of $\sim 10^5 - 10^7$), particularly useful for solar absorption.

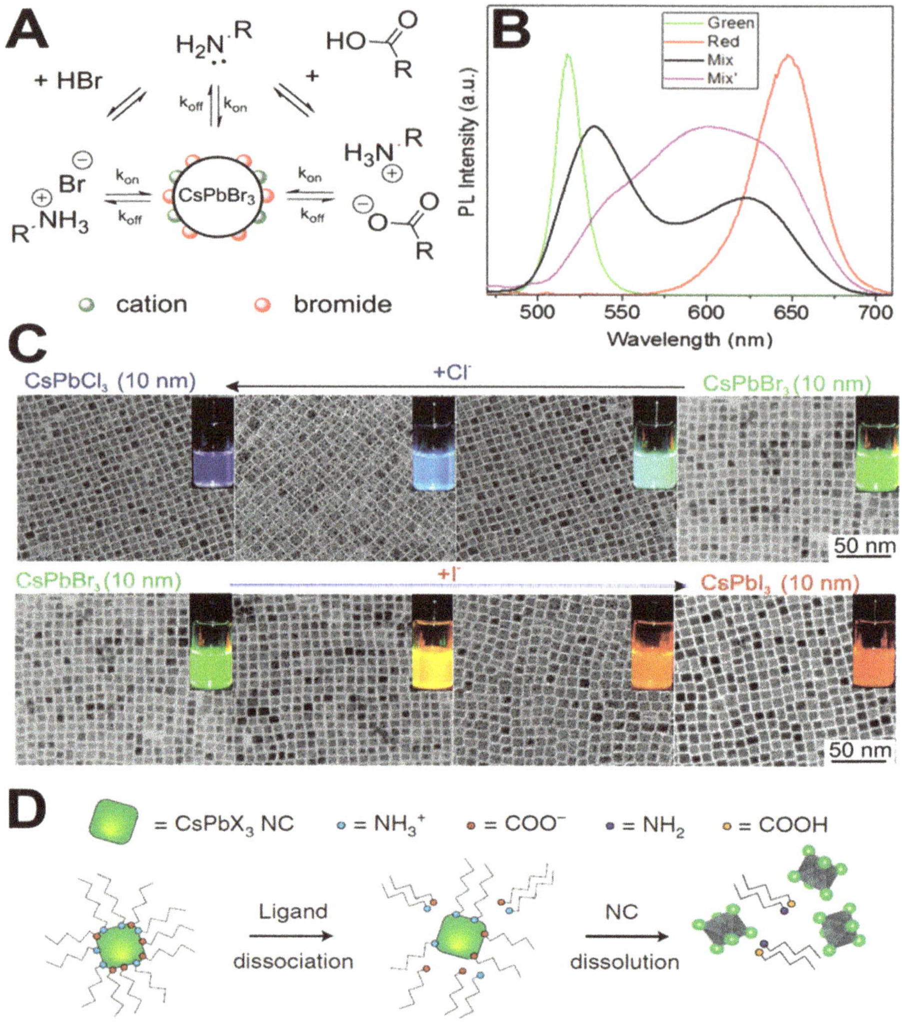

Figure 5.4. (A) Schematic representation of the dynamic surface of $CsPbBr_3$ PQDs. (B) PL spectra of green and red emissive PQDs. (C) Tuning the optical properties by treatment with various quantities of chloride or iodide anions. The figures show the evolution of TEM images and emission colours (under a UV lamp, l = 365 nm) upon forming mixed-halide $CsPb(Br/Cl)_3$ and $CsPb(Br/I)_3$ to fully exchanged $CsPbCl_3$ and $CsPbI_3$. (D) Schematic illustration that PQDs often lose their colloidal stability, or even structural integrity, due to the desorption of weakly bound ligands. Reprinted from [32] with permission from Royal Society of Chemistry.

5.1.3.2 Charge carrier multiplication—multiple exciton generation

There are many factors that limit solar cell efficiencies and hot charge carrier loss is an important performance-limiting mechanism in single-junction solar cells [6]. Hot carriers, electrons and holes that are excited with photon energies larger than the bandgap energy, usually rapidly relax to their respective band edges via phonon emission. In quantum-confined materials such as CQDs, the relative scarcity of available states can slow this process, possibly providing a pathway for capturing the

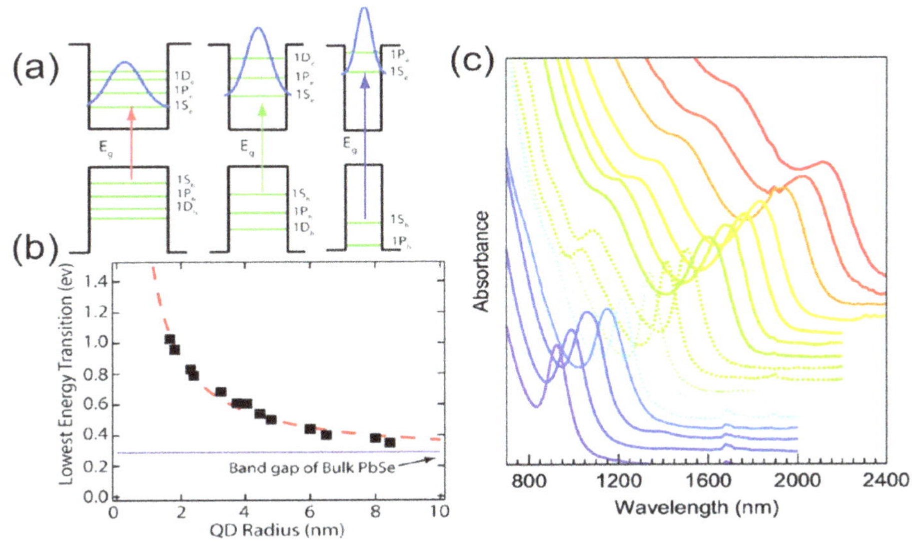

Figure 5.5. Quantum size effect. (a) Schematic demonstrating increase in quantum confinement with decreasing size of PbSe QDs. (b) Variation of the bandgap of PbSe QDs as a function of their size. (c) Red-shifted absorption spectra showing excitonic maxima of PbS QDs with sizes ranging from 3 to 10 nm [36]. Reprinted with permission from [36]. Copyright 2020 American Chemical Society.

energy in excess of the fundamental excitonic transition. This extra energy could potentially be harvested using either hot carrier solar cell concepts or MEG, which makes CQD solar cells attractive for improving efficiency beyond the Shockley–Queisser limit. MEG in QDs is an example of carrier multiplication (CM), where the absorption of a single photon generates multiple electron–hole pairs (excitons) [5, 6, 37, 38]. A high-energy, 'hot' electron or hole relaxes within the same band by creating new electron–hole pairs. This occurs via one or more impact ionization events whereby a pre-existing VB electron is excited to the CB via an Auger-type collision with a high-energy carrier. In an ideal MEG solar cell, the external quantum efficiency (EQE) would exceed 100% for energies greater than twice the bandgap energy because high-energy photons could produce more than one exciton. If a threshold of twice the bandgap energy could be realized for impact ionization in a CQD solar cell, the theoretical maximum PCE would shift to above 60% [6].

The primary criterion for MEG to occur is that the absorption photon energy is above the bandgap energy and at least one of the charge carriers created by absorption of a photon has sufficient energy to promote another electron across the bandgap, creating a second exciton [5]. Additionally, the multiexcitons produced within the QDs must be separated prior to Auger recombination, and the free-charge carriers must be transported to electron- and hole-accepting contacts. The first solar cells that utilized CM were reported by Semonin and co-workers [38] in 2011. As shown in figure 5.6(a), the devices using the 0.72 eV E_g of the PbSe CQDs achieved an EQE over 114% and an internal quantum efficiency (IQE) over 130% under 3 E_g excitation energy.

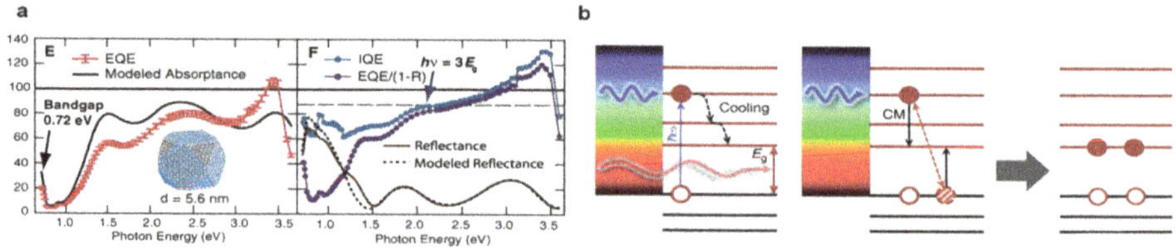

Figure 5.6. Carrier multiplication. (a) The measured EQE and IQE values of over 100% in CQD-SCs. (b) Schematic of energy losses in a device from cooling of hot carriers and schematic of the CM process. Reprinted from [37] with permission from Royal Society of Chemistry.

5.1.3.3 Charge carrier mobility

Charge carrier mobility in CQDs refers to how easily charge carriers can move through the CQD film under the influence of an electric field. While CQDs offer promising properties for optoelectronic devices, their mobility can be a limiting factor, particularly in applications like solar cells. Therefore, the transport of carriers including free-charge carriers and excitons within QD thin films is of particular interest for the better understanding of fundamental carrier transport dynamics and energy loss mechanisms in QD systems and optimization of efficiency in CQD-SCs [39]. As such, coupling of CQDs, especially along specific lattice planes, is essential in carrier transport and enhancing carrier mobility and diffusion length, leading to improved performance of CQD solar cells [40]. An important factor is the influence of interdot linking ligands on the CQD electronic energy band structure and carrier transport behaviour. Typically, short molecular lengths of linking ligands induce strong interdot coupling strengths, thereby forming an extended continuous energy band structure, which may result in an almost continuous carrier transport behaviour in CQD systems with proper doping and negligible spatial and energy disorders. Carrier transport between localized states in CQDs that contain photoexcited electrons and holes is quantitatively modelled through their tunnelling behaviour under a one-dimensional potential approximation in the presence of an applied electric field [39]. Overall, the interdot coupling strength that increases with reduced ligand-determined interdot distance has a significant influence on carrier transport physics. Through the enhancement of the interdot coupling strength, it has been found that the electrical properties of CQD systems evolve from a Coulomb blockade dominated insulating regime to a hopping conduction dominated semi-conducting regime in CQDs. Strong interdot coupling strength can also assist exciton dissociation into free electron and hole charge carriers during the tunnelling or hopping processes.

Carrier mobility can be improved in different ways. This can include exchanging long-chain organic ligands with short-chain organic or inorganic ligands which are essential for enhancing the coupling between QDs. Additionally, surface ligand exchange has become a significant process for passivating surface defects in CQDs which minimize trapping states and for improving the mobility of carriers and the performance of CQD solar cells. For example, the introduction of halide ligands has been shown to enhance carrier mobility compared with that of purely organic ligand systems [41]. Other approaches involved passivating with organic/inorganic hybrid passivation strategies [42]. It has also been shown that preferential carrier mobility can be achieved in anisotropic nanostructures, such as nanorods and tetrapods [37].

5.1.3.4 Trap density

Parts of this subsection have been reprinted from [39], with the permission of AIP Publishing.

The CQD surface is a highly dynamic region coordinated by ligands and can have drastic effects on the QD properties including the charge carrier characteristics and optical properties. These effects are typically ascribed to the lower coordination of surface atoms, also referred to as *dangling bonds*, compared to bulk atoms, which may lead to localized electronic states or highly reactive sites [43]. This results in the formation of shallow, or deep, mid-gap states as *surface traps* providing pathways for non-radiative exciton recombination, detrimental for QD-based optoelectronic applications. In this respect, the concentration of these electronic traps, described as trap density must be substantially eliminated since they can significantly impact the performance of QD-based devices like solar cells by reducing efficiency and stability. High trap density can lead to a decrease in PL intensity, a reduction in charge carrier mobility, and a decrease in the overall efficiency of QD-based devices. Uncapped bonds on the surface of QDs created during the ligand exchange process also induce defect trapping states which can act as a recombination center for the photo-generated carriers and also reduce the quasi-Fermi level splitting range under illumination, leading to lower V_{oc} [5]. Hence, the reduction of trap density is important for the improvement of solar cell performance.

The most exploited approach to eliminating surface traps is with core–shell QDs, which involves the growth of a (thick) shell of a wider bandgap material around the photoactive core, thus moving surface defects to the outer region of the inorganic shell [43]. This results in the exciton localized in the core, thus leading to almost unitary PLQY, as a signature of the low probability of trapping photogenerated charge carriers in the outermost surface sites. However, core–shell heterostructures have drawbacks in that charge carrier localization in the core hinders transport in core–shell QD solids. Outside of surface-related traps, it is established that the composition of CQD also leads to trap formation. In particular, nanocrystalline perovskites show PLQYs close to unity, whereas bare cadmium chalcogenides rarely exceed 10%, indicating strikingly different influence of trap states [44]. Whereas it has been assumed that full ligand coverage is necessary for perfectly passivated CQDs, it has been shown that for CQDs such as cadmium and lead chalcogenides, the composition and the crystal structure play a major role in the susceptibility of surfaces to defects and their forming trap states, as illustrated from computational studies in figure 5.7.

5.1.3.5 Dopant effects

One of the most effective strategies to tune the bandgap of CQDs for photo-absorption is by changing their constituent stoichiometries and by dopant inclusion. The inclusion of dopants increases the number of charge carriers and the density of

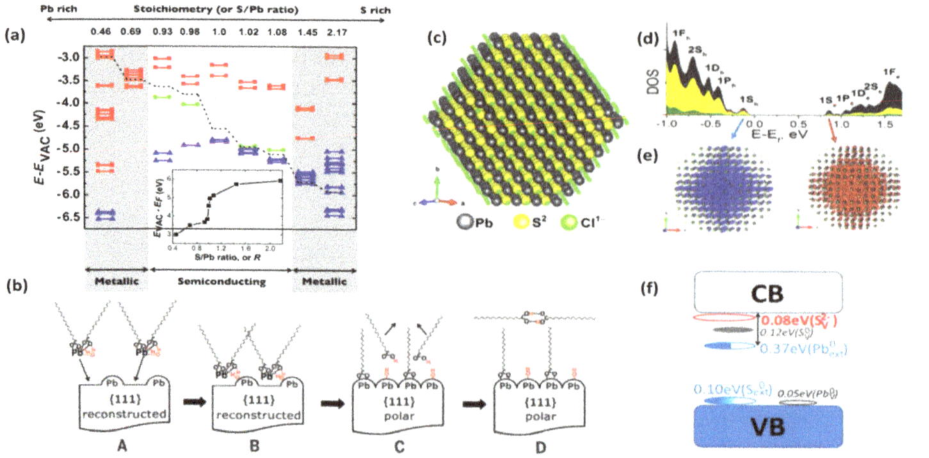

Figure 5.7. (a) Calculation of states for a bare PbS QD upon variation of the stoichiometry. (b) The polar (111) facet of as-synthesized CQDs is covered alternately by oleate and hydroxyl anions, the latter of which have formed during synthesis. Many computational approaches assume the bare (100) facets to be inert and use small Cl^- anions to passivate the polar (111) facets (c). The projected density of states (DOS) of a particle as shown in (d) is given in (c), highlighting the delocalized nature of the valence and conduction levels in (e). Neutral defects, although unlikely to form in solution, give rise to several trap levels (shallow for S and deep for Pb) (f). Reprinted with permission from [44]. Copyright 2020 AIP Publishing.

states within the electronic band structure thereby enhancing conductivity (diffusion length of charge carriers) and broadening the spectral response of the doped material. The creation of new energy states within the band structure shifts the Fermi level. That is, inclusion of acceptor impurity atoms increases the hole concentration and shifts the Fermi level closer to the VB edge resulting in a *p*-type material. Conversely, donor impurities shift the Fermi level closer to the CB as it increases electron concentration, resulting in an *n*-type material. An impurity can also be classified as shallow or deep based on the proximity to the edges of the CB minimum and VB maximum and the respective energy needed for ionization.

Lanthanide and transition metal ions have been effectively used to enhance optical properties and increase conductivity in the host QD [45–47]. In particular, they tend to increase the charge carrier lifetime which improves the charge separation processes, crucial for CQD-SC [48]. It has been widely reported that tunability of optical properties can be realized by varying the composition of the host QD and or the concentration of the dopant ion. For example, Cu-doped ZnS/$Zn_{1-x}Cd_xS$ QD alloys can be tuned from blue to red [49]. Zhang and co-workers reported the ability to control the emission properties by adjusting Cu concentration within CdS QDs without which NIR emission would not be possible [50]. Manganese(II) and nickel(II) are two of the most utilized metal dopants and as a result are established routes to modify material optoelectronic properties of QDs [51–54]. The energy transfer between the host and dopant can be very efficient. This is seen clearly in Mn-doped ZnS where there is direct coupling between the *d*–*d* states of the metal ion and that of the host with a characteristic emission at about

590 nm due to triplet-to-singlet electronic transition ($^4T_1 \rightarrow {}^6A_1$). The energy emitted by the dopant is red-shifted from that of the bandgap energy of the host, therefore self-absorption or quenching interferences by the host QD do not occur. Similarly, nickel-doped QDs have demonstrated enhanced and/or additional desirable features when compared to the undoped counterpart. To that end, Ni-doped ZnS, CdS, CdZnS, ZnSe and CdTeSe QDs have been prepared [54–56]. Yang and co-workers reported a red shift from 450 to 520 nm in Ni-doped ZnS QDs which is very likely due to the dopant [56].

5.2 Strategies for engineering colloidal quantum dots

Device efficiencies of CQD-SCs are related to the quality of the QD's chemical and electronic properties which are dependent on a number of factors including the nanoparticle's size distribution, shape control, crystallinity, phase purity, defect chemistry, stability, and surface quality. Engineering CQDs involves a range of strategies focussed on controlling their nucleation, growth and surface chemistry through manipulating their surface ligands, and assembly to tailor their properties for specific applications.

5.2.1 Synthesis: nucleation and growth control

For CQD-SCs, a critical factor is the synthesis of QDs with a highly monodisperse, narrow size distribution and uniform shape for which there are well understood mechanistic factors for achieving. The formation of monodisperse CQDs typically involves two steps: a rapid nucleation followed by a slow growth. The former is dependent on factors of temperature, interfacial tension, and degree of supersaturation in solution, and growth is dependent on effects of focussing and defocussing of particle size growth rate and size distribution [6]. In diffusion-controlled growth, a focussing model involves larger particles growing more slowly than smaller particles because the particle growth rate is inversely proportional to its radius. Here, narrower size distributions can be obtained provided secondary nucleation is avoided and all particles are growing. By contrast, the defocussing effect also known as Ostwald ripening is based on the solubility of the particles as a function of their size in which smaller particles dissolve and larger particles continue to grow due to the higher chemical potential of the smaller particles.

Colloidal synthesis of semiconductor nanoparticles has been established as an effective route for the fine control of size, morphology, composition, structure and optical properties [20, 57–59]. This is achieved through adjusting various parameters including surface passivation via ligands or core–shell formation, reaction/growth temperature, reaction/growth time and the nature of ligands and precursors employed. Synthesis of nanoparticles can be carried out via aqueous or organic media. However, organic based approaches tend to produce more monodispersed, non-agglomerated and phase-controlled nanoparticles including QDs.

Factors of time and temperature are important in nanoparticle synthesis, and their appropriate control is paramount to the synthesis of high-quality nanoparticles. For example, in the synthesis of ternary and quaternary chalcogenide nanoparticles at low temperatures, incomplete reactions occur which lead to undesirable binary phases. For example, Dilena and co-workers demonstrated the importance of reaction temperature on the phase purity of CIGS nanoparticles [60]. At low temperatures, CuS was predominant but with higher temperatures, more phase pure CIGS nanoparticles were grown. Another important factor to note is that increased temperature results in the higher rate of monomer addition to growing particles and thus it is expected that particle size will increase. In some cases, morphology changes may accompany an increase in size, as demonstrated by Zhou and co-workers [61]. They showed that CZTS nanoparticles underwent time-dependent morphology changes over a 24 h period.

During colloidal growth, to compensate for the high surface-to-volume ratio of CQDs, surface-passivating reagents known as surfactants or ligands are used. The majority of established CQD synthetic routes use surfactant molecules comprised of a long hydrocarbon tail and a polar head; such surfactants include oleic acid, oleylamine, trioctylphosphine oxide, or dodecane- thiol. These surfactants can tune the reactivity of the precursors, improving control over the nucleation and growth rate [6]. Also, capping ligands are effectively used to control the size of nanoparticles. In this respect, bulky ligands provide sufficient steric hindrance producing smaller nanoparticles whilst compact ligands give the reverse effect [62, 63]. Bulkier ligands tend to produce nanoparticles that are not agglomerated, however, they can provide an insulating barrier affecting charge transfer mechanisms. Thus, the length of the ligands used in nanoparticle synthesis is an important factor when one considers its application in optoelectronic devices, where the efficient charge transport between nanoparticles and in the thin film is critical for optimum device performance [64, 65]. This was demonstrated by Bhaumik and co-workers [65] where longer chain length thiol ligands resulted in a wider interdot spacing and as such, there was a reduction in the conducted energy of a light-emitting diode (LED) made from nanocrystalline thin films of the QDs.

Capping ligands have also been shown to affect growth processes and the morphology of the nanoparticles, as shown in transmission electron microscopic images of $CuInS_2$ nanoparticles in figure 5.8. In this case, Zhong and co-workers [66] demonstrated that with the addition of oleic acid, a mixture of triangular and

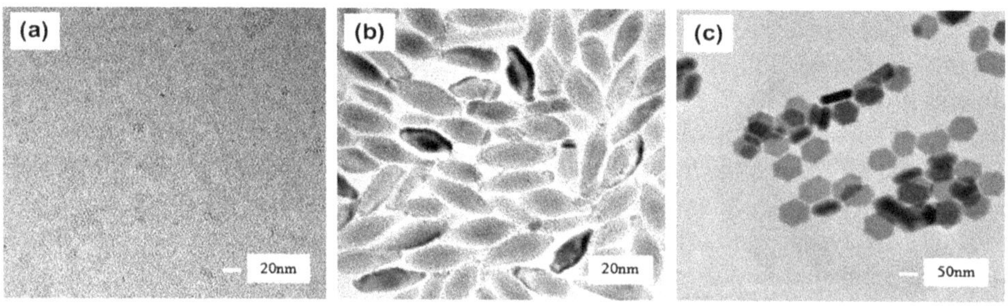

Figure 5.8. TEM images of CIS nanoparticles with (a) tri-*n*-octylamine (TOCA) (b) di-octylamine (DOCA) and (c) tri-*n*-octylphosphine (TOP). Reprinted with permission from [69]. Copyright 2012 Elsevier.

spherical shaped CIS nanoparticles were obtained, whilst without oleic acid nanorods were formed. The addition of oleylamine was reported to lower the decomposition temperatures in multicomponent materials, thus lowering the presence of unwanted phases [67]. There are numerous reports where capping ligands are used to control the structure of the as-synthesized nanoparticles [68].

One of the critical features of nanocrystals is the formation of surface defects through surface dangling bonds. These surface defects manifest themselves as surface electronic states which are undesirable mid-bandgap states which act as electrons and holes traps negatively affecting recombination processes. These surface defects are passivated by organic ligands during synthesis via covalent bonding which saturate the dangling bonds on the surface of the particle. Consequently, the type of ligand utilized whether hydrophilic or hydrophobic is based on the intended application. For instance, hydrophilic ligands would be deemed extremely necessary if nanoparticles were to be employed for bioimaging or even for photo-absorption and emission. An alternative to passivation by ligands is the growth of a shell over the nanoparticle core in place of a ligand; a material of the same or different type (shell) is grown on the surface of the nanoparticles which bonds with the unsaturated species on the surface of the core.

5.2.2 Synthetic methods

Size- and shape-controlled synthesis can be achieved using the low-cost, high-quality thermal decomposition method. This colloidal synthesis, also known as arrested precipitation has been the most widely used to reliably and controllably synthesize a wide range of metal, binary and composite nanoparticles [62, 70, 71]. This is because it effectively allows for the separation of nucleation and growth steps by thermal decomposition which can be accomplished using either the hot-injection or the heat-up approach [6]. The hot-injection method involves the rapid injection of precursor (s) into a hot, high boiling point surfactant, as shown in figure 5.9(a). The injection temperature is critical, as it regulates the decomposition of the precursor(s). Upon injection, nucleation is initiated due to induced supersaturation. Injecting a room temperature solution will decrease the overall reaction temperature, terminating the nucleation stage and commencing the growth stage. In the non-injection heat-up method, as shown in figure 5.9(b), the two-step mechanism is achieved through the steady heating of a mixture of precursor(s) and ligand. Following nanoparticle growth, purification of the as-synthesized CQD is essential in the overall fabrication process. This is because, the presence of unreacted precursors or excess surfactants can limit charge transfer and reduce the efficiency of the solar cells. Typically, CQD purification takes place through sequential precipitation and dispersion in a non-solvent and solvent, respectively. One drawback attributed to the purification protocol is that this can reduce the surface passivation by ligand which may result in reduced CQD photo- luminescence quantum yield due to the surface defect states present. This stripping of surface ligand can also affect the colloidal stability of the nanoparticles.

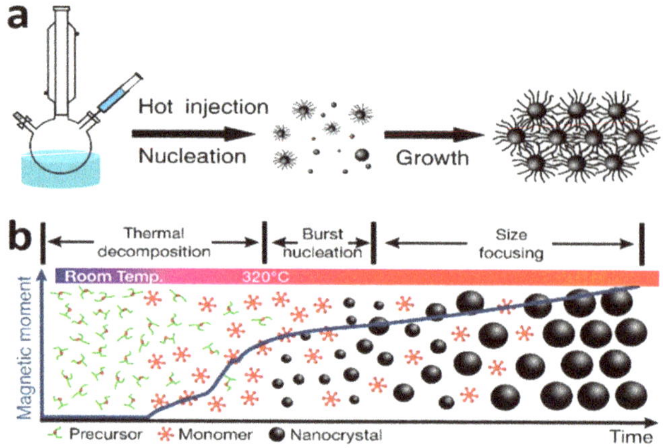

Figure 5.9. Schematic of (a) hot-injection synthesis, with nucleation occurring upon the injection of a room temperature precursor into a high-temperature surfactant, followed by reaction cooling leading to the growth phase. (b) Non-injection heat-up method [6]. Reprinted with permission from [6]. Copyright 2015 American Chemical Society

5.2.3 Strategies for doping colloidal nanoparticles

One strategy to alter the bandgap in cation- or anion-alloyed CQDs is by changing their constituent stoichiometries, in particular via doping. The relationship between the bandgap and the mole fraction of each component is nonlinear and is governed by a bowing parameter dependent on the two binary materials [6]. Tuning the bandgap of alloyed CQDs via stoichiometry and manipulating the defect chemistry enables tunable absorption and emission particularly relevant for spectral adaption of QDs solar cells. An essential benefit of impurity ions is their influence on the charge carrier separation and recombination dynamics to enhance the PCE of solar cells. Dopant inclusion in CQDs has been experimentally challenging for some time, however, by using the colloidal route, several doping strategies have emerged that can be employed, such as the use of single-source precursors, nucleation doping, growth doping and cation diffusion [72]. The single-source precursor route has the dopant in the required mole ratio in relation to host atoms and therefore would not require the addition of several sources to the reaction mixture. Nucleation doping occurs when both dopant and host material precursors are added simultaneously with the dopant nucleating first, resulting in a dopant–host core–shell nanoparticle where the dopant is center doped. In addition to this, growth doping occurs when the dopant is added while the nanoparticle is in the growth stage; following dopant deposition on the nanoparticle surface, an isocrystalline (host) or heterocrystalline shell is grown to further trap the ion. In cation diffusion, the as-synthesized nanoparticle undergoes ion diffusion with the dopant atoms.

Growth doping is one of the most common methods for doping, providing fairly good control over dopant incorporation and successful addition of internally doped nanocrystalline systems [72, 73]. Growth doping procedures involving isocrystalline growth can be carried out whether using a binary or high multicomponent host material (ternary, quaternary systems). For example, Pradhan and co-workers demonstrated isocrystalline growth for Cu-doped ZnSe nanocrystals where after ZnSe growth and subsequent Cu impurity addition, Zn and Se precursors were added to the system to form a ZnSe shell. Also, Zn and Mn dopants in ternary systems such as $AgInS_2$ and $CuInS_2$, respectively, have been demonstrated [48, 49]. For example, Taylor and co-workers [23] reported indium-rich and copper-deficient silver doped-$CuInS_2$ QDs with tunable bandgaps between 1.60 and 1.81 eV and their Ag^+ ion-dependent wurtzite to chalcopyrite phase transformation. They exhibited broad PL emission via a dual radiative pathway with long decay lifetimes, τ_1 (0.68–2.11 ms) and τ_2 (3.37–7.38 ms), implicating DAP transitions of indium interstitials, and/or indium–copper antisites, to copper vacancies, and with their long radiative emission lifetimes suggesting their potential utility in QD solar cells. Heterocrystalline growth strategies generally include the addition of similar transition metal impurities, however, a different shell material is utilized as seen for Cu and Mn doping of CdSe/ZnS and CdS/ZnS core–shell structures [74, 75].

5.2.4 Surface passivation of colloidal nanoparticles

As previously mentioned, surface passivation of colloidal nanoparticles has a profound effect on the optical and electrical properties for applications in solar cells and is therefore critical for nanoparticle engineering. Typically, during colloidal growth, long-chain ligands such as oleic acid, oleylamine and dodecanethiol are effective in controlling the growth of the nanoparticles. Whilst this is desirable, often these long-chain ligands exhibit insulating properties that hinder dot-to-dot carrier transport [37]. This renders low collection efficiency of photogenerated charge carriers in CQD-SCs. Additionally, these long-chain ligands along with the surface defects which are not effectively eliminated, act as non-radiative recombination centers of photogenerated charge carriers, thereby decreasing the V_{oc} and FF values of the CQD-SCs. To circumvent these issues, two primary approaches have been used: (1) ligand exchange with ligands that are less bulky and that can more effectively passivate these defects, (2) halide passivation, and (3) passivating surface defects by adding a shell layer which improves charge carrier recombination. Overall, ligand exchange shortens the distances between the CQDs and increases their packing density, thereby resulting in more efficient charge transport in the CQD film. This can be done in solution or in solid-state. Solution-phase ligand exchange involves mixing CQDs capped with long alkyl chains dispersed in a nonpolar solvent and short hydrophilic organic ligands dissolved in a polar solvent [76]. The CQDs become encapsulated with the short hydrophilic ligands and transfer into the polar solvent. Solid-state ligand exchange involves CQDs with long alkyl chain ligands which are deposited onto the substrate by spin-coating for example, after which a solution containing short organic ligands is deposited on the film. Following a short soaking time and the removal of the solution by spin-coating, a protic solvent is cast onto the film to wash away the exchanged ligands as well as the excess of new short ligands. Furthermore, as shown in figure 5.10, an atomic ligand with halide ions can passivate the defect states of a CQD and control its energy levels. For example, Cao and co-workers carried out solution-phase ligand exchange by mixed

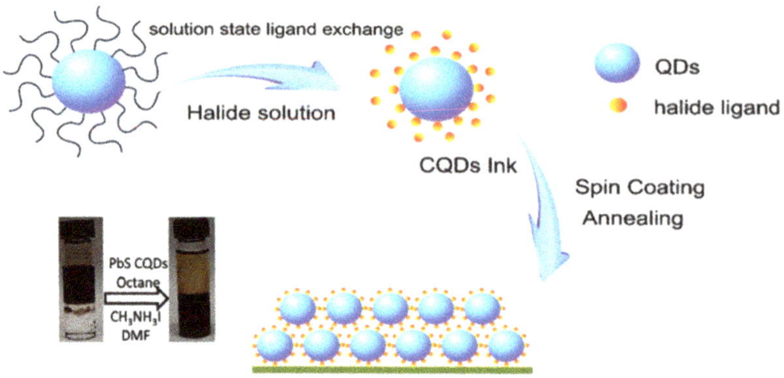

Figure 5.10. Schematic diagram of solution phase ligand exchange with a halide solution. Long insulating OA chains are efficiently exchanged to short halide ligands. Reprinted from [76] with permission from Royal Society of Chemistry.

halides for $AgBiS_2$ CQDs. This approach, involving AgCl and other halides, effectively suppressed surface defects, significantly improving optoelectronic characteristics and enhanced the photodetector performance [77].

As stated previously, the core–shell architecture in which the surface of the core QDs is passivated by an inorganic layer shell is an effective approach to reduce the undesirable carrier recombination through reducing the density of surface defects/traps and prolonging emission lifetime. Core–shell CQDs also display enhanced chemical, thermal and photo-chemical/physical stability toward the surrounding environment compared to bare CQDs [78]. In the synthesis of core–shell CQDs the choice of shell and its thickness in relation to the core material and its size is critical for band alignment and confinement potential. Of note is that a thick shell layer of a type I core–shell CQD impedes exciton dissociation, charge extraction, and transport. For example, Zamkov and co-workers showed that having a CdS shell on PbS QD improves passivation and that the distance between PbS QDs and hence the thickness of the CdS layer impacts carrier mobility and device performance [79]. Watt and co-workers [80] also showed that CQD-SCs employing PbS–CdSe core–shell CQDs had higher V_{oc} and device efficiency but lower J_{sc}. This was due to the effective surface passivation by the CdS shell layer and the confinement of the core electronic energies from the surface chemistry. However, to mitigate the lower J_{sc} they showed that it was possible to alter the energy barrier imposed by the shell material by the use of halide ligands and that the extent of passivation could be regulated by synergistic use of both halide and bifunctional ethane-1,2-dithiol ligands and were able to boost the photocurrent without losing the effect of shell passivation.

The approach to synthesizing core–shell QDs primarily involves first the synthesis of colloidal core QDs via co-precipitation or hot-injection approaches, followed by purification and re-dispersal in organic solvent for the shell growth [78]. Shell growth has commonly been conducted via several approaches. For example, cation exchange involving injection of a mixture of cationic and anionic precursor at the growth temperature and via a successive ion layer adsorption and reaction (SILAR) method involving alternative injection of cationic and anionic precursors. In the cation exchange approach, the cationic precursor of the shell material is introduced during the shell growth, which gradually replaces the cation in core QDs. The overall size of resulting core–shell QDs does not change significantly compared to the starting core QDs. However, in the SILAR approach, the thickness of the shell grown over the core QDs, can be controlled by changing the added amount of shell precursors, growth time and growth temperature.

5.3 Cell architecture, device physics and performance

The device structure of CQD-SCs is the most important factor that affects charge generation and transport, and ultimately performance. As shown for different types in figure 5.11, the evolution of device structures has mainly focussed on extending the depletion region in the CQD layer and optimizing the energy level alignment for efficient charge extraction to the electrodes. Various types of devices involve the original Schottky CQD-SC, depleted heterojunction CQD-SC, depleted bulk heterojunction (BHJ) CQD-SC, and CQD-sensitized solar cell, as the main examples, which will be described herein.

5.3.1 Schottky CQD solar cell

Schottky CQD-SCs have emerged over the last two decades as a promising type of solar cell that utilizes a simple Schottky junction structure with solution-processed QDs for low-cost fabrication, offering advantages like efficient carrier extraction and air stability. Built upon the architecture of the simple and well established Schottky junction, early progress in CQD PV device performance was achieved through engineering of the readily fabricated Schottky CQD solar cell [2]. The device comprises a thin film of CQDs that is ohmically contacted on one side via a transparent conductive oxide (TCO) typically indium tin oxide (ITO) and on the other side using a low work function metal such as aluminum and magnesium at which a significant built-in potential is established relative to the *p*-type CQD film which leads to the charge extraction of photogenerated carriers to each electrode [37]. Figure 5.6(a) shows the device architecture along with a corresponding spatial band diagram. There are several factors that affect the performance of these devices. In theory, the height of the Schottky barrier is estimated by the difference in the work function of each electrode. However, Sargent and co-workers demonstrated that the Schottky barrier heights are normally less than the difference in the work functions of each electrode [81]. It has also been established that the work function difference between metal electrodes (Ca, Mg, Ag, Al and Au) resulted in different V_{oc} values as well as different Schottky barriers [82]. Additionally, carrier transport is improved by the insertion of a buffer layer between the electrodes and CQD layer enabled by an increased energy barrier. This was demonstrated from work by Sargent and co-workers [83] where the air stability and PCE of solution-processed Schottky junction PbS QD solar cells were dramatically improved by the insertion of 0.8 nm LiF between the PbS CQD film and the Al contact. Despite

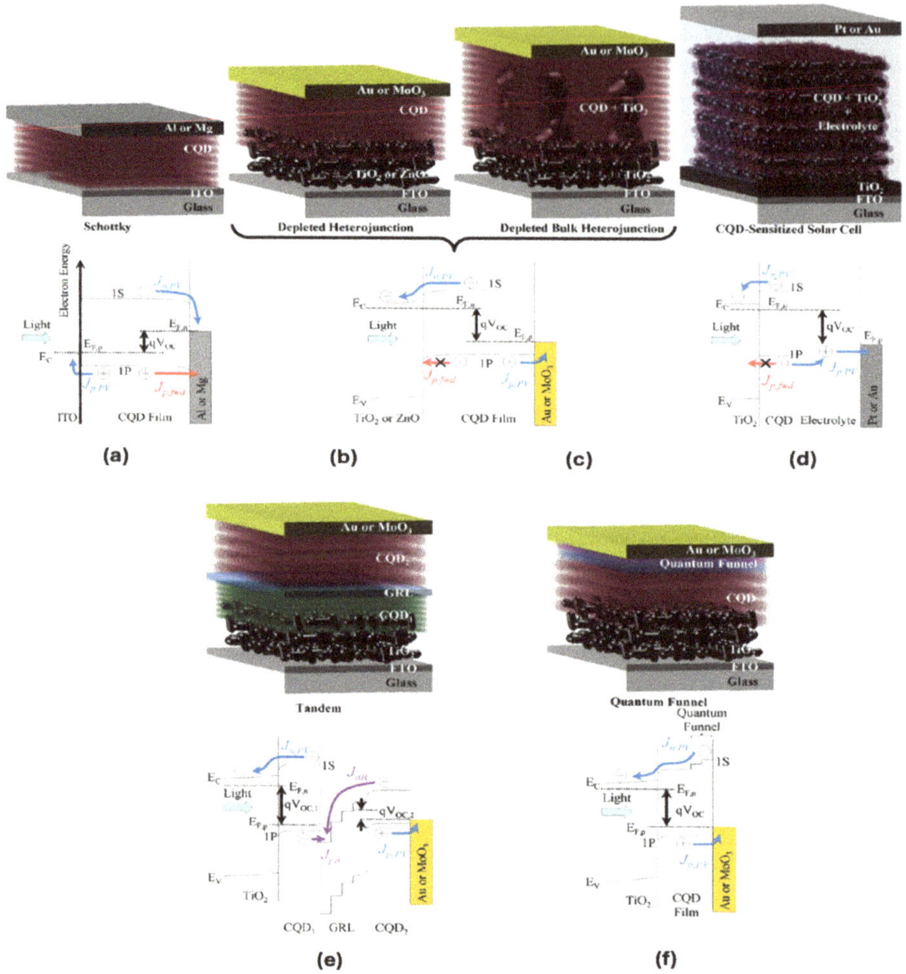

Figure 5.11. Single-junction CQD-SC architectures with band diagrams: (a) Schottky CQD-SC; (b) Depleted heterojunction CQD-SC; (c) Depleted BHJ CQD-SC; (d) CQD-sensitized solar cell. Multijunction CQD-SC architectures with band diagrams: (e) Tandem CQD-SC using a graded recombination layer (GRL); (f) Quantum funnel solar cell. Adapted with permission from [2]. Copyright 2011 American Chemical Society.

simple fabrication, Schottky CQD-SCs have some limitations, including: (1) low air stability, (2) V_{oc} deficiency, (3) inefficient carrier extraction, and (4) poor harvesting of short-wavelength light [84]. However, these limitations can be overcome by using an *n*-type, transparent metal oxide layer to create a front junction with the *p*-type QD layer.

In 2014, Jeong and co-workers [85] introduced a novel architecture, the inverted Schottky CQD-SC, which consisted of a thin film of PbS CQDs sandwiched between a low-work-function, TCO (L_ϕ-TCO) and a high-work-function metal anode. On L_ϕ-TCO substrates, which were generated by coating a thin layer of polyethylenimine (PEI) onto FTO—a series of inverted Schottky CQD-SCs with varied PbS

CQD sizes and QD layer thicknesses. A Schottky junction, of about 180 nm in width, was formed at the front TCO contact exhibiting a PCE of 3.8% retained over several weeks of air exposure.

5.3.2 Depleted heterojunction CQD solar cell

Heterojunction CQD-SCs leverage the unique tunable optical properties of QDs and heterojunction solar cell architectures to enhance light absorption and charge separation, offering potential for improved solar energy conversion efficiency. They have emerged out of the limited PCEs of Schottky CQD-SCs and have shown improved performance with developments in bandgap engineering and interfacial dynamics for improved charge extraction. In particular, depleted heterojunction (DH) architecture entails a wide bandgap semiconductor such as TiO_2 or ZnO with CQDs [2]. In contrast with CQD-sensitized SCs, and in resemblance with Schottky CQD cells, DH devices exploit transport of electrons and holes through a many-layer CQD film. These devices were developed to overcome one of the main limitations of the Schottky CQD-SCs by introducing a junction between the electron transport layer (ETL) such as wide bandgap oxides (TiO_2 or ZnO) and the CQD [37]. Here, the photogenerated charge carriers are created in the junction region and can be efficiently extracted by the ETLs. Furthermore, the ETLs block hole transport from the CQD layers to the cathode, thereby resulting in reduced non-radiative electron–hole recombination.

The development of DH-CQD-SCs has involved a range of strategies to improve their performance. For example, early studies focused on energy level alignment between the CQD layers and charge transport layers (CTLs) by controlling the CQD diameter or thickness. For example, Choi and co-workers [86] fabricated from PbSe CQD films sandwiched between layers of ZnO nanoparticles and PEDOT:PSS as electron and hole transporting elements, respectively. They demonstrated that charge extraction from the CQD active layer was driven by a photoinduced chemical potential energy gradient at the nanostructured heterojunction establishing a direct correlation between interfacial energy level offsets and PV device performance. Also, a comparison between DH and Schottky-type SCs revealed that higher V_{oc} and *FF* values can be obtained by incorporating specific CTLs [87]. Early studies by Sargent and co-workers [88] showed that bandgap tuning based on the sizes of PbS CQDs was very effective in improving the junction kinetics at interfaces for DH-CQD-SCs fabricated in TCO. These devices exhibited PCEs of 5.1% and *FF*s of 60% enabled through the combination of efficient hole blocking of the heterojunction and very small minority carrier density (depletion) in the wide bandgap layer. Importantly, strategies focused on ETLs involving extending or modifying the structure of the depletion region have been used to optimize performance. These include: (1) adding mesoporous TiO_2 which elongates the depletion region between the ETL and the CQD layer, improving charge extraction at the junction region; (2) applying thicker CQD layers which results in enhanced light absorption; (3) adding mesoporous TiO_2 that has been modified to a more periodic pillar structure for improved charge collection; (4) adding ZnO nanowires to enlarge the surface area of the thick depletion region; and (5) chemical modification of the ZnO surface with ligands to reduce its surface trap sites and improve recombination [37].

5.3.3 Bulk heterojunction CQD solar cell

One of the limitations of the Schottky and depleted heterojunction CQD-SCs is that the depletion region in the junction area is not thick enough for electronic transport. As such, depleted BHJ CQD devices (DBJ-CQD-SCs) were developed to overcome this limitation and combine elements of QD and BHJ solar cell architectures. For these, the *n*-type wide bandgap semiconductor and CQD film form an interpenetrating depletion region of a nanoporous architecture. This is usually accomplished by structuring the TiO_2 or ZnO and infiltrating the CQDs into the structured electrode. The structured interface allows for the extension of the depletion region and the addition of more absorbing CQD material, improving absorption with enhanced depletion for driving the separation of electron–hole pairs and enhances charge extraction efficiency [6]. The first device was reported in 2012 by Konstantatos and co-workers [89] based on *n*-type Bi_2S_3 nanocrystals and *p*-type PbS QDs, demonstrating a more than a threefold improvement in performance compared to their bilayer analogue, as a result of suppressed recombination. Recently, Tabernig and co-workers [90] designed an optically resonant BHJ solar cell involving ZnO nanoparticles to create a nanohole template in an ETL infiltrated with PbS QDs to form a nanopatterned DH. For these devices shown in figure 5.12, optical simulations showed that the absorption per unit volume in the cylindrical QD absorber layer was enhanced by 19.5% compared to a planar reference. This is achieved for a square array of QD nanopillars of 330 nm height and 320 nm diameter, with a pitch of 500 nm on top of a residual QD layer of 70 nm, surrounded by ZnO. The patterned cell showed a J_{sc} of 3.2 mA cm^{-1} [2] and higher V_{oc} than that of the counterpart planar cell with lower *FF* and a PCE of 11.6% in comparison to the thin reference (11.2%).

The main drawback with BHJ CQD-SCs is related to the increased interfacial area which increases bimolecular recombination. In particular, owing to the increase of CQD/metal oxide interface area, BHJ CQD-SCs are more susceptible to non-radiative recombination at interfaces because of interfacial defects, leading to larger

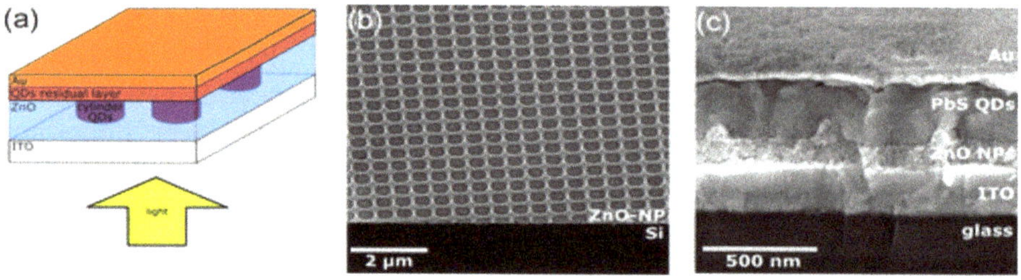

Figure 5.12. (a) Patterned *p*–*i*–*n* for a PbS QD solar cell heterojunction structure. (b) SEM perspective of SCIL-patterned ZnO nanoparticle layer on top of a silicon substrate. (c) SEM cross-section image of the patterned cell geometry. Reprinted from [90] CC BY 4.0.

V_{oc} loss. Also, the increase in electron extraction rate further exacerbates the imbalance between electron and hole extraction [91]. This results in lower device built-in voltage due to the need to manage the CQD-metal oxide CB offsets to prevent back-recombination [6]. The drawbacks associated with heterojunctions can, in principle, be overcome by using a homojunction-like architecture composed of *p*- and *n*-type CQDs [6]. These so-called nano-heterojunctions of quantum junction CQD-SCs were demonstrated using widely tuned CQD bandgaps of 0.6–1.6 eV. Further optimization of the device structure and ligand strategies produced competitive PCEs. Recently, Sargent and co-workers [92] developed CQD nano-heterojunction solar cells that exhibited extended carrier transport length, enabling efficient IR light harvesting. They employed an in-solution doping strategy for large-diameter CQDs to address the complex interplay between (100) facets and doping agents, enabling control of CQD doping, energetic configuration, and size homogeneity. The hetero-offset between *n*-type CQDs and *p*-type CQDs influenced the transfer of electrons and holes into distinct carrier extraction pathways from active layers exceeding thicknesses of 700 nm without compromising V_{oc} and *FF* and accomplishing >90% charge extraction efficiency across the UV-to-IR range (350–1400 nm).

5.3.4 CQD-sensitized solar cell

Efforts to replace molecular absorber dyes in dye-sensitized solar cells (DSSCs) with semiconductor nanoparticle absorbers resulted in the development of QD-sensitized solar cells. CQD-sensitized solar cells (CQDS-SCs) have several advantages such as low cost, bandgap tunability, and high absorption coefficients. The configuration of CQDS-SCs is very similar to that of DSSCs in that they contain a photoanode of a mesoporous structure based on QDs and a counter electrode (CE). In the QDSSC, the excitonic absorber absorbs solar radiation leading to electron–hole generation in the absorber [93]. In these devices, a nanostructured, large surface area, wide bandgap semiconductor electrode is used as the electron transport material (ETM). The advantage of the large surface area of the electrode is that it allows for sufficient deposition of the excitonic sensitizer to absorb incident solar radiation. TiO_2 is one of the most widely investigated and used ETMs among other wide bandgap semiconductors such as ZnO and SnO_2. A redox couple is used as a hole transport material (HTM) such as the iodide and polysulfide redox systems. In the device, an excitonic QD absorbs incident solar radiation, leading to exciton generation which is dissociated at ETM/absorber interface and electron is transferred to the ETM, whilst the hole is subsequently transferred to the HTM.

Despite the simplicity and cost-effectiveness of solution-processing device fabrication and the potential of CQDS-SCs, some significant problems such as fast recombination of charges can significantly impede their performance. In particular, there are usually three types of interfacial paths that support the process of recombination in the device: (1) QDs/electrolyte interface; (2) photoanode/electrolyte interface; and (3) photoanode/QD interface [94]. Additionally, the inherent property of the QDs also facilitates various pathways for recombination such as:

(1) charge recombination in QDs; (2) photoexcited electron trapped in a surface defect of the photoanode or QDs; (3) back transfer of an electron from the CB to a hole in the VB of the QD; and (4) recombination of photoexcited electrons present in the photoanode or QDs with the oxidized species of an electrolyte. One strategy to reduce recombination is the surface passivation of the photoanode or QDs by using wide bandgap semiconducting materials. For example, Bein and co-workers [95] prepared QDSSCs with a PbS QD absorber layer from thin mesoporous TiO_2 layers by the SILAR method with spiro-OMeTAD as the organic *p*-type HTM. They passivated the surface of the PbS QDs with the tripeptide l-glutathione (GSH) and showed that this treatment more than doubled the short-circuit current and PCE. This improved performance was attributed to enhanced charge injection from PbS QDs into the CB of TiO_2 through increased exciton recombination lifetime.

Another main reason for the intermediate PCEs of CQDS-SCs is due to the low QD loading (i.e., low QD coverage) on the photoanode surface and the concomitant insufficient light-harvesting capacity as well as charge recombination loss [96]. In this respect, a high QD loading on the photoanode surface is essential for high photocurrent and high PV performance. High QD loading can reduce the necessary thickness of a sensitized photoanode to capture all incident solar photons. A thin photoanode means a short transportation path for photoexcited electrons through the photoanode to the current collector plate and consequently a reduced possibility for undesirable charge recombination. Furthermore, a high QD loading corresponds to a decrease in the proportion of the uncovered TiO_2 surface directly exposed to the electrolyte and less possibility of photogenerated electrons captured by the redox couple in the electrolyte, thereby benefiting the PV performance.

5.3.5 Tandem/multijunction CQD solar cell

In the last decade, tandem device structures using numerous arrangements of photoactive materials for complementary absorption have been studied. A multijunction solar cell or tandem solar cell is a configuration of two or more sub-cells, which can convert sunlight into electricity with minimal energy losses [97]. In particular, tandem/multijunction CQD solar cells exploit facile layering of size-effect-tuned QD layers having multiple, different, optimally chosen bandgaps [2]. These tandem cells typically employ two stacked DH cells, where the upper metal contact of the bottom cell and lower transparent contact of the top cell are replaced using a transparent, carefully engineered recombination layer. The recombination layer enables holes generated in the bottom cell to recombine with electrons generated in the top cell, maintaining charge neutrality in the overall device and facilitating current matching. The first report of a multijunction CQD-SC was by Sargent and co-workers [98] involving layers of bandgap size-tuned CQD PbS. The cell consisted of a graded recombination layer to provide a progression of work functions from the hole-accepting electrode in the bottom cell to the electron-accepting electrode in the top cell, allowing matched electron and hole to meet and recombine. The tandem solar cell exhibited an V_{oc} of 1.06 V, equal to the sum of the two single-junction devices, and a solar PCE up to 4.2%.

5.3.6 Tandem perovskite CQD solar cell

Metal halide perovskite solar cells have shown tremendous potential to become a low-cost alternative for front-cells in tandem solar cells, and also for conventional photovoltaics. Di and co-workers [99] were the first to model and fabricate a tandem cell design based on a halide perovskite top cell and a chalcogenide CQD bottom cell, where both materials provide bandgap tunability and solution processability. For these, a theoretical efficiency of 43% was calculated for tandem cells with bandgap combinations of 1.55 eV (perovskite) and 1.0 eV (CQDs) under 1-sun illumination. Their solution-processed monolithic perovskite/CQD tandem solar cell, showing evidence for sub-cell voltage addition exhibited a PCE of 29.7% and paved the way for the development of these types of cells.

There are different configurations of these tandem solar cells including 2-terminal (2T) and 4-terminal (4T) tandem configurations [100]. Of these, the 4T tandem devices seem to be more attractive due to no current-matching limitation and therefore enabling easier manufacturing processes [97]. However, constructing a successful 4T tandem solar cell entails many challenges. These involve the fabrication of an efficient, highly transparent and stable perovskite front sub-cell, the development of a spectrally matched CQD back cell with strong IR absorption and the design and modification of a semi-transparent interlayer between them. To address this issue, Johansson and co-workers [97] fabricated 4T tandem solar cells using a methylammonium lead iodide ($MAPbI_3$) perovskite solar cell as the front cell and a lead sulfide CQD-SC as the back cell. To improve the infrared transparency of the perovskite sub-cell, different modifications of the tandem interlayer, at the interface between the sub-cells were tested. This included the incorporation of a semi-transparent thin gold electrode on the $MAPbI_3$ solar cell, followed by adding a molybdenum(VI) oxide (MoO_3) layer. These interlayer modifications resulted in an increase of the IR transmittance to the back cell and improved the optical stability, compared to that in the reference devices.

Prior to this, Sargent and co-workers [101] reported a solution-processed 4T tandem solar cell comprising a perovskite front cell and a CQD back cell which exhibited a PCE exceeding 20%. The front semi-transparent perovskite solar cell employed a dielectric–metal–dielectric (DMD) electrode constructed from a metal film (silver/gold) sandwiched between dielectric (MoO_3) layers and the back cell was based on an IR CQD absorber layer complementary to the IR transmittance of the semi-transparent perovskite front cell. Recently, Singh and Rani [102] reported a cost-effective 2 T perovskite-CQD monolithic tandem solar cell consisting of wide bandgap all-inorganic perovskite $CsPbI_3$ top sub-cell and narrow bandgap PbS QDs bottom sub-cell. For optimization of the PCE of the tandem structure, variations of absorber layer thickness, total defect density, interfacial defect density and back metal contact were carried out. Optimized devices designed with the 530 nm/450 nm thick absorber layers of top/bottom sub-cells resulted in J_{sc} of 19.28 mA cm^{-2}, V_{oc} of 1.89 V, and PCE of 27.32%. It was found that performance of the device was very sensitive to the total defect density of the bottom sub-cell. Owing to the high PCEs, solution-processed hybrid tandem PV combining these technologies offer to contribute to higher-efficiency solar cells for next-generation flexible PV devices.

5.3.7 CQD-organic hybrid solar cell

The development of organic/inorganic hybrid nanocomposite systems that enable efficient solar energy conversion has been an important pursuit. These CQD-organic hybrid solar cells are constructed by blending QDs with semiconducting polymers and organic small molecules in which they generate charge carriers and transfer them to electrodes [103]. In particular, fullerene (C_{60}) and its derivatives (e.g., [6,6]-phenyl-C_{61}-butyric acid methyl ester ($PC_{61}BM$)) have long been employed in the fabrication of these CQD/organic solar cells because of their excellent electron-accepting ability [104]. The essential criterion for these types of solar cells is that the QDs and organic molecules have appropriately aligned energy levels in order to promote charge carrier separation at their interface [103]. Their mixing also leads to smaller domain size than the exciton diffusion length (10–20 nm) of conjugated polymers to suppress exciton recombination. Ideally, the inter-penetrating heterostructure of QDs and polymer enables maximized interface area for charge separation and respective electron and hole transport channels. For example, early work by Bang and Kamat [104] involved the development of CdSe–C_{60} nanocomposite solar cells by chemically linking CdSe CQDs with thiol-functionalized C_{60}. In particular, CQD-polymer BHJ structures have been widely investigated involving PDPPTPD, PDTPBT, P3HT, PDTPQx and si-PCPDTBT polymers in combination with various CQDs [105]. On the basis of these hybrid devices and with the goal of improving performance, recent developments by Sargent and wo-workers [105] involved preparing a hybrid architecture involving small molecules into the CQD-organic stacked structure. The small molecule complemented CQD absorption and created an exciton cascade with the host polymer, thus enabling efficient energy transfer and also promoting exciton dissociation at heterointerfaces. The resulting hybrid solar cells exhibited PCEs of 13.1% and retained over 80% of their initial PCE after 150 h of continuous operation.

Despite these attractive attributes of CQD-organic hybrid solar cells with competitive PCEs, suppressing charge accumulation at photoactive/charge transporting layer interfaces, such as CQD-organic HTL interfaces, has been a challenge to overcome. To address this, Lee and co-workers [106] systematically investigated how ZnO/CQD interfaces affect the charge extraction properties. To do this they used a library of cinnamic acid (CA) groups that are well-matched with the ETL region among various ligands and employed an interfacial layer for the ETL using several types of CA ligands, which had fine energy level tunability and a superior electron transport property. They showed that devices, as represented in figure 5.13, where the ETL with its conduction band situated between ZnO and CQD interface, established a cascading band alignment with significantly reducing potential barriers, facilitating charge transfer by suppressing charge accumulation and reducing interfacial recombination.

The development of HTMs for solution-processed solar cells has involved an array of organic *p*-type materials which have been successfully incorporated into

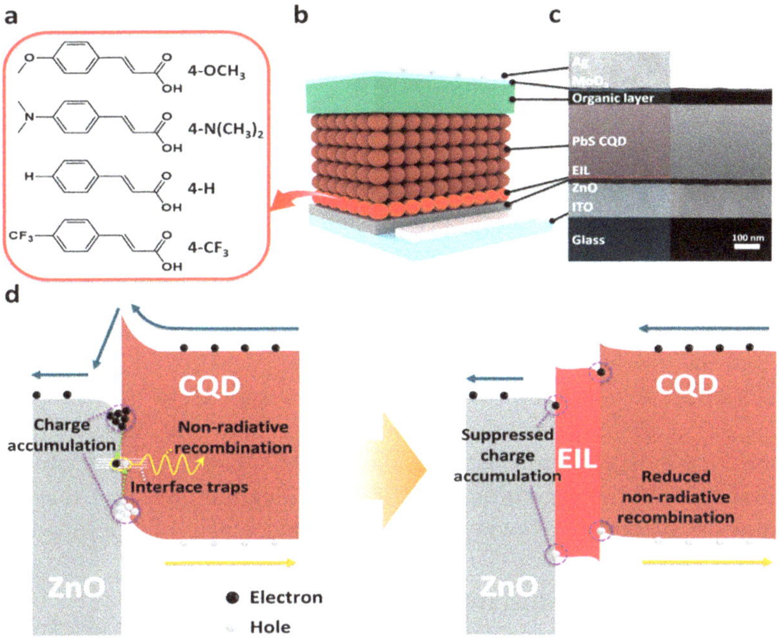

Figure 5.13. Overview of a CQD/organic hybrid solar cell with an EIL (ETL). (a) Chemical structures of CA ligands. (b) Schematic illustration of CQD/organic hybrid solar cells. (c) TEM cross-sectional image of a CQD/organic hybrid solar cell. (d) Proposed interface engineering scheme: insertion of an EIL between ZnO and CQD layers in a CQD/organic hybrid solar cell to suppress charge accumulation. Reprinted with permission from [106]. Copyright 2023 American Chemical Society.

CQD-polymer devices. Among these, π-conjugated polymers (π-CPs) are good candidates as highly processable HTMs owing to their good hole-accepting/transporting characteristics [107]. Critically, the HOMO level of the π-CPs must be higher than that of the CQD active layers to enable efficient hole transfer/electron block, and their hole mobility should be sufficiently high to prevent charge accumulation. In particular, the design of π-CPs with chemical structures of alternating electron-rich (donor, D) and electron-deficient (acceptor, A) moieties has been demonstrated as effective due to their tunable optical and electrical properties such as energy levels, bandgap, and charge mobility [108]. Examples of these π-CPs include those with benzo[1,2-b:4,5:b′]-dithiophene (BDT) and ethylhexyloxy (EHO)-substituted BDT (EHO-BDT) units and systems such as polythieno [3,4-b]-thiophene-*co*-benzodithiophene (PTB7) poly(3-hexylthiophene) (P3HT) as HTMs [107]. On this premise, Jeon and co-workers [107] developed CQD-polymer devices involving organic π-CP-based HTMs. In particular, a device using PBDTTPD-HT achieved PCE of 11.53% with good air-storage stability and at that time, was the highest reported PCE among CQD-SCs using organic HTMs, and even higher than the reported best solid-state ligand exchange-free CQD-SC using *p*-type CQD HTM. Further work by Cho and workers [107] involved CQD-SCs using triisopropylsilylethynyl (TIPS)-derivatized BDT-containing π-CPs which achieved a PCE of 13.03% which was substantially higher than those previously reported using *p*-type CQD HTM (11.33%) or π-CPs (11.25%) owing to the improved charge collection efficiency near the photoactive CQD layer/HTM interface.

5.4 Cell degradation and failure analysis

The stability of CQD-SCs exceeds 1100 h (T_{80}, under illumination) and has been increasing. However, stability under normal operating conditions, particularly under sunlight and ambient air, remains a challenge. Several factors contribute to instability, including surface trap states of CQDs, degradation due to moisture and oxygen, and the inherent instability of certain QDs like perovskite QDs. There has been much effort towards improving stability through CQD surface passivation, optimizing device architecture, and developing new materials. Discussed herein are the effects of the most common cell degradation of CQD-SCs sources which includes oxygen and humidity degradation, light-induced degradation, and thermal degradation, and the strategies for overcoming these device degradation processes.

5.4.1 Evaluation of degradation mechanisms and analysis

5.4.1.1 Oxygen degradation

Although the efficiency of CQD-SCs has improved significantly during recent years, the air stability still remains a major obstacle to their commercialization [109]. This is because exposure to oxygen can lead to the oxidation of the QDs which can introduce defects and trap states, reducing the efficiency of charge transport and thus lowering the device performance. For example, lead chalcogenide QDs are prone to degradation in air due to their large surface-to-volume ratios and high surface energies. It has been reported that the surfaces of Pb-based CQDs such as PbS and PbSe oxidize upon reaction with oxygen from ambient air, forming lead oxide (PbO), lead sulfite ($PbSO_3$) or lead sulfate ($PbSO_4$) where the ratio of the latter two degradation products depends on the size of the QDs [37, 110]. For PbSe QDs, spontaneous thermally activated oxidation has been linked to the formation of a thin outer shell which effectively reduces the size of the PbSe 'core' which increases quantum confinement, causing shifts of the PL band and the absorption onset to higher energies [111]. The exposure of QD solutions to air also causes rapid PL quenching attributed to enhanced carrier trapping induced by mid-bandgap states from adsorption of oxygen onto the nanocrystal surface and non-radiative interband recombination. Early work by Wood and co-workers [112] involved investigation of trap states in PbS QD-based thin films and determined their influence on the electronic behaviour of Schottky-type devices. They provided evidence that the traps interact efficiently with both the VBs and CBs and can act as recombination centers in the solar cell. As shown in figure 5.14, they did this by leveraging the change in measured trap densities as a function of time and showed device aging in air where oxidation results in a shift in the Fermi level and trap occupation rather than in the creation of additional traps. Additionally, CQDs are prone to oxygen degradation during the ligand exchange process, since detached ligands leave empty

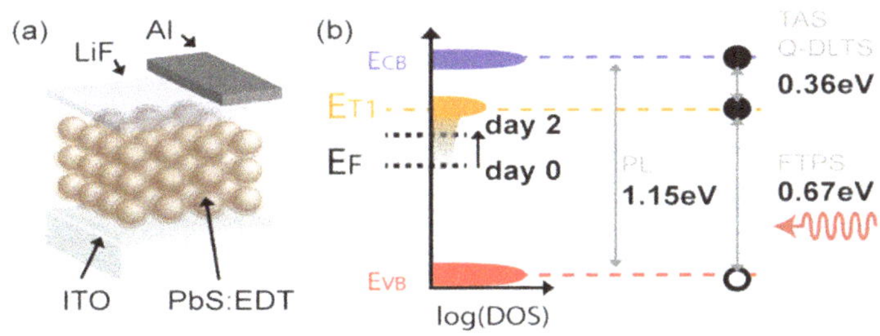

Figure 5.14. Schematic of: (a) Schottky junction solar cell. (b) Density of states in the bandgap of the PbS-ethanedithiol (EDT) semiconductor showing change in Fermi level and occupation of trap states over time. Reprinted with permission from [112]. Copyright 2013 American Chemical Society.

sites which are exposed to oxygen which now acts as a defect state [37]. This oxygen-induced quenching site provides additional trap states for photogenerated charge carriers, thereby decreasing the device performance and stability.

Beygi and co-workers [109] investigated the air exposure oxidation mechanisms of PbS QD thin films and solar cells. The QDs were passivated with organic and inorganic ligands, butylamine (BA), mercaptopropionic acid (MPA), tetrabutylammonium iodide (TBAI), methylammonium iodide (MAI) and the perovskite, methylammonium lead triiodide ($MAPbI_3$). They reported that the films upon exposure to air exhibited blue-shift and emission quenching due to rapid oxidation of QD thin films involving the formation of $PbSO_3$ and $PbSO_4$ on the (100) facets of the PbS QDs. However, though $MAPbI_3$ treatment led to the complete passivation of QDs in the air, the perovskite shelling partially oxidized to PbO and $PbCO_3$. Fabricated *p–n* and *p–i–n* structured solar cells with PbS QDs showed increased PCE for those passivated with the inorganic layer with high stability measured as 1% PCE loss after 500 h of storage in the air.

5.4.1.2 Humidity degradation

Parts of this section have been reproduced from [113]. CC BY 4.0.

The surface-dominant nature of CQDs makes them extremely surface sensitive towards ambient humidity, which has significant impacts on their optical properties and overall stability, leading to degradation of solar cells. Studies have shown that water molecules can hydroxylate the solid surface via dissociative chemisorption during material synthesis or ambient device fabrication [113]. Early work from Wang and co-workers [114] demonstrated that the dissociated water molecules could cause surface hydroxylation during PbS CQDs synthetic process. The interposition of the hydroxyl anions (OH^-) between steric oleic acid (OA) ligands effectively releases the surface energy of the (111) facet, preserving facet stabilization. However, the OH^- ions are difficult to remove by conventional ligand exchange due to their strong bonding with surface Pb^{2+} ions. With this understanding, Konstantatos and co-workers [115] showed that suppression of the hydroxide on the PbS QD surface through appropriate thermal annealing and a halide salt for the ligand exchange reduced mid-gap states, diminished charge recombination and improved both the PV performance and photostability of QD solar cells. Recently, Liu and co-workers [113] carried out an extensive experimental and theoretical investigation involving *in situ* temperature-dependent x-ray photoelectron spectroscopy (XPS), x-ray absorption spectroscopy (XAS), and density functional theory (DFT) calculations on the effect of hydrogen-bonded water on hydroxylated PbS CQDs. They confirmed that the adsorbed water could govern the temperature-dependent change of CQD surface chemistry through water desorption, hydroxylation, and dehydration process. Also, they showed that exposure to ambient water during device processing enhances the surface hydroxylation and water adsorption, which subsequently aids the fusion of adjacent CQDs which are responsible for mid-bandgap trap states and energetic disorder, affecting performance of CQD solar cells. Critically, they linked the thermal instability of PbS CQD solar cells to the continuous fusion, as well as surface iodine loss and migration during thermal aging. They demonstrated that densely packed CQD arrays could spatially extrude the ambient water and thus reduce surface hydroxylates, leading to improved device performance and thermal stability.

5.4.1.3 Light-induced degradation

CQD-SCs are susceptible to light soaking, which is the degradation of device performance over time due to prolonged exposure to light, particularly under high-intensity conditions like direct sunlight. Light soaking in CQD solar cells can involve various mechanisms, including: (1) surface oxidation of CQDs, altering their electronic properties and affecting charge transport; (2) interface states where the interface between the CQDs layer and other materials in the solar cell can degrade, leading to increased recombination of charge carriers; and (3) material disordering, where the crystalline structure or packing of the CQDs can become disordered or undergo phase transitions, impacting their ability to absorb light and affecting charge transport. Critically, light exposure also influences oxygen and humidity-related degradations in many CQDs and their solar cells.

It has been reported that in colloidal PbSe QDs, UV irradiation can speed up the oxidation-induced blue-shift of the first excitonic peak when the measurement is made in ambient air [116]. In particular, no UV-induced blue-shift was observed in a nitrogen atmosphere, suggesting that UV exposure does not cause degradation of PbSe QDs on its own, but rather accelerates the process of oxygen-induced degradation. It has also been reported that the emission peak of solutions of CdSe/ZnS core–shell QDs illuminated in air for a continuous period of time undergo a blue-shift and photobleaching. In a nitrogen atmosphere, the blue-shift is absent while photobleaching occurs after much longer times (10–15 min) [117]. The investigations of this light-induced photo-oxidation points to a number of effects related to loss of ligand passivation, where defect sites at their surface from dangling bonds due to incomplete ligand coverage or grain boundaries may act as trap centers for photo-oxidation [116]. Also, the broadening of the PL spectra can occur due to the gradual loss of ligands and the subsequent QD agglomeration in solution. Likewise, other II–VI core QDs such as CdSe and CdTe and III–V QDs such as InP also show decreased PL signal due to agglomeration or photo-oxidation, with CdTe QDs being most sensitive in this respect [116]. Additionally, the hydroxides on CQD surfaces have been reported to be unstable under light exposure [37]. The most widely used ligand exchange processes with EDT and TBAI were reported to lead to hydroxide species on Pb-terminated (111) surface facets which exhibit the quenching of photogenerated electrons and holes, thus increasing the non-radiative recombination rate, which is critical for device stability under illumination.

Additionally, it has been found that the decomposition of PQDs such as $CsPbI_3$ PQDs does not occur in a pure oxygen atmosphere until light and water are involved [118]. For these the compositional instability is attributed to reactive oxygen species initiated from oxygen, light and surface defect states, which has been reported is two orders of magnitude slower than their thin-film or bulk counterparts, owing to their unique surface chemistries.

Light-soaking induced structure degradation of QD ink-based solar cells has been investigated by several authors. For example, in an *operando* study by Xu and co-workers [119] structural degradation of the active layer of a PbS CQD solar cell during the device operation—illumination was observed as illustrated in figure 5.15. This involved spontaneous decrease of the QD interdot distance with an increase in the spatial disorder in the active layer (PbX_2–PbS QDs, X = I and Br). The structure disorder-induced broadening of the energy state distribution was cited to be responsible for the decrease in V_{oc} leading to the device degradation.

With respect to degradation from ETL, critical for electron charge extraction, Kirmani and co-workers [120] showed that PbS CQD-SCs (*n*–*i*–*p*) employing a solution phase halide ligand exchange and a >100 nm thick ZnO nanoparticle ETL

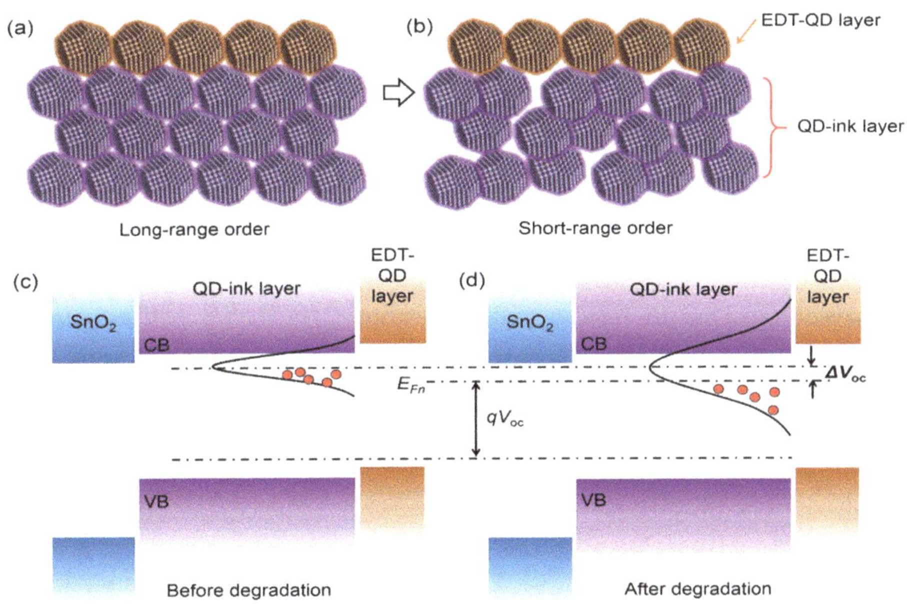

Figure 5.15. Schematic of the structure in the QD layers, (a) having long-range order before and (b) short-range order after the degradation. Energy state distribution in the QD-ink layer (c) before and (d) after degradation. Reprinted from [119] with permission from Royal Society of Chemistry.

suffer from degradation under exposure to UV-light and compromised long-term stability, owing to parasitic UV absorption and carrier recombination in ZnO. They were able to suppress this with an ultrathin (ca. 20 nm), quantum-confined, solution-processed In_2O_3/ZnO bilayer ETL which led to CQD-SCs with enhanced UV-resilience and long-term stability.

5.4.1.4 Thermal degradation

CQD-SCs degrade with increasing temperature due to several factors including reduced quantum yield, decreased optical efficiency, and changes in charge transport. Higher temperatures can also lead to phase transitions and decomposition of the QD materials, affecting their performance. Importantly, the degradation caused by thermal influence seems inevitable since QD solar cell processing and operation requires elevated temperatures [118]. In particular, exposure to heat could also reduce the emission intensity of QDs, which is partly related to the removal of surface ligands at elevated temperatures [116]. It has been shown that thick protective shells can help to enhance their thermal stability.

CQD-SC degradation also arises from the interplay of oxygen or moisture and thermally induced evolution of surface chemical environment, which significantly influence the nanostructures, carrier dynamics, and trap behaviours in the devices. For example, Ma and co-workers [113] reported that for PbS CQDs adsorbed water could govern the temperature-dependent change of CQD surface chemistry through water desorption, hydroxylation, and dehydration. These effects resulted in inter-band trap states and energetic disorder introduced by fused CQDs which impeded the performance of CQD solar cells. In particular, they showed that thermal instability of PbS CQD-SCs may have originated from the continuous CQD fusion, as well as surface iodine loss and migration during thermal aging. Also, it has been reported that the formation of lead sulfite or sulfate at the CQD surface from oxidation could be accelerated at elevated temperatures [37]. The increased lead sulfite and sulfate amounts induce additional trap states that lie below the CBs (0.1 and 0.3 eV for sulfite and sulfate, respectively), which can cause degradation of charge extraction in the solar cell.

The influence of temperature on phase transition of perovskite bulk and QD materials is reported to have a profound impact on device performance. For example, of the perovskite phases of $CsPbX_3$, the cubic α-$CsPbI_3$ black phase, which is ideal for strong light absorption in the visible range, is thermodynamically unfavourable [116]. This black phase degrades into an undesired wider bandgap orthorhombic yellow δ-phase quickly under ambient conditions which makes device fabrication very challenging since the δ-phase tends to form during annealing and exposure to humid ambient conditions. Recently, Zhao and co-workers [118] investigated the relationship of crystal structure, morphology, and optical properties and the thermal behaviour for $Cs_xFA_{1-x}PbI_3$ PQDs with a more complete understanding of the concurrent effects of QD chemical composition and surface ligand binding. They showed that the thermal degradation process of Cs-rich PQDs was dominated by a phase transition from black α-phase ($\sim$30 °C–130 °C) to yellow δ-phase ($\sim$130 °C–300 °C) and then finally to black α-phase ($\sim$300 °C–450 °C), while FA-rich PQDs were mainly decomposed into PbI_2, showing similar or even slightly higher thermal stability than Cs-rich PQDs. They attributed this difference of degradation behaviour not only to the variation of chemical composition, but also the extra hydrogen bonds formed between the surface-exposed FA cations of PQDs and the carboxylic or amino groups of ligands, which also led to an increasing binding energy of OA/OAm ligands with the increase in the FA/Cs ratio. There was also grain growth to produce perovskite bulks at elevated temperature.

5.5 Conclusion, challenges and future prospects

CQD-SCs have emerged as strong candidates for next-generation PV because of the development in materials and device engineering, leading to competitive PCEs upwards of 25%. Fundamentally, advantageous CQD materials properties such as quantum confinement effects allowing for size- and shape-tunable bandgaps, large intrinsic dipole moments ideal for rapid charge carrier separation and extraction, energy level alignment via surface treatment, multiexciton generation, spectral adaptation encompassing utilization of infrared light, limited photobleaching and overall stability have made CQD-SCs particularly attractive. Importantly, ease of synthesis of CQDs and their ease of solution processing to fabricate CQD-SCs make them attractive and advantageous with strong potential to enter the flexible portable PV market in the near future. In terms of optimizing performance, improvements in device engineering have led to an enlarged depletion region and efficient charge extraction of photogenerated charges with the development of various device architectures.

Notwithstanding the advances in CQD-SCs, there are still issues that need to be tackled to make them more competitive in terms of performance. For instance, to reduce the trap density in QDs, improved surface passivation strategies have to be developed, including new and hybrid ligands. In particular, the hydroxide components on the CQD surfaces should be effectively passivated via ligand exchange or surface treatments in order to reduce surface trap states and also to improve stability in humid environments. For the enhancement of charge transport in QDs films, packing density has to be improved by creating a large-area self-assembly process and also unintentional impurities on the QDs and in the films have to be considerably reduced. Aspects of charge injection and collection at the electrodes need to be improved. This can be tackled with engineering the band structure of top and bottom contacts, electron and hole transport layers that will have better alignment with the QDs' band structure. Also, the interfacial junctions between each cell constituent layer should be carefully investigated because the defect states at these junctions are problematic since they act as non-radiative recombination centers during operation, thereby causing energy losses.

Other critical issues that need to be of focus include the device stability and environmental toxicity of the materials with the operational lifetime of the devices required to be improved. Such issues should involve more effective device encapsulation to exclude environmental degradation sources, such as oxygen and humidity, which will improve the stability of the light-unstable components in the

CQD devices. Also, with the improvements in other device systems and materials, such as perovskite and organic materials and solar cells, there are opportunities for advancing multijunction architectures for improved performance. With these, the goal is to create high-efficiency devices through improved spectral utilization and minimal loss associated with photocarrier thermalization. In particular, tandem and triple-junction all-CQD solar cells which make use of tunable bandgaps through QD size variation are useful to maximize spectral capture efficiency and boost device efficiency.

Notwithstanding these challenges, the field of CQD PV is rapidly growing and certainly has a tremendous potential to capture a large sector in the solar energy marketplace. This is enabled by advancements in colloidal synthesis and printing technologies which have paved the way for more cost-effective production methods. Also, the development of non-toxic alternatives to cadmium and lead-based CQDs has involved materials like copper indium sulfide and zinc-based QDs, which offer similar performance without the environmental risks. Critically, optimized and novel cell architectures and interface engineering have improved charge extraction. Some designs incorporate specially designed transport layers to facilitate the movement of electrons and holes to their respective electrodes. With further advances in these areas of development, CQD-SCs will realize their tremendous potential and wide-scale application.

References

[1] Sargent E H 2012 Colloidal quantum dot solar cells *Nat. Photonics* **6** 133–5

[2] Kramer I J and Sargent E H 2011 Colloidal quantum dot photovoltaics: a path forward *ACS Nano* **5** 8506–14

[3] Ekimov A I, Efros A L and Onushchenko A A 1985 Quantum size effect in semiconductor microcrystals *Solid State Commun.* **56** 921–4

[4] Brus L 1986 Electronic wave functions in semiconductor clusters: experiment and theory *J. Phys. Chem.* **90** 2555–60

[5] Taylor R A and Ramasamy K 2017 Colloidal quantum dots solar cells *Nanoscience* **vol 4** ed P J Thomas and N Revaprasadu (London: The Royal Society of Chemistry)

[6] Carey G H, Abdelhady A L, Ning Z, Thon S M, Bakr O M and Sargent E H 2015 Colloidal quantum dot solar cells *Chem. Rev.* **115** 12732–63

[7] Jasim K E 2015 Quantum dots solar cells *Solar Cells—New Approaches and Reviews* ed L A Kosyachenko (London: IntechOpen)

[8] Khan Z U, Khan L U, Brito H F, Gidlund M, Malta O L and Di Mascio P 2023 Colloidal quantum dots as an emerging vast platform and versatile sensitizer for singlet molecular oxygen generation *ACS Omega* **8** 34328–53

[9] Zhang L, Xiang W and Zhang J 2020 Thick-shell core/shell quantum dots *Core/Shell Quantum Dots: Synthesis, Properties and Devices* ed X M Tong and Z Wang (Cham: Springer International Publishing) pp 197–218

[10] Ghosh Chaudhuri R and Paria S 2012 Core/shell nanoparticles: classes, properties, synthesis mechanisms, characterization, and applications *Chem. Rev.* **112** 2373–433

[11] Ji B, Koley S, Slobodkin I, Remennik S and Banin U 2020 ZnSe/ZnS core/shell quantum dots with superior optical properties through thermodynamic shell growth *Nano Lett.* **20** 2387–95

[12] Ivanov S A, Piryatinski A, Nanda J, Tretiak S, Zavadil K R, Wallace W O, Werder D and Klimov V I 2007 Type-II core/shell CdS/ZnSe nanocrystals: synthesis, electronic structures, and spectroscopic properties *J. Am. Chem. Soc.* **129** 11708–19

[13] Wu K, Liang G, Kong D, Chen J, Chen Z, Shan X, McBride J R and Lian T 2016 Quasi-type II $CuInS_2$/CdS core/shell quantum dots *Chem. Sci.* **7** 1238–44

[14] Toufanian R, Piryatinski A, Mahler A H, Iyer R, Hollingsworth J A and Dennis A M 2018 Bandgap engineering of indium phosphide-based core/shell heterostructures through shell composition and thickness *Front. Chem.* **6** 567

[15] Wu L, Li Y, Liu G-Q and Yu S-H 2024 Polytypic metal chalcogenide nanocrystals *Chem. Soc. Rev.* **53** 9832–73

[16] Coughlan C, Ibáñez M, Dobrozhan O, Singh A, Cabot A and Ryan K M 2017 Compound copper chalcogenide nanocrystals *Chem. Rev.* **117** 5865–6109

[17] Hamanaka Y, Ogawa T, Tsuzuki M and Kuzuya T 2011 Photoluminescence properties and its origin of $AgInS_2$ quantum dots with chalcopyrite structure *J. Phys. Chem. C.* **115** 1786–92

[18] Uematsu T, Wajima K, Sharma D K, Hirata S, Yamamoto T, Kameyama T, Vacha M, Torimoto T and Kuwabata S 2018 Narrow band-edge photoluminescence from AgInS2 semiconductor nanoparticles by the formation of amorphous III–VI semiconductor shells *NPG Asia Mater.* **10** 713–26

[19] Afzaal M and O'Brien P 2006 Recent developments in II–VI and III–VI semiconductors and their applications in solar cells *J. Mater. Chem.* **16** 1597–602

[20] Ramasamy K, Malik M A, Revaprasadu N and O'Brien P 2013 Routes to nanostructured inorganic materials with potential for solar energy applications *Chem. Mater.* **25** 3551–69

[21] Zhao C, Li Z, Tang T, Sun J, Zhan W, Xu B, Sun H, Jiang H, Liu K, Qu S *et al* 2021 Novel III-V semiconductor epitaxy for optoelectronic devices through two-dimensional materials *Prog. Quantum Electron.* **76** 100313

[22] Gazis T A, Cartlidge A J and Matthews P D 2023 Colloidal III–V quantum dots: a synthetic perspective *J. Mater. Chem.* C **11** 3926–35

[23] Ming S-K, Taylor R A, McNaughter P D, Lewis D J and O'Brien P 2022 Tunable structural and optical properties of $AgxCuyInS_2$ colloidal quantum dots *New J. Chem.* **46** 18899–910

[24] Ming S-K, Taylor R A, McNaughter P D, Lewis D J, Leontiadou M A and O'Brien P 2021 Tunable structural and optical properties of $CuInS_2$ colloidal quantum dots as photovoltaic absorbers *RSC Adv.* **11** 21351–8

[25] Hong S P, Park H K, Oh J H, Yang H and Do Y R 2012 Comparisons of the structural and optical properties of o-$AgInS_2$, t-$AgInS_2$, and c-$AgIn_5S_8$ nanocrystals and their solid-solution nanocrystals with ZnS *J. Mater. Chem.* **22** 18939–49

[26] Kipkorir A, Murray S and Kamat P V 2024 How effective are sub-bandgap states in $AgInS_2$ quantum dots for electron transfer? *Chem. Mater.* **36** 4591–9

[27] Shi W, Khabibullin A R and Woods L M 2020 Exploring phase stability and properties of I-II2-III-VI4 Quaternary Chalcogenides *Adv. Theory Simul.* **3** 2000041

[28] Ling X, Yuan J and Ma W 2022 The rise of colloidal lead halide perovskite quantum dot solar cells *Acc.Mater. Res.* **3** 866–78

[29] Shen J and Zhu Q 2022 Stability strategies of perovskite quantum dots and their extended applications in extreme environment: a review *Mater. Res. Bull.* **156** 111987

[30] Shi J, Li F, Yuan J, Ling X, Zhou S, Qian Y and Ma W 2019 Efficient and stable $CsPbI_3$ perovskite quantum dots enabled by *in situ* ytterbium doping for photovoltaic applications *J. Mater. Chem.* A **7** 20936–44

[31] Zou J, Li M, Zhang X and Zheng W 2022 Perovskite quantum dots: synthesis, applications, prospects, and challenges *J. Appl. Phys.* **132** 220901

[32] Wei Y, Cheng Z and Lin J 2019 An overview on enhancing the stability of lead halide perovskite quantum dots and their applications in phosphor-converted LEDs *Chem. Soc. Rev.* **48** 310–50

[33] De Roo J, Ibáñez M, Geiregat P, Nedelcu G, Walravens W, Maes J, Martins J C, Van Driessche I, Kovalenko M V and Hens Z 2016 Highly dynamic ligand binding and light absorption coefficient of cesium lead bromide perovskite nanocrystals *ACS Nano* **10** 2071–81

[34] Li Q, Wu K, Zhu H, Yang Y, He S and Lian T 2024 Charge transfer from quantum-confined 0D, 1D, and 2D nanocrystals *Chem. Rev.* **124** 5695–763

[35] Murray C B, Norris D J and Bawendi M G 1993 Synthesis and characterization of nearly monodisperse CdE (E = sulfur, selenium, tellurium) semiconductor nanocrystallites *J. Am. Chem. Soc.* **115** 8706–15

[36] Kirmani A R, Luther J M, Abolhasani M and Amassian A 2020 Colloidal quantum dot photovoltaics: current progress and path to gigawatt scale enabled by smart manufacturing *ACS Energy Lett.* **5** 3069–100

[37] Lee H, Song H-J, Shim M and Lee C 2020 Towards the commercialization of colloidal quantum dot solar cells: perspectives on device structures and manufacturing *Energy Environ. Sci.* **13** 404–31

[38] Semonin O E, Luther J M, Choi S, Chen H-Y, Gao J, Nozik A J and Beard M C 2011 Peak external photocurrent quantum efficiency exceeding 100% via MEG in a quantum dot solar cell *Science* **334** 1530–3

[39] Hu L and Mandelis A 2021 Advanced characterization methods of carrier transport in quantum dot photovoltaic solar cells *J. Appl. Phys.* **129** 091101

[40] Li Q, Deng L, Du Y, Chang S, Wen S, Qin R, Wang J, Feng W, Gu B and Liu H 2025 Interfacial coupling enables high carrier mobility in PbS colloidal quantum dot photodetectors *Nano Res.* **18** 94907223

[41] Tang J, Kemp K W, Hoogland S, Jeong K S, Liu H, Levina L, Furukawa M, Wang X, Debnath R, Cha D *et al* 2011 Colloidal-quantum-dot photovoltaics using atomic-ligand passivation *Nat. Mater.* **10** 765–71

[42] Hong J, Hou B, Lim J, Pak S, Kim B-S, Cho Y, Lee J, Lee Y-W, Giraud P, Lee S *et al* 2016 Enhanced charge carrier transport properties in colloidal quantum dot solar cells via organic and inorganic hybrid surface passivation *J. Mater. Chem.* A **4** 18769–75

[43] Giansante C and Infante I 2017 Surface traps in colloidal quantum dots: a combined experimental and theoretical perspective *J. Phys. Chem. Lett.* **8** 5209–15

[44] Kahmann S and Loi M A 2020 Trap states in lead chalcogenide colloidal quantum dots—origin, impact, and remedies *Appl. Phys. Rev.* **7** 041305

45.

Rakov N, Guimarães R B and Maciel G S 2011 Strong infrared-to-visible frequency upconversion in Er^{3+}-doped Sr_2CeO_4 powders *J. Lumin.* **131** 342–6

[46] Akhtar M S, Malik M A, Alghamdi Y G, Ahmad K S, Riaz S and Naseem S 2015 Chemical bath deposition of Fe-doped ZnS thin films: investigations of their ferromagnetic and half-metallic properties *Mater. Sci. Semicond. Process.* **39** 283–91
[47] Pradhan N and Sarma D D 2011 Advances in light-emitting doped semiconductor nanocrystals *J. Phys. Chem. Lett.* **2** 2818–26
[48] Cao S, Li C, Wang L, Shang M, Wei G, Zheng J and Yang W 2014 Long-lived and well-resolved Mn^{2+} ion emissions in CuInS–ZnS quantum dots *Sci. Rep.* **4** 7510
[49] Liu S and Su X 2013 The synthesis and application of doped semiconductor nanocrystals *Anal. Methods* **5** 4541–8
[50] Zhang F, He X-W, Li W-Y and Zhang Y-K 2012 One-pot aqueous synthesis of composition-tunable near-infrared emitting Cu-doped CdS quantum dots as fluorescence imaging probes in living cells *J. Mater. Chem.* **22** 22250–7
[51] Makhal A, Sarkar S and Pal S K 2012 Protein-mediated synthesis of nanosized Mn-doped ZnS: a multifunctional, UV-durable bio-nanocomposite *Inorg. Chem.* **51** 10203–10
[52] Karmakar R, Neogi S K, Banerjee A and Bandyopadhyay S 2012 Structural; morphological; optical and magnetic properties of Mn doped ferromagnetic ZnO thin film *Appl. Surf. Sci.* **263** 671–7
[53] Manikandan A, Hema E, Durka M, Amutha Selvi M, Alagesan T and Arul Antony S 2015 Mn^{2+}-doped NiS ($Mn_xNi_{1-x}S$: $x = 0.0$, 0.3 and 0.5) nanocrystals: structural, morphological, opto-magnetic and photocatalytic properties *J. Inorg. Organomet. Polym.* **25** 804–15
[54] Jana S, Srivastava B B, Jana S, Bose R and Pradhan N 2012 Multifunctional doped semiconductor nanocrystals *J. Phys. Chem. Lett.* **3** 2535–40
[55] Singh N, Charan S, Sanjiv K, Huang S H, Hsiao Y C, Kuo C W, Chien F C, Lee T C and Chen P 2012 Synthesis of tunable and multifunctional Ni-doped near-infrared QDs for cancer cell targeting and cellular sorting *Bioconjug. Chem.* **23** 421–30
[56] Yang P, Lü M, Xu D, Yuan D, Song C, Liu S and Cheng X 2003 Luminescence characteristics of ZnS nanoparticles co-doped with Ni^{2+} and Mn^{2+} *Opt. Mater.* **24** 497–502
[57] Castro S, Bailey S, Banger K, Hepp A and Raffaelle R 2004 Synthesis and characterization of colloidal CuInS2 nanoparticles from a molecular single-source precursor *J. Phys. Chem.* **108** 12429–35
[58] van Embden J, Chesman A S R and Jasieniak J J 2015 The heat-up synthesis of colloidal nanocrystals *Chem. Mater.* **27** 2246–85
[59] Gaponik N, Talapin D V, Rogach A L, Hoppe K, Shevchenko E V, Kornowski A, Eychmüller A and Weller H 2002 Thiol-capping of CdTe nanocrystals: an alternative to organometallic synthetic routes *J. Phys. Chem.* B **106** 7177–85
[60] Dilena E, Xie Y, Brescia R, Prato M, Maserati L, Krahne R, Paolella A, Bertoni G, Povia M, Moreels I *et al* 2013 $CuIn_xGa_{1-x}S_2$ nanocrystals with tunable composition and band gap synthesized via a phosphine-free and scalable procedure *Chem. Mater.* **25** 3180–7
[61] Zhou Y-L, Zhou W-H, Li M, Du Y-F and Wu S-X 2011 Hierarchical Cu_2ZnSnS_4 particles for a low-cost solar cell: morphology control and growth mechanism *J. Phys. Chem.* C **115** 19632–9
[62] Murray C B, Sun S, Gaschler W, Doyle H, Betley T A and Kagan C R 2001 Colloidal synthesis of nanocrystals and nanocrystal superlattices *IBM J. Res. Dev.* **45** 47–56
[63] Chang J and Waclawik E R 2014 Colloidal semiconductor nanocrystals: controlled synthesis and surface chemistry in organic media *RSC Adv.* **4** 23505–27

[64] Xu F, Gerlein L F, Ma X, Haughn C R, Doty M F and Cloutier S G 2015 Impact of different surface ligands on the optical properties of PbS quantum dot solids *Materials* **8** 1858–70

[65] Bhaumik S and Pal A J 2013 Quantum dot light-emitting diodes in the visible region: energy level of ligands and their role in controlling interdot spacing and device performance *J. Phys. Chem.* C **117** 25390–6

[66] Zhong H, Zhou Y, Ye M, He Y, Ye J, Yang C and Li Y 2008 Controlled synthesis and optical properties of colloidal ternary chalcogenide $CuInS_2$ nanocrystals *Chem. Mater.* **20** 6434–43

[67] Khare A, Wills A W, Ammerman L M, Norris D J and Aydil E S 2011 Size control and quantum confinement in Cu_2ZnSnS_4 nanocrystals *Chem. Commun.* **47** 11721–3

[68] Pan D, An L, Sun Z, Hou W, Yang Y, Yang Z and Lu Y 2008 Synthesis of Cu–In–S ternary nanocrystals with tunable structure and composition *J. Am. Chem. Soc.* **130** 5620–1

[69] Kuzuya T, Hamanaka Y, Itoh K, Kino T, Sumiyama K, Fukunaka Y and Hirai S 2012 Phase control and its mechanism of $CuInS_2$ nanoparticles *J. Colloid Interface Sci.* **388** 137–43

[70] Kolny-Olesiak J and Weller H 2013 Synthesis and application of colloidal $CuInS_2$ semiconductor nanocrystals *ACS Appl. Mater. Interfaces* **5** 12221–37

[71] Fengcong G, Shouqin T, Baoshun L, Dehua X and Xiujian Z 2014 Oleic acid assisted formation mechanism of $CuInS_2$ nanocrystals with tunable structures *RSC Adv.* **4** 36875–81

[72] Buonsanti R and Milliron D J 2013 Chemistry of doped colloidal nanocrystals *Chem. Mater.* **25** 1305–17

[73] Radovanovic P V and Gamelin D R 2003 High-temperature ferromagnetism in Ni-doped ZnO aggregates prepared from colloidal diluted magnetic semiconductor quantum dots *Phys. Rev. Lett.* **91** 157202

[74] Bear J C, Hollingsworth N, McNaughter P D, Mayes A G, Ward M B, Nann T, Hogarth G and Parkin I P 2014 Copper-doped CdSe/ZnS quantum dots: controllable photo-activated copper(I) cation storage and release vectors for catalysis *Angew. Chem. Int. Ed.* **53** 1598–601

[75] Yang Y, Chen O, Angerhofer A and Cao Y C 2006 Radial-position-controlled doping in CdS/ZnS core/shell nanocrystals *J. Am. Chem. Soc.* **128** 12428–9

[76] Wang R, Shang Y, Kanjanaboos P, Zhou W, Ning Z and Sargent E H 2016 Colloidal quantum dot ligand engineering for high performance solar cells *Energy Environ. Sci.* **9** 1130–43

[77] Yuan D, Han Z, Cao F, Liu X, Liu M, Zhang L, Cao S, Li J, Zeng T, Chen Y *et al* 2024 Mixed halide passivation of $AgBiS_2$ quantum dots for high-performance photodetectors *ACS Appl. Electron. Mater.* **6** 8455–62

[78] Selopal G S, Zhao H, Wang Z M and Rosei F 2020 Core/shell quantum dots solar cells *Adv. Funct. Mater.* **30** 1908762

[79] Kinder E, Moroz P, Diederich G, Johnson A, Kirsanova M, Nemchinov A, O'Connor T, Roth D and Zamkov M 2011 Fabrication of all-inorganic nanocrystal solids through matrix encapsulation of nanocrystal arrays *J. Am. Chem. Soc.* **133** 20488–99

[80] Neo D C J, Cheng C, Stranks S D, Fairclough S M, Kim J S, Kirkland A I, Smith J M, Snaith H J, Assender H E and Watt A A R 2014 Influence of shell thickness and surface passivation on PbS/CdS core/shell colloidal quantum dot solar cells *Chem. Mater.* **26** 4004–13

[81] Clifford J P, Johnston K W, Levina L and Sargent E H 2007 Schottky barriers to colloidal quantum dot films *Appl. Phys. Lett.* **91** 253117

[82] Luther J M, Law M, Beard M C, Song Q, Reese M O, Ellingson R J and Nozik A J 2008 Schottky solar cells based on colloidal nanocrystal films *Nano Lett.* **8** 3488–92

[83] Tang J, Wang X, Brzozowski L, Barkhouse D A R, Debnath R, Levina L and Sargent E H 2010 Schottky quantum dot solar cells stable in air under solar illumination *Adv. Mater.* **22** 1398–402

[84] Mai V-T, Duong N-H and Mai X-D 2019 Boosting the current density in inverted Schottky PbS quantum dot solar cells with conjugated electrolyte *Mater. Lett.* **249** 37–40

[85] Mai X-D, An H J, Song J H, Jang J, Kim S and Jeong S 2014 Inverted Schottky quantum dot solar cells with enhanced carrier extraction and air-stability *J. Mater. Chem.* A **2** 20799–805

[86] Choi J J *et al* 2009 PbSe nanocrystal excitonic solar cells *Nano Lett.* **9** 3749–55

[87] Zhao N, Osedach T P, Chang L-Y, Geyer S M, Wanger D, Binda M T, Arango A C, Bawendi M G and Bulovic V 2010 Colloidal PbS quantum ot solar cells with high fill factor *ACS Nano* **4** 3743–52

[88] Pattantyus-Abraham A G *et al* 2010 Depleted-heterojunction colloidal quantum dot solar cells *ACS Nano* **4** 3374–80

[89] Rath A K, Bernechea M, Martinez L, de Arquer F P G, Osmond J and Konstantatos G 2012 Solution-processed inorganic bulk nano-heterojunctions and their application to solar cells *Nat. Photonics* **6** 529–34

[90] Tabernig S W, Yuan L, Cordaro A, Teh Z L, Gao Y, Patterson R J, Pusch A, Huang S and Polman A 2022 Optically resonant bulk heterojunction PbS quantum dot solar cell *ACS Nano* **16** 13750–60

[91] Ding C, Zhang L, Shen Q and Ding,. L 2021 Colloidal quantum-dot bulk-heterojunction solar cells *J. Semiconduct.* **42** 110203

[92] Choi M-J *et al* 2020 Colloidal quantum dot bulk heterojunction solids with near-unity charge extraction efficiency *Adv. Sci.* **7** 2000894

[93] Sahu A, Garg A and Dixit A 2020 A review on quantum dot sensitized solar cells: past, present and future towards carrier multiplication with a possibility for higher efficiency *Sol. Energy* **203** 210–39

[94] Basit M A, Aanish Ali M, Masroor Z, Tariq Z and Bang J H 2023 Quantum dot-sensitized solar cells: a review on interfacial engineering strategies for boosting efficiency *J. Ind. Eng. Chem.* **120** 1–26

[95] Jumabekov A N, Cordes N, Siegler T D, Docampo P, Ivanova A, Fominykh K, Medina D D, Peter L M and Bein T 2016 Passivation of PbS quantum dot surface with l-glutathione in solid-state quantum-dot-sensitized solar cells *ACS Appl. Mater. Interfaces* **8** 4600–7

[96] Song H, Lin Y, Zhang Z, Rao H, Wang W, Fang Y, Pan Z and Zhong X 2021 Improving the efficiency of quantum dot sensitized solar cells beyond 15% via secondary deposition *J. Am. Chem. Soc.* **143** 4790–800

[97] Andruszkiewicz A, Zhang X, Johansson M B, Yuan L and Johansson E M J 2021 Perovskite and quantum dot tandem solar cells with interlayer modification for improved optical semitransparency and stability *Nanoscale* **13** 6234–40

[98] Wang X, Koleilat G I, Tang J, Liu H, Kramer I J, Debnath R, Brzozowski L, Barkhouse D A R, Levina L, Hoogland S *et al* 2011 Tandem colloidal quantum dot solar cells employing a graded recombination layer *Nat. Photonics* **5** 480–4

[99] Karani A, Yang L, Bai S, Futscher M H, Snaith H J, Ehrler B, Greenham N C and Di D 2018 Perovskite/colloidal quantum dot tandem solar cells: theoretical modeling and monolithic structure *ACS Energy Lett.* **3** 869–74
[100] Raza E and Ahmad Z 2022 Review on two-terminal and four-terminal crystalline-silicon/perovskite tandem solar cells; progress, challenges, and future perspectives *Energy Rep.* **8** 5820–51
[101] Manekkathodi A *et al* 2019 Solution-processed perovskite-colloidal quantum dot tandem solar cells for photon collection beyond 1000 nm *J. Mater. Chem.* A **7** 26020–8
[102] Singh R and Rani M 2024 Optimization of perovskite/colloidal quantum dot monolithic tandem solar cell to enhance device performance via solar cell capacitance simulator 1D *Physica Status Solidi b* **261** 2300475
[103] Du Z and Ma D 2025 Recent progress in I–III–VI colloidal quantum dots-integrated solar cells *Curr. Opin. Colloid Interface Sci.* **75** 101890
[104] Bang J H and Kamat P V 2011 CdSe quantum dot–fullerene hybrid nanocomposite for solar energy conversion: electron transfer and photoelectro chemistry *ACS Nano* **5** 9421–7
[105] Baek S-W *et al* 2019 Efficient hybrid colloidal quantum dot/organic solar cells mediated by near-infrared sensitizing small molecules *Nat. Energy* **4** 969–76
[106] Lee J *et al* 2023 Unlocking the potential of colloidal quantum dot/organic hybrid solar cells: band tunable interfacial layer approach *ACS Appl. Mater. Interfaces* **15** 39408–16
[107] Mubarok M A, Aqoma H, Wibowo F T A, Lee W, Kim H M, Ryu D Y, Jeon J-W and Jang S-Y 2020 Molecular engineering in hole transport π-conjugated polymers to enable high efficiency colloidal quantum dot solar cells *Adv. Energy Mater.* **10** 1902933
[108] Liu C, Wang K, Gong X and Heeger A J 2016 Low bandgap semiconducting polymers for polymeric photovoltaics *Chem. Soc. Rev.* **45** 4825–46
[109] Beygi H, Sajjadi S A, Babakhani A, Young J F and van Veggel F C J M 2019 Air exposure oxidation and photooxidation of solution-phase treated PbS quantum dot thin films and solar cells *Sol. Energy Mater. Sol. Cells* **203** 110163
[110] Becker-Koch D, Albaladejo-Siguan M, Lami V, Paulus F, Xiang H, Chen Z and Vaynzof Y 2020 Ligand dependent oxidation dictates the performance evolution of high efficiency PbS quantum dot solar cells *Sustain. Energy Fuels* **4** 108–15
[111] Sykora M, Koposov A Y, McGuire J A, Schulze R K, Tretiak O, Pietryga J M and Klimov V I 2010 Effect of air exposure on surface properties, electronic structure, and carrier relaxation in PbSe nanocrystals *ACS Nano* **4** 2021–34
[112] Bozyigit D, Volk S, Yarema O and Wood V 2013 Quantification of deep traps in nanocrystal solids, their electronic properties, and their influence on device behavior *Nano Lett.* **13** 5284–8
[113] Shi G *et al* 2021 The effect of water on colloidal quantum dot solar cells *Nat. Commun.* **12** 4381
[114] Zherebetskyy D, Scheele M, Zhang Y, Bronstein N, Thompson C, Britt D, Salmeron M, Alivisatos P and Wang L-W 2014 Hydroxylation of the surface of PbS nanocrystals passivated with oleic acid *Science* **344** 1380–4
[115] Cao Y, Stavrinadis A, Lasanta T, So D and Konstantatos G 2016 The role of surface passivation for efficient and photostable PbS quantum dot solar cells *Nat. Energy* **1** 16035
[116] Albaladejo-Siguan M, Baird E C, Becker-Koch D, Li Y, Rogach A L and Vaynzof Y 2021 Stability of quantum dot solar cells: a matter of (life) time *Adv. Energy Mater.* **11** 2003457

[117] van Sark W G J H M, Frederix P L T M, van den Heuvel D J, Bol A A, van Lingen J N J, de Mello Donegá C, Gerritsen H C and Meijerink A 2002 Time-resolved fluorescence spectroscopy study on the photophysical behavior of quantum dots *J. Fluoresc.* **12** 69–76

[118] Wang S, Zhao Q, Hazarika A, Li S, Wu Y, Zhai Y, Chen X, Luther J M and Li G 2023 Thermal tolerance of perovskite quantum dots dependent on A-site cation and surface ligand *Nat. Commun.* **14** 2216

[119] Chen W *et al* 2021 Operando structure degradation study of PbS quantum dot solar cells *Energy Environ. Sci.* **14** 3420–9

[120] Kirmani A R *et al* 2020 Colloidal quantum dot photovoltaics using ultrathin, solution-processed bilayer In_2O_3/ZnO electron transport layers with improved stability *ACS Appl. Energy Mater.* **3** 5135–41

IOP Publishing

Solution-Processed Solar Cells
Materials and device engineering
Richard A Taylor and Karthik Ramasamy

Chapter 6

Dye-sensitized solar cells

Dye-sensitized solar cells (DSSCs) are solution-processed thin film solar cells that utilize dye molecules to absorb photons and generate electricity. They are based on a semiconductor formed between a photo-sensitized (dye molecule) anode and an electrolyte, a photoelectrochemical system. DSSCs are low-cost alternatives to traditional silicon solar cells and offer flexibility and transparency particularly useful for wearable technology, building-integrated photovoltaics (BIPV), indoor lighting, and portable electronic devices. Typical DSSCs utilize organic (natural and synthetic) dyes and inorganic, mainly ruthenium complexes as a photosensitizer. In addition to their low cost and relative non-toxicity, the simple preparation and ease of production of DSSCs using conventional roll-to-roll printing techniques makes them attractive alternative solar cells. However, there are limitations associated with these devices. It has proven difficult to eliminate a number of expensive materials, notably platinum and ruthenium dye sensitizers, and the liquid electrolyte presents serious stability issues making them challenging for use in some environments. Notwithstanding, DSSCs are attractive replacements for existing technologies in 'low density' applications like rooftop solar collectors, where the mechanical robustness and light weight of the glass-less collector is a major advantage. Currently, their performance stands at around 19% efficiency and they are expected to become more competitive in the next few years. Though they may not be as attractive for large-scale deployments where higher-cost and higher-efficiency cells are more viable, small increases in the DSSC conversion efficiency might make them viable in alternative applications.

6.1 Working principles of dye-sensitized solar cells

The seminal work at the basis of the development of the DSSC was reported in 1991 by O'Regan and Grätzel [1]. The so-called Grätzel cell, primarily a photoelectrochemical device was based on the sensitization of a wide bandgap metal oxide semiconductor, TiO_2, with a ruthenium-based metallo-organic dye. Unlike *p–n*

 © IOP Publishing Ltd 2025. All rights, including for text and data mining (TDM), artificial intelligence (AI) training, and similar technologies, are reserved.

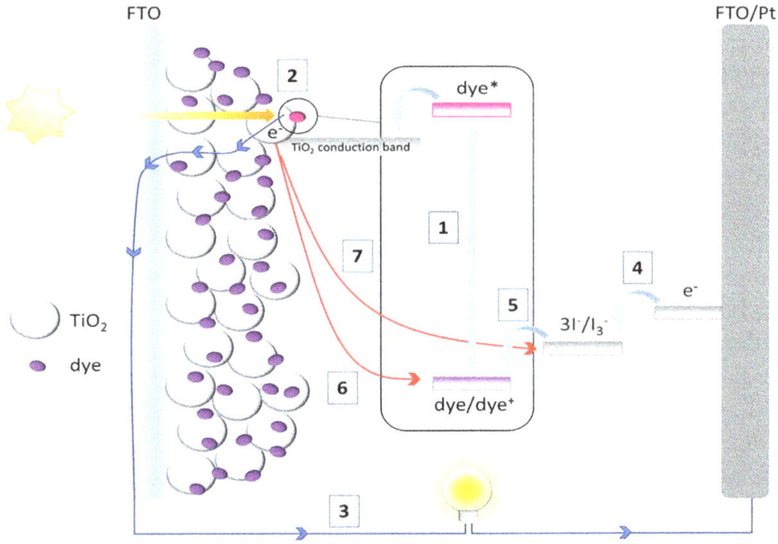

Figure 6.1. Architecture and operating principle of the typical DSSC; Grätzel cell [2]. Reprinted from [2] CC BY 4.0.

junction cells, DSSCs operate on the principle of a photosynthetic plant cell with a dye-sensitizer molecule mimicking chlorophyll, involving several steps. As shown in figure 6.1 [2] the first step [1] involves the absorption of a photon by the sensitizer molecule, conventionally a ruthenium-based metallo-organic complex, in which an electron is excited from the ground state—highest occupied molecular orbital (HOMO) into the excited state—lowest unoccupied molecular orbital (LUMO), leaving behind a hole. Similar to a typical *p–n* junction, the electron and hole must be effectively separated for charge extraction, circumventing their recombination. The second step [2] involves fast injection of the electron from the LUMO state of the excited sensitizer into the lower energy conduction band (CB) of the layered semiconductor, leaving the sensitizer in the oxidized state. This electron transport layer (ETL), which is typically nanostructured TiO_2, allows for gradient diffusion of the injected electron through its mesoporous structure. The ETL which is interfaced with the working electrode (WE) consists of dye molecules embedded through chelation into the TiO_2 matrix via immersion in a dye solution. This interface allows for collection of electrons at the WE, a transparent conductive oxide (TCO) electrode/anode (typically, fluorine tin oxide—FTO) and enables current generation (step 3). In the other side of the device, the hole transport layer (HTL)—electrolyte component interfaces the counter electrode (CE)—cathode, facilitating migration of excited holes (steps 4 and 5). The HTL enables transport of holes from excited dye molecules to the cathode via a redox couple that reduces the oxidized dye sensitizer with the extracted electron, regenerating the original dye and completing the circuit. The first dye-sensitized cells created by Grätzel and co-workers [1] utilized an iodide–triiodide (I^-/I_3^-) redox couple in which triiodide ions diffuse to the cathode then become reduced back to iodide ions, and the holes are transported from dye molecules to the CE. Overall, developments in these cells involved the use of a range of new HTL materials, including new liquid electrolytes, solid-state conjugated polymers and quasi-solid polymer-electrolyte composites.

For DSSCs, the efficiency which depends on the open-circuit voltage, V_{oc} correlates with the difference in electrochemical potential between the HTL and ETL, ascribed to the difference in energy between the Fermi energy of electrons in the illuminated TiO_2 (roughly approximated as the CB edge) and the redox potential of the liquid electrolyte. In the case of the more reliable solid-state DSSCs, the difference between the CB minimum of the TiO_2 and the HOMO level of the HTL material determines the V_{oc}. Similar to other PV cells, there are some undesirable processes that limit the V_{oc} and overall conversion efficiency in these cells. These include, charge recombination of the injected electrons with the oxidized sensitizer (step 6—figure 6.1) or with the oxidized state of the redox couple (step 7) which competes with the injection of electrons into the semiconductor for available photocurrent [2]. Although these cells are not yet marketable due to their relatively low efficiencies compared to *p–n* junction cells, improvements in each of the three main components (dye, ETL and HTL), as well as improved stability and lifetime, will enable Grätzel cells and their more advanced counterparts such as the solid-state devices, achieve their full technological potential across a range of aforementioned applications.

6.2 Structure and properties of dye sensitizer materials

Parts of this section have been reproduced from [2]. CC BY 4.0.

The dye sensitizer of the photoanode is one of the most important components of a DSSC. Its main role is the absorption of solar photons and injection of the photoexcited electrons into the CB of the *n*-type semiconductor. Invariably, the sensitizer controls the breadth of the solar spectrum used and the quantum yield for electron injection, and its properties have a considerable effect on the light-harvesting efficiency and the overall power conversion efficiency (PCE) of the DSSC. Accordingly, based on their structure and chemical characteristics, dye molecules must fulfil several key requirements as effective sensitizers for DSSCs which include [2]:

1. They should have a wide range of optical absorption in the visible–NIR region with high molar absorption coefficients, desirable for efficient solar photons harvesting, especially since a thin layer of the ETL is required.
2. Since they must be easily anchored and embedded into the semiconductor layer, their molecular structure should be characterized by peripheral anchoring groups, typically acidic in nature, such as carboxylic or silyl, for strong bonding interaction. This enables them to sufficiently passivate the surface of the semiconductor electrode, reducing undesirable interfacial recombination.
3. The excited state LUMO energy level should be higher than the CB edge of the *n*-type semiconductor for efficient, ultrafast electron injection.

4. The HOMO level should be lower than the energy level of the redox mediator to promote dye regeneration.
5. They should feature high photostability to resist degradation from continuous light exposure, along with thermal and electrochemical stability.
6. For more advanced cells, they are required to be inexpensive, easily synthesized and environmentally benign.

A wide variety of photosensitizers for DSSC applications have been studied and can be divided into inorganic-based and organic dyes according to their molecular structures. Of these, the main classes have been ruthenium(II) polypyridyl complexes, zinc(II) porphyrin derivatives, and organic (i.e. metal-free) dyes. Organic dyes are particularly attractive since they are easy to synthesize, their properties can be easily tuned, and they are less expensive compared to ruthenium-based sensitizers.

6.2.1 Ruthenium-based dyes

Figure 6.2 shows a selection of known bipyridyl-based ruthenium(II) dye complexes used as sensitizers in the photoanodes of DSSCs. Dyes, namely N3, N719, CYC-B11, and the Black Dye are commercially available and remain common benchmarks in DSSC applications with PCEs > 11% [3]. The DSSC device developed by Grätzel and co-workers incorporated the first ruthenium-based dye (dye 1—Black

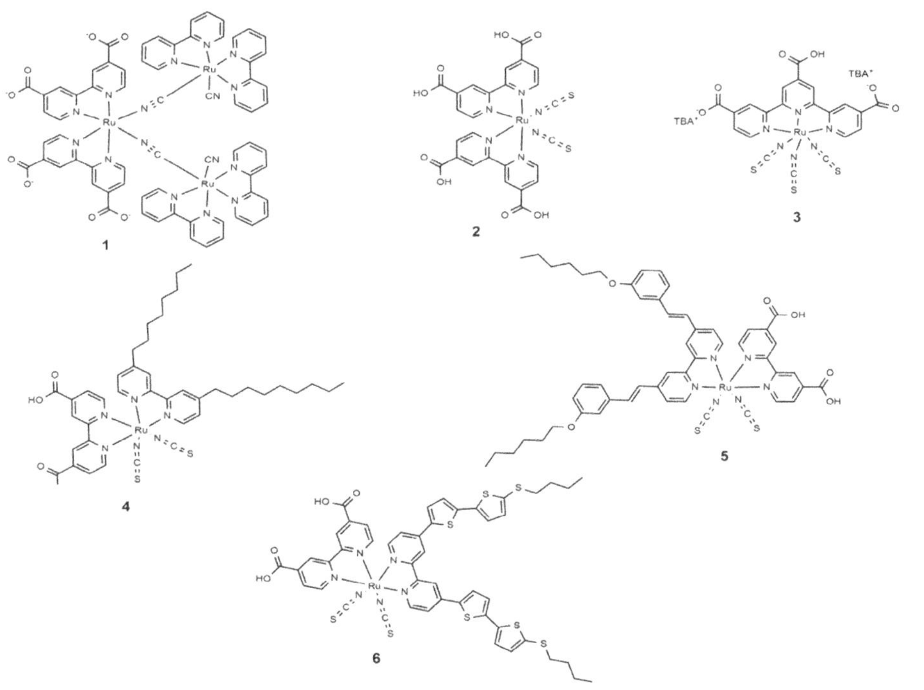

Figure 6.2. Selected examples of a range of ruthenium-based dyes of different molecular structures [3].

dye), a trinuclear bipyridyl-based complex. The cell which had an efficiency of 7.12% was improved to an efficiency of 10% with the use of its mononuclear counterpart (dye 2—N3) [1]. Subsequently, the primary metallo-organic dyes of heavy transition metals of ruthenium, osmium and iridium have been widely used as inorganic dyes in DSSCs because of their long excited-state lifetime, highly efficient metal-to-ligand charge transfer (MLCT) and strong redox potential [4]. These dyes are of the general formula, $ML_2(X)_2$, where M = metal, typically of octahedral geometry, L = ligand such as 2,2′-bipyridyl-4,4′-dicarboxylic acid, and X = halide, cyanide, thiocyanate, acetyl acetonate, thiocarbamate, or water substituents. Of these, ruthenium(II) polypyridyl complexes have been extensively incorporated in DSSCs with the best efficiencies due to their thermal and chemical stability, and wide absorption range from visible to NIR. Their dominant MLCT transitions with moderate absorption coefficient (<18 000 M^{-1} cm^{-1}) make them particularly useful as sensitizers. Their absorption and electrochemical properties can be engineered by suitable modification of their molecular structures for variability of absorption across the visible–NIR spectral range. In particular, the bipyridyl moieties can be modified to produce carboxylate, phosphate and polynuclear-type ruthenium dyes.

As an example, dye N3, is characterized with two thiocyanate groups compared to its pre-congener dye 1, resulting in a red-shift in the absorption spectrum up to 800 nm which was at the basis of its J_{sc} of 18 mA cm^{-2}, and an overall efficiency of 10% in the cell [2]. The dye also exhibited a short excited-state lifetime of 60 ns leading to fast charge injection. This was attributed to the degree of protonation resulting from photoexcitation which enhances adsorption of the dye to the surface of TiO_2, facilitating fast electron injection from its excited state into the TiO_2 CB, resulting in high photocurrents. Simultaneously, there is also a positive shift of the CB edge which, in turn, lowers the V_{oc}. This understanding ensured that a delicate balance between these two competing factors is considered in designing sensitizers for example, dye N719 (dye 3), which led to device efficiencies of 11.2%.

Strategies towards the development of ruthenium bipyridyl-based dyes has involved extending the π-conjugation network in order to tune the LUMO and increase their molar absorption coefficient, and to improve directionality in the excited state. For example, dyes such as N945, Z910, K19 among others, exhibited enhanced properties in addition to being more electrochemically and thermally stable [3]. Also, thiophene derivatives with and without conjugation have been developed, with those without, setting comparatively higher DSSC efficiencies of 11.3%–11.5% [2]. Other strategies involved increasing the hydrophobicity with alkyl chains that repel water and triiodide from the TiO_2 surface, thereby improving anchoring and enhancing stability. For example, a device consisting of the amphiphilic ruthenium complex, K19 (dye 5—figure 6.2) achieved good efficiency of ~8% and remarkable long-term stability at 55 °C under one sun illumination exposure [5]. The high stability of the device, in which a co-adsorbent (hexadecyl-malonic acid) was grafted onto TiO_2 together with the sensitizers was due to formation of a hydrophobic layer which hindered the desorption of the dye, whilst forming an insulating barrier between the semiconductor and the electrolyte, thereby reducing the dark current and increasing V_{oc}. Another key strategy has been to include different anchoring groups to improve binding to the surface of the TiO_2. Typical carboxylic groups lose their binding strength at pH > 9 which causes desorption of the dye from the surface. Much of this strategy has incorporated phosphonic acid groups since their binding strength is greater than that of carboxylic acids. For example, use of the ruthenium complex, Z955 with a phosphonic acid group resulted in a DSSC with improved efficiency of 8% accompanied by good stability under prolonged light exposure for about 1000 h at AM 1.5 °C and 55 °C [6]. Also, ruthenium dyes with thiophene functional groups have also exhibited optimized light-harvesting potential due to the electron rich characteristics of these groups. For example, dye CYC-B11 (dye 6) characterized with an electron rich thioalkyl group with the bithiophene moiety, afforded a highly stable device of 11.5% efficiency [6]. This was mainly attributed to a high molar extinction coefficient (2.42×10^4 M^{-1} cm^{-1}) which resulted in the use of a thinner TiO_2 layer and caused a decreased dark current leakage and increased V_{oc}. Other routes towards functionalizing polypyridyl ruthenium dyes involved incorporating ancillary ligands with highly polarizable donor antennae or hole transport chromophores, or thiocyanate-free groups, for improved efficiency up to 11% [7].

6.2.2 Porphyrin-based dyes

Photosensitizers containing a conjugated system of pyrroles complexed with metal ions have become very important for DSSCs because of their intense optical absorptions, *vis* the Soret band between 400 and 500 nm and the Q-band between 550 and 750 nm. These functionalized porphyrin macrocycles which rival ruthenium-based pyrrole dyes allow for tunable absorption in such a way that panchromatic dyes suitable for DSSC applications can be obtained. An important breakthrough reported by Tachibana and co-workers [8] showed that the *tetrakis* (4-carboxyphenyl)porphyrin dye exhibits very fast electron injection into the TiO_2 semiconductor, similar to that of ruthenium-based photosensitizers, and established the basis for the design and study of various porphyrin-based sensitizer molecules for DSSCs. Overall, porphyrins are attractive since they exhibit strong absorption and emission in the visible region, have a long excited state lifetime (>1 ns), ultrafast electron injection rate (femtosecond), millisecond timescale electron recombination rate, and tunable redox potentials [4]. In addition to these, the strategy involving the use of porphyrins has been to exploit the large molar absorptivity between the Soret and Q spectral bands.

One of the issues associated with charge injection is poor photoanode coverage due to aggregation of dye molecules. As such, strategies in designing porphyrin dye sensitizers have involved reducing their aggregation on the semiconductor surface which led to a series of zinc(II) based tetraphenyl porphyrins with lower symmetry. These involved for example, substitution with styril carboxylic acid in the β-pyrrolic position, in which the phenyl moiety acting as donor and the carboxylic acid as acceptor, results in a red-shift of the Q-band. As forerunner examples, dye sensitizers, 8 and 9 shown in figure 6.3, yielded efficiencies of 4.1% and 4.8%, respectively, in DSSCs [2]. Additionally, increasing surface anchoring has been

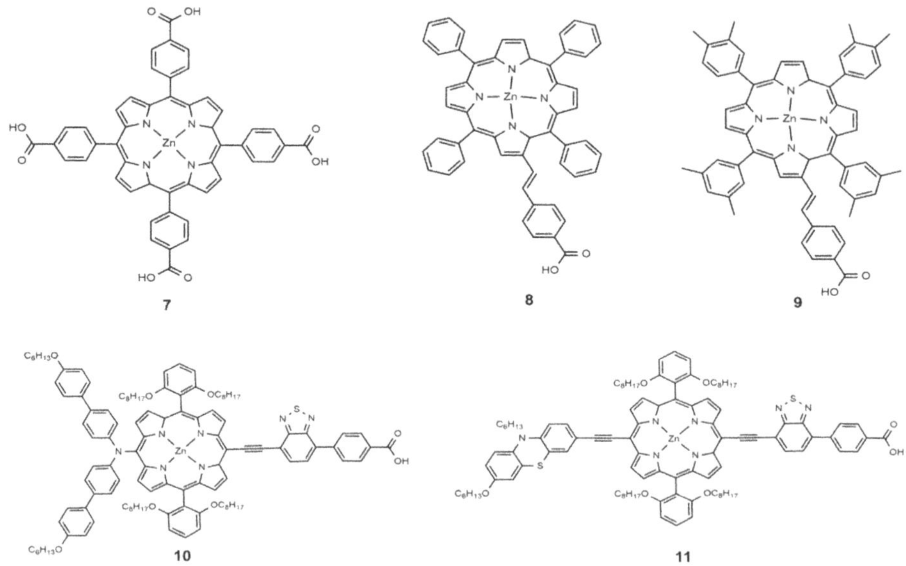

Figure 6.3. Selected examples of a range of porphyrin-based dyes of different molecular structures.[3].

achieved with electron withdrawing cyano functional groups and ethylenic phenyl carboxylic acids, which not only stabilize the anchoring but extends conjugation for increased charge injection.

Perhaps the most impactful strategy with these types of dyes has been employing the *push—pull*, donor (D)–acceptor (A) motif linked via a π-conjugated bridge, giving a typically elongated D–π–A structure. In these, the macrocycle acts as a π-conjugated spacer group between strong electron D and A groups, which extends optical absorption and harvests a larger fraction of solar photons. Notable examples included zinc-based dyes in which diarylamino substituents act as strong electron donors and an ethynyl-carboxy moiety acting as an acceptor. These types of dyes, examples of which are shown in figure 6.3, have yielded competitive conversion efficiencies above 10% relative to ruthenium-based dyes. For many of these, the increase in the number of the diarylamino groups widens, and slightly blue-shifts the Soret band, with a new band at 450–500 nm and the Q-band red-shifting. Improved PCEs using this mechanism have also been achieved by incorporating a benzothiadiazole (BTD) group near the benzoic acid anchor for example, to create the dye 10. This improved absorption in the 500–650 nm range resulted in a PCE of 12.8% from a single-dye DSSC device with a J_{sc} of 18.5 mA cm^{-2} using a cobalt-based electrolyte [3]. Overall, the D–π–A molecular structure has become the basis for the increased efficiency beyond 13% for dyes with highly functionalized π-conjugated electron withdrawing and donating groups.

Notwithstanding the aforementioned desirable attributes, one of the main drawbacks of porphyrin-based sensitizers relates to reduced absorption between 500 and 600 nm. An effective approach to overcome this has been to use two or more different sensitizers with complementary absorption properties to improve the

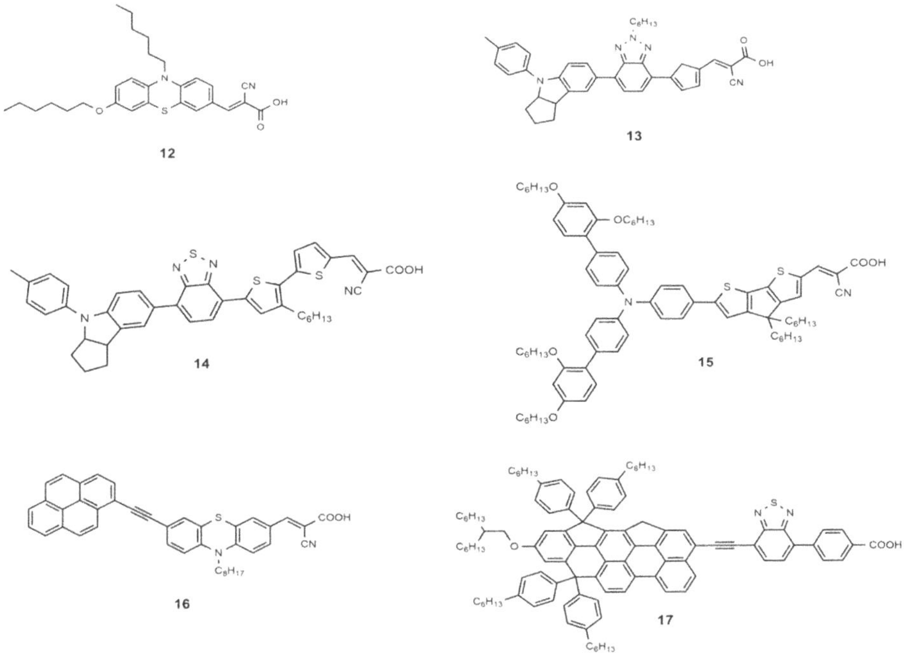

Figure 6.4. Selected examples of a range of organic based dyes of different molecular structures [3].

conversion efficiency of the DSSC device. Co-sensitization, as it is described, is primarily focussed on increasing the absorption coefficient and reducing the detrimental charge recombination through better coverage of the semiconductor surface. This approach can be achieved primarily in two ways [1]; by quantitative mixing of two sensitizer solutions, or [2] via sequential adsorption of the two different dye solutions [2]. As reported by Xie and co-workers [9], co-sensitization achieved by sequential adsorption of phenothiazine-containing porphyrin sensitizer dye 11 (figure 6.3) and the organic dye 13 (figure 6.4) afforded an efficiency of 11.5%, relative to 7.8% by a device sensitized only with dye 12 (figure 6.4). This was attributed to the complementary absorptions of the dyes, where dye 12 had an intense and broad band at around 510 nm, and dye 11 exhibited the typical Soret band (465 nm) and Q-band (683 nm).

To further improve efficiencies, a collection of strategies has emerged with respect to porphyrin dye design. These include reducing aggregation through novel constructs, improving spectral response across the visible–NIR region via building block incorporation, co-linking of chromophores, and design of supra-molecular assemblies towards tailored aggregation such as aggregate-induced red-shifting of the absorption spectrum [3]. With respect to the linear donor–porphyrin–acceptor design with meso-substituted de-aggregating groups, common methods for extending the absorption range focus on adding donor groups, fusing non-amine donor groups for π-extended donor groups, or adding acceptor groups to promote lower energy intramolecular charge transfer. Additionally, the use of a π-extended donor group has shown promise for improving device performances, for example through introduction of an anthracene group between an amine donor and porphyrin, which resulted in a red-shift of both the Soret and Q-bands relative to a donor without anthracene.

6.2.3 Organic dyes

Although ruthenium-based dyes are highly effective sensitizers, they are not particularly suitable for cost-effective and environmentally friendly PV due to their cost and availability of ruthenium [4]. Consequently, the strategy has been to develop inexpensive and environmentally friendly alternatives, including a range of functional organic dyes. An important advantage of many of these organic dyes over ruthenium-based ones is that their absorption coefficient is typically one order of magnitude higher, which enables a thinner TiO_2 layer, critical for charge carrier extraction and diffusion. The strategy has been to ensure larger photocurrent responses from photo-absorption across the visible–NIR region in which there is a sufficiently positive HOMO level relative to the CE's redox potential, and sufficiently negative LUMO than the CB edge level of the TiO_2 semiconductor, to ensure efficient charge injection and separation. Typical classes of these dyes include, tetrahydroquinolines, pyrolidines, di- and tri-phenylamines (DPA and TPA), coumarins, indolines, carbazoles (CBZ), phenothiazines (PTZ), phenoxazines (POZ), hemicyanines, merocyanines, squaraines, perylenes, anthraquinones, boradiazaindacenes (BODIPY), oligothiophenes, and polymeric dyes. A notable limitation associated with these dyes is their lower thermal stability which is a primary factor for device degradation.

The prototypal organic dye features a molecular design of an elongated donor–π-spacer–acceptor (D–π–A) structure, similar to what was described for the porphyrin-based dyes but without the metal [2]. The electron D unit includes functionalities such as indoline, triarylamine, coumarin, and fluorine, whereas electron A units are typically carboxylic acid, cyanoacrylic acid, and rhodamine functionalities with the π-spacer groups including conjugated polyene and oligothiophene. Typically, the electron acceptor also acts as the anchoring group to the semiconductor surface, as illustrated in figure 6.4 [10], and this architecture enables tunable absorption across a wide range. Effectively, a higher photo-induced electron transfer from the donor to acceptor through the π-spacer to the CB of the TiO_2 layer can result from shifting the HOMO and LUMO energy levels of the donor and acceptor groups, respectively. This is done by extending the π-conjugation either via the methine unit or by introducing aromatic rings such as benzene, thiophene, and furan or by adding either electron donating or withdrawing groups. The objective of these structural modulations is to effectively separate the photo-induced charge, thereby limiting their recombination between the semiconductor and oxidized dye. Overall, the rational design of these organic sensitizers has been focussed on investigating different electron donors and π-linkers and, to a lesser extent, electron-withdrawing anchoring moieties. The aim of these efforts has been to obtain systems with adequate optical absorption, which extend particularly into the NIR spectral region, well-aligned electronic levels to minimize losses, and high stability of the device.

Various strategies have been employed in developing organic dye sensitizers, and one alternative to the D–π–A structure involves the insertion of an electron-poor moiety into the π-bridge, thus creating a D–A–π–A structure, as illustrated in figure 6.5. This type of structure has become particularly attractive, especially to achieve high extinction coefficients, broaden the absorption range and improve the photo/thermal stability of organic sensitizers. One common example explored is the incorporation of a benzothiadiazole group in the π-bridge [2]. As shown in figure 6.5, the electron-poor benzothiadiazole group is inserted between an indoline donor and a bithiophene bearing the final acceptor (anchoring) cyanoacrylic moiety (dye 14—figure 6.4). This resulted in a red-shift of the absorption wavelength and an efficiency of 9.04% of the DSSC with a J_{sc} of 18.9 mA cm^{-2}. Several other DSSCs using this benzothiadiazole moiety and their derivatives with variable π-linkers, showing promising efficiencies, high current density and outstanding stability, and retaining 90%–95% of the initial efficiency after continuous light exposure for 1000 h at relatively high temperatures have been designed. Other design strategies have

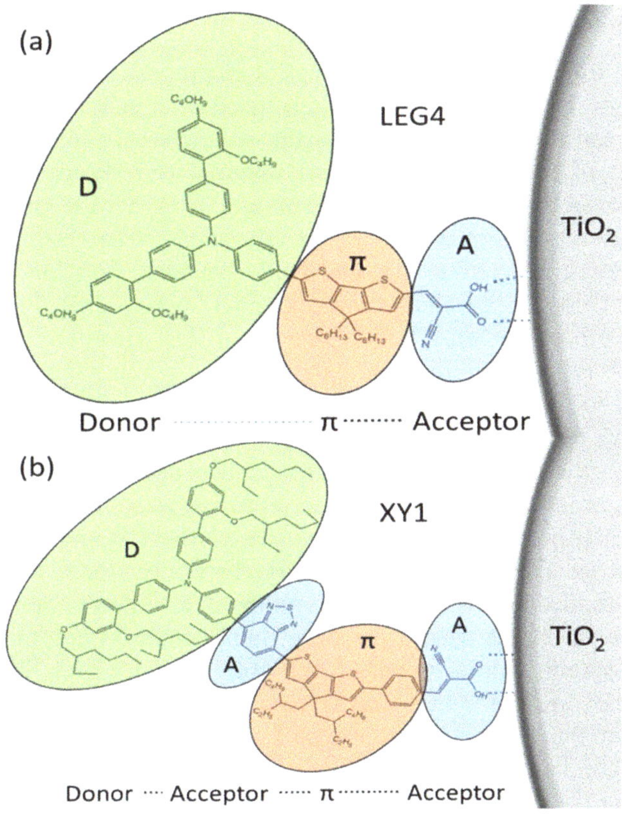

Figure 6.5. D–π–A/D–A–π–A structures of the organic dyes (a) LEG4 and (b) XY1 [10]. Reprinted with permission from reference [10]. Copyright 2018 Royal Society of Chemistry. CC-BY-NC 4.0.

included the use of a rigid aromatic system in the molecular backbone in order to increase the overall molecular planarity and favour a strong electronic connection between the donor and the acceptor moieties. These have resulted in improved device performances of PCEs exceeding 12%, and high J_{sc} values with examples such as dye 16 and 17, which have donor and acceptor groups linked via an alkynyl which extends conjugation and improves the absorption coefficient.

Improving the anchoring capability of the dye to the semiconductor is essential to increasing absorption coefficient and reducing aggregation as is already established. Several anchoring groups utilizing pyridine, phosphonic acid, benzoic acid, tetrazole and triazole derivatives with cyanoacrylic-like anchoring groups have been reported as the best performing organic sensitizers [2]. However, in recent years, the alkoxysilyl moiety as an alternative anchoring group has been investigated in a variety of dye-sensitizers to show better performances for devices with over 14% efficiency in those with co-sensitization. Additional strategies to increase molar absorption coefficients (1.0×10^5 M^{-1} cm^{-1}) and reduce aggregation have involved incorporating two anchoring groups, for example, bridging two mono-anchoring dye molecules.

6.2.4 Quantum dot sensitizers

Quantum dots (QDs) due to their exceptional properties have been effectively employed as sensitizers in quantum dots-sensitized solar cells (QDSSCs). For use in QDSSCs, QD sensitizers should possess the characteristics of a higher CB edge relative to that of the electron transport material (ETM) for effective electron injection, a narrow bandgap to absorb sunlight over a wide range of the solar spectrum, a high absorption coefficient for efficient light harvesting, good stability towards light, heat, and electrolyte, and facile synthesis with low toxicity. Of the range of QDs, those of the binary chalcogenides such as InP, InAs, CdS, CdSe, CdTe, PbS have been the most widely employed [11]. However, the key problem with binary QDs is that it is difficult to balance the narrower bandgap and higher CB edge important for band alignment and charge injection in the ETM.

Of the various types of QDs, the unique properties of core–shell QDs have made them versatile for electro-optical and electrochemical applications. Heterostructured core–shell QDs are composed of one material acting as a shell passivating the core material, and their properties are based on the relative CB and valence band (VB) edge alignment of the core and shell [12]. Typically, these are made with a semiconductor-semiconductor core–shell material to improve photoconductivity efficiency, decrease charge carrier response time, enhance photoluminescence and increase photostability, and therefore suitable as sensitizers in QDSSCs. Of particular interest are the binary and ternary chalcogenide semiconductor alloys, which can be of the Type I, reverse Type I and Type II structures [12, 13]. The inorganic semiconductor shell enables complete and long-lasting isolation of the core atoms from the surrounding environment, and the choice of the core and shell materials depends on two parameters: band offset alignment and the lattice mismatch between the core and shell. Furthermore, in core–shell QDs, the band edge alignment of the core and shell enables tuning of the light-absorption range, charge separation, and the recombination processes in QDSSCs [11]. Since the first report by Lee and co-workers in 2009 there have been several involving the use of core–shell sensitizers in QDSSCs [11]. Of note, are the promising outcomes from the work conducted by Zhong's group. For example, they employed type-I-structured CdSe(Te)-CdS QDs in QDSSCs, which exhibited a PCE of 9.48%, which was distinctly higher than that for the CdSe(Te) QD counterpart (8.02%). They attributed this to suppressed charge-recombination rates at the QD/TiO_2/electrolyte interface due to the reduced trapping state defects in the QDs afforded by the passivating shell layer [11, 14]. They also employed type-II core–shell QDs, ZnTe-CdSe, in QDSSCs which exhibited a good PCE of 7.17%. Here they attributed this improvement to the much larger CB offset in comparison to CdTe/CdSe [11, 15].

Of the chalcogenides, multinary-alloyed QD sensitizers are attractive alternatives to binary ones due to their composition-dependent tunable properties and higher chemical stability based on their stable lattice structure [11]. Of these, the I–III–VI, $AgInS_2$, $CuInS_2$ and $CuInSe_2$ have been attractive because they are more environmentally benign and possess high absorption coefficient with tunable narrow bandgap. There are several reports where they have been effectively employed as sensitizers. For example, Zhong's group developed a Zn–Cu–In–S-alloyed QDSSC exhibiting a PCE of 8.55% [16]. It was demonstrated that the alloyed structure is superior to that of the CIS–ZnS core–shell and pristine CIS QD sensitizers, due to suppressed charge recombination, as well enhanced electron-injection efficiency [11, 16]. Additionally, Zhong's group explored an alloyed Zn–Cu–In–Se (ZCISe) QD sensitizer possessing a narrow bandgap and high CB edge, achieving a PCE for a QDSSC of 11.61% [17]. They also reported even more impressive efficiency of 12.57% for QDSSC employing copper-deficient ZCISe (Zn–$CuInSe_2$) QD sensitizers and related this to defect related donor–acceptor pair (DAP) transitions in the QD [18].

Furthermore, the incorporation of dopant ions in QDs is a reliable means of tuning their electronic and photophysical properties and for effective use as sensitizers in QDSSCs [11]. Of the range of metal ion dopants, Mn^{2+} has been widely studied since their *d–d* transition which is spin- and orbital-forbidden, extending the long electron–hole recombination lifetime and is therefore useful for QDSSCs, since electron carriers can be more readily extracted for photocurrent [19]. For example, Zhong and co-workers incorporated Mn^{2+} dopants in $CdSe_xTe_{1-x}$ QD sensitizers with ZnS passivation shell, improving the PCE of the corresponding QDSSCs from 8.55% to 9.40% [20]. They attributed the improved performance to a suppression of the charge-recombination process in the QDSSCs caused by reduction of the trapping electronic states by the dopant ions in the QD sensitizers.

6.3 Photoanode development

Arguably, the most essential component of the DSSC is the photoanode, since it serves as the overall energy conversion unit of the device which influences the photovoltage and photocurrent. As shown in figure 6.1, the photoanode is essentially composed of: [1] a transparent conducting substrate [2]; a layer of semiconducting material—the ETL; and [3] embedded sensitizer molecules (*vide supra*) [2]. Importantly, desirable photoanodes should be capable of enabling high dye loading capacity, low internal resistances, low electron recombination rates and faster electron transport rates, resulting in high open-circuit voltage, current density and good cell efficiency. Typical candidates for the substrate include conducting glass such as indium tin oxide (ITO) or fluorine tin oxide (FTO) and conductive polymers. The conducting substrate is characterized as having high light transparency, high mechanical strength, and low surface resistivity for good cell performance. Typically, the ETL framework of DSSC photoanodes includes mesoporous, wide bandgap oxide semiconductor films such as TiO_2, ZnO, and SnO_2. Additionally, a light-scattering layer containing TiO_2 nanoparticles is typically deposited on top of the mesoporous layer which reflects transmitted light back into the active film to improve light absorption and the efficiency of the device. In addition to the wide bandgap, the semiconductor layer should feature good conductivity and stability, and appropriate morphological structures of high surface area and porosity to maximize dye sensitizer loading towards achieving efficient electron transport. In conventional cells, the film thickness ranges from several microns to 10 mm and composed of a three-dimensional (3D) network of randomly dispersed semiconductor micro/nanostructures and is usually layered using the conventional screen printing or doctor-blading techniques, and then sintered at around 400 °C–500 °C for compactness. The nature of coating depends on the physical properties (e.g. viscosity, nature of binder, solvent, etc) of the respective ink or paste, and as such, the performance of DSSCs is critically dependent on the preparation technique. However, the science behind controlling film thickness, microstructure, particle size, pore size and its distribution towards superior performance of the cell is still being uncovered. Consequently, these usually inhomogeneous films are characterized as a disordered network of numerous grain boundaries, which limits electron mobility and results in slow transport and recombination of photoexcited electrons. Such inherent problems associated with standard photoanode architectures have necessitated development of a variety of more effective nanostructured photoanodes of inorganic, organic, polymeric and hybrid materials and morphologies. Strategies for their development employ a variety of solution processing film preparation techniques including sol–gel, hydrothermal/solvothermal, electrochemical anodization, electrospinning and spray pyrolysis. They have resulted in a diverse array of nanostructured semiconductor photoanode materials such as TiO_2, ZnO, SnO_2 and Nb_2O_5 with improved properties. They include morphologies of nanorods, nanotubes, nanosheets, mesoporous structures and 3D hierarchical architectures, as well as graphene-based sheets and hybrid structures. As an advantage, the anisotropic dimensionality of these morphologies enhances electron transport and their injection rates. For example, it has been reported that 1D semiconductor nanostructures exhibit excellent charge transport properties and 3D mesoporous nano/microspheres possess better light-scattering properties due to their larger surface area (>100 $m^2\ g^{-1}$).

6.3.1 Binary metal oxides

Titanium dioxide, TiO_2 is the most widely used semiconductor photoanode in DSSCs and demonstrates the best device efficiencies in either liquid- or solid-state devices. Grätzel cells incorporate a highly mesoporous (pores of 2–50 nm) layer of TiO_2 or other metal oxide semiconductors deposited from a colloidal solution, followed by drying and sintering [1, 21]. This allows dyes to infiltrate the pores of the semiconductor, vastly increasing dye-loading and light-trapping capacity. As illustrated in figure 6.6, TiO_2 exists in several polymorphs of which the two more prominent ones are the metastable tetragonal anatase and rutile phases [22]. The bandgap of anatase is 3.2 eV, slightly wider than that of rutile of 3.0 eV with absorption around 390–400 nm. Rutile is the most stable bulk form but anatase is reported with enhanced photoconductivity and is more stable as a nanostructure [10]. TiO_2 has strong light absorption due to its high refractive index which results in efficient diffused scattering of the light within its mesoporous structure, thereby increasing photon interaction with the dye molecule. These mesoporous TiO_2 films also have a high internal surface area to support the monolayer of a dye sensitizer and coupled with their CB edge slightly below the LUMO level of many dye sensitizers, makes it an almost ideal photoanode for electron injection, transport and diffusion to the TCO. Furthermore, its high dielectric constant ($\varepsilon = 80$ for anatase) provides good electrostatic shielding of the injected electrons from recombination with the oxidized dye, an important attribute for enhanced performance. However, TiO_2 has a low intrinsic film conductivity, with nanocrystals not having built-in electric field.

One challenge associated with mesoporous TiO_2 films is that they are plagued with cracks and gaps that create electron traps which limit electron tunnelling between particles, thus retarding electron transport [21]. This leads to increased probability of recombination with holes from the HTL, limiting device efficiency.

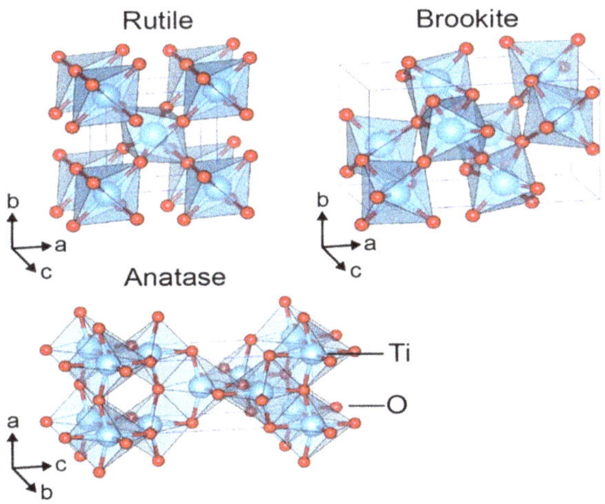

Figure 6.6. Crystal structures of TiO_2 rutile (tetragonal, $P4_2/mmm$), brookite (orthorhombic, Pbca) and anatase (tetragonal, $I4_1/amd$) polymorphs. Reprinted from [22] CC BY 4.0.

Consequently, to counter this limitation, alternative morphologies of the semiconductor layer to improve electron transport and cell performance have been pursued. For example, early work by Adachi and co-workers [23] resulted in efficiencies of 4.88% using disordered nanotube arrays, whilst Mor and co-workers [24] reported the use of highly-ordered TiO_2 nanotubes for improved electron transport. Though their devices exhibited low efficiency of around 2.9%, the nanotube layer was 360 nm thick. They noted that efficiency scaled linearly with nanotube length, suggesting that dye adsorption appears to be the limiting factor. Additionally, they hypothesized that with micron-length nanotubes, this system has a promising limiting efficiency of ~31%. Importantly, such highly-ordered nanotubes with increased surface area, unlike a mesoporous TiO_2 layer provide very fast, 1D electron transport, circumventing any electron traps between nanostructures, and increases the diffusion length of the semiconductor layer, prior to electron–hole recombination with the electrolyte of the HTL. The thrust towards variable morphology has shown that 2D nanostructures of TiO_2 such as nanosheets are superior materials relative to 1D nanostructures for DSSCs due to their higher surface area and porosity for dye loading. There are reports of DSSCs incorporating TiO_2 nanoribbons, nanoleaves and nanoflakes with efficiency up to 8%, with electron microscopy images of several examples shown in figure 6.7 [25]. In these, it is suggested that structural continuity of the nanocrystalline architecture plays a role in the electron transfer and cell efficiency.

Considering the cost associated with TiO_2, zinc oxide, ZnO emerged as its best oxide alternative as a photoanode ETL for DSSCs. As shown in table 6.1 [26], the bandgap of the ZnO is 3.3 eV with the energy levels of the CB edge similar to TiO_2. However, ZnO has some advantages over TiO_2 for application in DSSCs, including its low cost, higher electron carrier mobility and lifetime, photo-corrosion stability, low growth temperature, easy crystallization and anisotropic growth, including an abundance of easily synthesized nanostructured morphologies such as nanoparticles, nanowires, nanorods, and nanotubes. Overall, the diverse architectures of ZnO are possible through reliable solution-processed routes such as low-temperature solution phase, hydrothermal growth, electrodeposition, forced hydrolysis, as nanostructures, aggregates, layers and hierarchical structures. For example, in early work, Ko and co-workers [27] prepared branched ZnO nanowires hydrothermally. DSSCs made from these nanostructures displayed efficiencies up to 2.6% which depended on their hierarchy ranging from standing to branched nanowires. These structures afforded larger surface area, facilitating increased dye loading and photon absorption, with the branched nanostructures increasing the electron diffusion length and rate of electron transport. However, despite their promise as a photoanode, the PCEs of ZnO-based DSSCs are still lower than that of TiO_2-based DSSCs, with the highest record around 8.0%. The limitations are associated with their instability in ruthenium-based dyes which lowers electron injection rates, and instability in the acidic I^-/I_3^- electrolyte which decreases cell efficiency. Strategies to overcome the limitations of ZnO in the presence of corrosive redox I^-/I_3^- electrolyte have involved including layered double hydroxide. Other strategies to improve the efficiency of ZnO-based DSSCs also include designing of new dyes that are

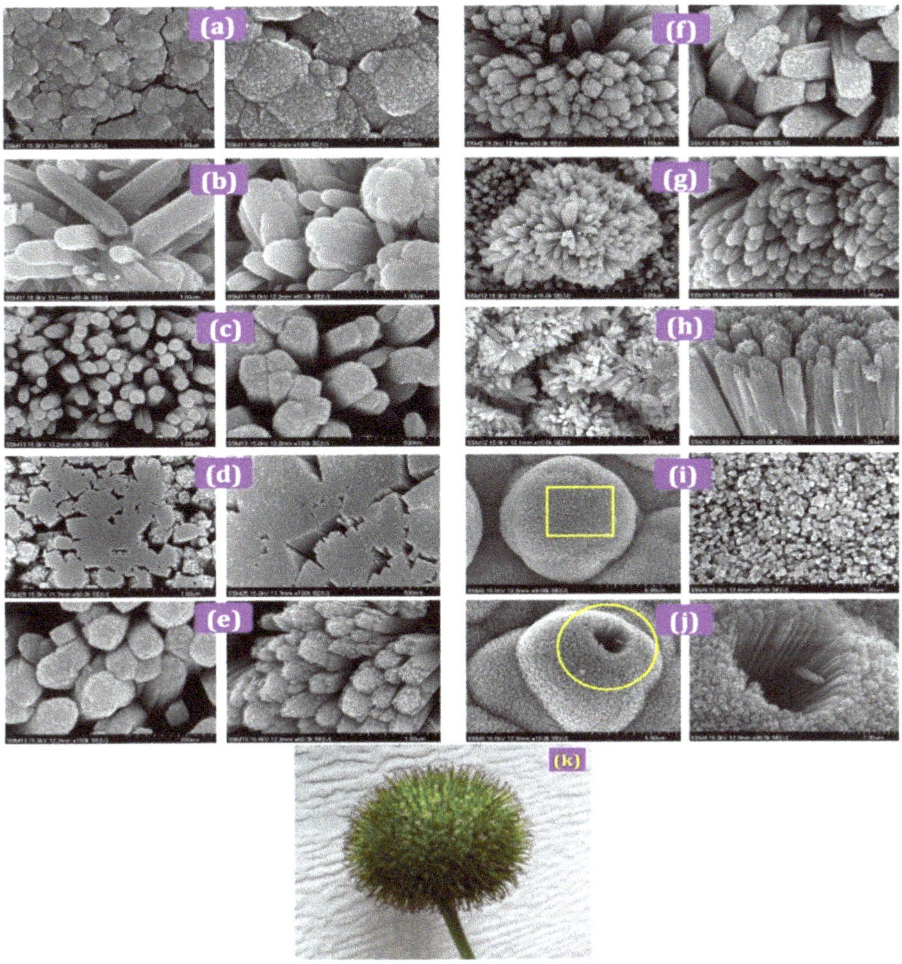

Figure 6.7. Electron micrographs of nanostructures of hydrothermally grown 1D/3D TiO_2 at different hydrolysis conditions. (a) T100, (b) T110, (c) T120, (d) T130, (e) T140, (f) T150, (g) T160, (h) T170, (i) T180 and (j) T190. The images on right hand show their respective highly magnified FESEM micrographs. (k) Photograph of a hollow platanus seed found in the nature that mimics the nanostructures of T190 sample. Reprinted from [25] with permission from Springer Nature.

suitable for the ZnO photoanode as well as replacing the liquid electrolyte with hole transport materials (HTMs) in solid-state devices. Notwithstanding these developments, the efficiencies of these ZnO-based devices have been poor.

There are several other oxides that have been investigated as alternative photoanodes for DSSCs with tin oxide (SnO_2) as a promising candidate due to its higher electron mobility and wider optical transparency than TiO_2 and ZnO. However, DSSCs employing SnO_2 show significantly lower conversion efficiencies, compared to TiO_2 and ZnO, due to their intrinsic limitations such as lower CB energy and isoelectric point, as shown in table 6.1 [26, 28]. Despite these limitations, SnO_2 is a

Table 6.1. Comparison of the structural and physical properties of ZnO and TiO_2 [26, 28].

Oxide material	TiO_2	ZnO	SnO_2
Crystal structure	Anatase (t) Rutile (t) Brookite (o)	Wurtzite (h) Zinc blende (c)	Rutile (t) CaCl2-type and α-PbO2-type (o)
Bandgap energy (eV)	3.2–3.3	3.4	3.6
CB energy min. versus vacuum (eV)	−4.1	−4.0	−5.0
VB energy min. versus vacuum (eV)	−7.3	−7.4	−8.6
Electron mobility ($cm^2\ V^{-1}\ s^{-1}$)	0.1–4	205–300 (b), 10^3 (s-NW)	∼100–200
Refractive index	2.5	2.0	2.0
Electron effective mass (m_e)	9	0.26	—
Relative dielectric constant	170	8.5	12.5
Electron diffusion coefficient ($cm^2\ s^{-1}$)	0.5 (b), 10^{-8}–10^{-4} (n)	5.2 (b)	—
Isoelectric point	6–7	∼9	4.5

Abbreviations: n (bulk), n (nano), s-NW (single nanowire), t (tetragonal), h (hexagonal), c (cubic), and o (orthorhombic).

material of choice in DSSCs due to its high optical transparency in the visible spectral range, high electron effective mass, m_e, and superior stability under UV light compared to TiO_2 and ZnO. The high photostability arises from its comparatively wider bandgap which reduces degradation of embedded dye molecules upon irradiation. However, renewed interest in SnO_2 as a photoanode material has been based on strategies to: [1] improve the dye loading capacity by increasing surface roughness and novel morphologies [2]; increase its Fermi energy level for better band alignment with the dye for electron injection; and [3] reduce the recombination by doping with transition metals. In response to these improvements, SnO_2-based DSSCs showed similar V_{oc} but superior J_{sc} relative to TiO_2 [28]. To that end, novel morphologies such as flowers availed increased surface area, yielding a remarkable V_{oc} of ∼700 mV with pure SnO_2, and nanotubes, porous nanofibers, core–shell and composite nanostructures provided high surface area and improved electron diffusion. Additionally, SnO_2 doped with various metal ions such as Zn^{2+}, Cd^{2+}, Ni^{2+}, Cu^{2+}, and Pb^{2+} improved V_{oc} due to their increased CB edge. Overall, such developments have enabled SnO_2 nanostructured-based DSSCs to exhibit efficiencies up to 9.5%.

In general, it has been established that the morphology of oxide nanostructures is critical for electron transport in the photoanode layer. As illustrated in figure 6.8, 1D nanostructures such as nanotubes provide a vertical pathway for electron transport and thus minimize electron loss [29]. However, the variable alignment of nanorods yields poor efficiencies because of the electron hopping mechanism which lowers

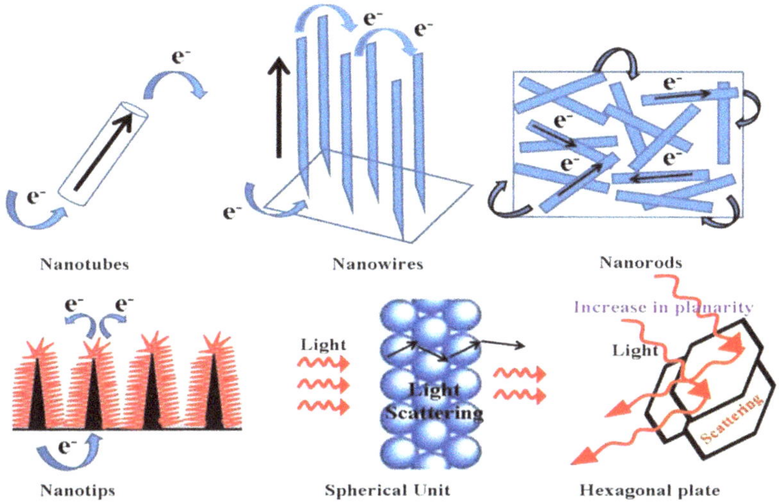

Figure 6.8. Electron transport mechanism of nanostructured photoanodes. Reprinted from [29], copyright 2016 with permission from Elsevier.

mobility, though they provide shorter distance for electron transport, facilitating faster electron accumulation. Although the 1D nanostructures have enhanced facile electronic transport thereby reducing charge recombination, 2D and 3D nanostructures on the other hand are useful for their sufficient surface area for dye loading to enhance photo-absorption. Notwithstanding this, they tend to not increase electron mobility due to extended defects and discontinuity in their architecture and yield lower conversion efficiencies.

6.3.2 Graphene-based materials

There is now tremendous interest in the use of graphene-based materials as photoanodes in DSSCs because of the fast electron transfer and good optoelectronic properties of the atomic thick carbon sheet. These graphene-based materials primarily include pristine graphene, graphene oxide (GO) nanosheets, reduced graphene oxide (rGO) nanosheets, carbon–graphene nanocomposites and vertically-oriented graphene sheets [30]. Importantly, graphene-based materials have the appropriate properties that make them attractive as photoanodes for DSSCs. These include ultrahigh theoretical surface area of 2600 $m^2\ g^{-1}$, high thermoconductivity of 5×10^3 W $(m\ K)^{-1}$, extraordinary electron mobility of $\sim$1500 $m^2\ (V\ s)^{-1}$ at room temperature and a current density of $\sim 1 \times 10^9$ A cm^{-2} due to graphene's long-range π-conjugation. Additionally, pristine graphene absorbs 2.3% of visible light with a very high extinction coefficient of about 10^8 mol cm^{-2} and tunable bandgap as nanoribbons or quantum dots across the entire solar light spectrum, attractive for photo-electron generation. Furthermore, the 2D morphology of graphene offers a conductive pathway for a high rate of electron transfer and reduced electron–hole recombination rate, essential characteristics of an excellent photoanode for high-performance DSSCs.

Of the many methods that have been employed to prepare graphene, solution exfoliation in organic and surfactant solutions, chemical vapour deposition (CVD), and reduction of GO are the most popular approaches because of their ease, low cost and low setup requirements, requisite for large-scale production [30]. However, large amounts of rGO are commonly synthesized via solution-processed methods accordingly. Firstly, producing GO using Hummer's method and secondly reducing these with agents such as hydrazine to obtain rGO sheets with sizes in the range of 1–10 μm. GO and rGO are commonly used in DSSCs because they can be easily functionalized to enhance electronic properties. However, the method of preparation has a strong effect on their properties especially with respect to defects which typically affect the chemical, electronic, optical and mechanical properties, limiting the performance of the DSSC. For example, intrinsic graphene is expected to have electron mobility as high as 200 000 cm^2 $(V\ s)^{-1}$ at room temperature but is much lower because of the various defects such as holes and grain boundaries associated with the preparation method. On the other hand, the ability to surface functionalize graphene materials can enhance chemical, electronic and transport properties for enhanced device performance. For example, ions introduced onto a graphene surface during GO synthesis can lead to higher surface resistivity of rGO compared to pristine graphene which may cause deterioration of electron transfer in the cell and/or lead to contamination of the main cell components. However, such surface modification alters the chemical properties and enhances the reactivity of graphene sheets which can enhance device performance. Additionally, some of these synthesis methods lack the control over layer thickness and it has been demonstrated that multilayer graphene generally has lower transparency.

As a recent example, Makal and Das [31] reported a facile one-step hydrothermal method to prepare reduced graphene oxide-laminated TiO_2–bronze (TiO_2–B) nanowire composites (TNWG), which contained two-dimensional graphene oxide nanosheets and TiO_2–B nanowires. DSSCs based on a TNWG hybrid photoanode with a reduced graphene oxide content of 8 wt% demonstrated an overall light-to-electricity conversion efficiency of 4.95%, accompanied by a short-circuit current density of 10.41 mA cm^{-2}, an open-circuit voltage of 0.71 V, and a fill factor of 67%, which were much higher than those of DSSC made with a pure TiO_2–B NW-based photoanode. The overall improvement in photovoltaic performance was associated to the intense visible light absorption and enhanced dye adsorption because of the increased surface area of the composite, together with faster electron transport due to reduced carrier recombination.

6.3.3 Doped and hybrid materials

Notwithstanding some of the performance characteristics of the aforementioned photoanodes, they are still limited in their electron transport capabilities. In addition to modifying the morphology of TiO_2 and ZnO, modifying the surface of TiO_2 by doping, preparing semiconductor hybrid bilayers and multilayers, creating hybrids with other materials such as carbon nanotubes (CNTs), GO and reduced graphene oxide (rGO) have been explored as routes towards designing photoanodes that enhance electron transport and prevent charge recombination. In particular, doping metal oxides with suitable ions has been investigated as a means to increase electron lifetimes and reduce recombination. Dopant ions introduce energy states which modify the bandgap by adjusting the position of the VBs and CBs of the absorber material. In effect, introducing cations as dopants in metal oxide photoanodes exerts larger dipole moment for electron transfer. It is noteworthy that TiO_2 has been doped with several ions resulting in notably high DSSC conversion efficiencies. For example, it was reported that doping of strontium into TiO_2 decreases the CB edge, thereby increasing the efficiency of electron–hole separation at the interface, and thus reduces the electron/electrolyte recombination rate resulting in an increase in photo-current and open-circuit voltage [29]. An efficiency of 9.1% was achieved with tungsten ion doping, attributed to the formation of sub-bandgap electronic states in TiO_2. Tong and co-workers [32] introduced an intermediate band into the mesoporous TiO_2 backbone of DSSCs to take full advantage of the sunlight and enhance the PCE by adding a nominal trace amount W-doped TiO_2 nanocrystalline films. A notable improvement of the device performance was obtained when N-type W-doped TiO_2 films were applied as the photoanode of DSSCs. The J_{sc} increased from 12.40 to 15.10 mA cm^{-2}, and the conversion efficiency increased from 6.64% to 7.42% when nominal 50 ppm W-doped TiO_2 was adopted. Also, Latini and co-workers tested several DSSCs with TiO_2 (anatase) photoanodes at different scandium doping both under solar simulator and in the dark [33]. The maximum efficiency of 9.6% was found at 0.2 at.% of scandium in anatase, which is 6.7% higher with respect to the DSSCs with pure anatase. Likewise, a promising efficiency of 8.4% for chromium-doped TiO_2 was due to lowering of the VBs and CBs of TiO_2 [34]. Overall, increased conversion efficiencies were due to enhanced electronic coupling of the LUMO orbital of the sensitizer dye to the lowered metal oxide CBs. Also, dopant ions introduce sub-bandgap electronic trapping states resulting in faster charge injection than recombination.

Hybrid photoanodes are also attractive alternatives to the conventional ones, primarily those composed of ZnO and TiO_2 due to the instability of ZnO in acid dyes. Indeed, photoanodes composed of two or more materials have attracted considerable attention due to the obvious advantages of combining different materials [35], for example, in the case of the high electron transport rate of ZnO, and the high electron injection efficiency of TiO_2 from ruthenium-based dyes. One of the earliest reports involved a photoanode of a ZnO/TiO_2 composite film fabricated on FTO transparent conductive glass substrate using solution-processed techniques such as electrophoretic deposition, screen printing and colloidal spray coating [35]. ZnO tetrapods were prepared via thermal evaporation and ZnO nanorods were obtained via hydrothermal growth. The best power conversion of 1.87%, which corresponded to the laminated TiO_2/ZnO/TiO_2 structure prepared via screen printing was attributed to the morphology of the semiconductor layers, which limited charge transport.

Other strategies have employed various semiconductor oxide structures to serve as a top scattering layer of a bilayer structured photoanode. These included TiO_2, ZnO, SnO_2, and CeO_2 as hollow spherical microspheres, nanocrystalline spherical aggregates, and tailored spherical aggregates of nanosheets or nanotubes, spherical voids, and composite structures. Of these layers, CeO_2 has attracted increasing attention because it has a high refractive index, good transmission for visible light, as well as strong adhesion and high stability [36]. In this respect, cubic shaped CeO_2 nanoparticles offered strong light scattering for DSSCs due to their high refractive index and exposed mirror-like facets [37]. PCEs of DSSCs with photoanode composed of cubic shaped CeO_2 nanoparticles as the top layer were improved by 17.8% with limited dye loading capacity. Alternatively, Song and co-workers [36] reported a photoanode composed of a scattering layer of hydrothermally prepared porous flower-like CeO_2 microspheres, and a thin TiO_2 film deposited by atomic layer deposition (ALD) resulted in a DSSC of efficiency of 9.86%, 31% higher than its conventional counterpart. This was attributed to improvement in the interconnection of particles and electrical contact between the bilayer and conducting TiO_2 film, which effectively reduced the electron recombination and facilitated electron transport, thus enhancing the charge collection efficiency of the device.

Despite the promising properties of graphene-based materials in photoanodes, as highlighted in the foregoing, their performances have not been impressive. For example, early work involving their use as TCO, prepared by dip coating from a hot GO-solution resulted in devices with a PCE of 0.26%, and was mainly attributed to the lower conductivity and optical transparency [30]. Even though their use as sensitizers is also an attractive prospect, perhaps the most effective means of employing graphene-based materials in photoanodes is as additives to the semiconductor layer. The effect of this is to increase electron transfer and significantly reduce electron–hole recombination rates in the oxide material layer. In this regard, graphene nanosheets are very useful since they create a much larger surface area to anchor oxide nanoparticles, and the photogenerated electrons can be captured and transferred more efficiently via intermolecular interactions of charge transfer, physisorption, or electrostatic binding with oxides. Early work on this by Yang and co-workers [38] demonstrated that the short-circuit current improved from 3.35 to 11.26 mA cm^{-2} and PCE from 0.58% to 5.01% by introducing graphene additives into the TiO_2 photoanode of their DSSCs. Here, rGO sheets which ranged from 0.2 to 2 μm in size were dispersed into TiO_2 P25 polymer-supported graphene water/ethanol solutions for photoanode fabrication. In another case, graphene was combined with TiO_2 nanofibers to increase the surface area in a photoanode incorporated into a cell with a short-circuit current of 16.2 mA cm^{-2}, an open-circuit voltage of 0.71 V, a fill factor of 0.66, and an efficiency of 7.6%, compared with pure TiO_2 nanofibers at 13.9 mA cm^{-2}, 0.71 V, 0.63, and 6.3%, respectively [39]. More recent studies have involved constructing TiO_2 photoanodes with a gradient graphene content. Typically, rGO prepared under hydrothermal conditions of other solution methods were used for layers of graphene/TiO_2 composites with different graphene content by mixing with TiO_2. For example, Singh and co-workers [40] employed the modified Hummers' method for the synthesis of GO. Thin films coated on corning glass substrates as well as synthesized GO powder were gradually reduced by the vacuum annealing process at different temperatures. DSSCs fabricated with an optimized photoanode having a mesoporous TiO_2 mixed with GO nanostructures reduced at 150 °C showed the best performance with a J_{sc} of 16.90 mA cm^{-2}, V_{oc} of 739 mV, and FF of 70.5%, with a PCE of 8.80%. They demonstrated that the incorporation of rGO reduced at the optimum level can play an essential role in improving the DSSC performance.

6.4 Electrolyte development

As the hole transport medium of the DSSC, the redox electrolyte functions as the component to transfer electrons from the CE, regenerating the oxidized dye sensitizer to ground state. Alternatively, it can be seen as transporting holes from the excited dye molecules to the cathode, reducing the oxidized dye sensitizer with the extracted electron, regenerating the original dye and completing the circuit. As such, it is a crucial component for determining the overall performance of the solar cell, including the J_{sc}, V_{oc}, *FF*, and its stability [3]. The original Grätzel DSSC incorporates an electrolyte comprising an iodide/triodide (I^-/I_3^-) redox couple composed of tetrapropylammonium iodide, iodine as active components, ethylene carbonate and acetonitrile as solvents. The relatively good efficiency of 7%–8% was attributed to the excellent mobility of the liquid electrolyte and its ability to make contact with, and rapidly regenerate the dye inside the ETL. This HTL has been the most widely used redox shuttle because of its fast oxidation of I^- ions at the photoanode/electrolyte interface for efficient dye regeneration and slow reduction of I_3^- ions at the electrolyte/CE interface for high carrier collection, excellent infiltration, relatively high stability, overall low cost and easy preparation. These typical electrolyte HTMs have five main components, including [1], a redox couple, such as I^-/I_3^-, Br^-/Br_2^-, SCN^-/ SCN_2 and Co(II)/Co(III) [2], solvent [3], additives [4], ionic liquids, and [5] counter cations, and should exhibit the following characteristics.

1. The hole carriers must be injected efficiently to regenerate the oxidized dye.
2. They should permit fast diffusion of charge carriers, enhance conductivity, and create effective contact between the working and CEs.
3. The absorption spectrum of an electrolyte should not overlap with that of the dye.
4. They should possess long-term chemical, thermal, optical and electrochemical stability to prevent dye degradation.
5. They should be non-corrosive with other DSSC components.

Crucially, the solubility and ionic mobility of a redox couple in the organic medium, the driving force for the dye regeneration, and fast electron transfer kinetics with a minimal overpotential at the CE render an effective redox electrolyte. For DSSCs, electrolytes are classified as liquid, quasi-solid and solid-state, due to their physical characteristics, which significantly impact the efficiency and stability. The fundamental differences between the various charge transport materials are the charge mobility and mechanism. For liquid electrolytes the primary transport mode is ionic conductivity, however in polymeric and solid-state HTMs the mechanism can be a combination of ionic and electronic transport, or a predominantly electronic [3].

6.4.1 Liquid electrolytes

To date, the PCE of I^-/I_3^- redox electrolyte-based DSSCs has been around 11%. The I^-/I_3^- redox couple fulfils several requirements for an ideal electrolyte, and it was for several decades the benchmark for research and industry because of several advantages including; a suitable redox potential for many dyes, small molecular size for high diffusion, good solubility in a wide range of solvents at high concentration for high conductivity, and good stability [3]. However, there are several limitations associated with the electrolyte's application due to absorption of visible light at 430 nm, corrosion of the noble metal CE, high volatility responsible for dye desorption, and poor long-term stability stemming from iodine diffusion. As a result, several alternative electrolytes have been investigated, including Co(II/III) polypyridyl complexes, ferrocenium/ferrocene (Fc/Fc^+) couple, Cu(I/II) complex, and thiolate/disulfide couples [41]. Notably, DSSCs with the Co(II/III) polypyridyl redox couple displayed an impressive PCE of 12.3% with a high V_{oc} of 0.935 V [41, 42]. This was attributed to slow diffusion of the bulky complex into the photoanode films, and fast recombination of electrons with the oxidized redox species. However, issues with stability rendered these Co(II/III) complex-based electrolytes a challenge for DSSCs. Importantly, the electrolyte solvent plays a critical role in the cell's efficiency. Typically, organic solvents should have high dielectric constants, examples of which include acetonitrile (ACN), 3-methoxypropionitrile (MePN), propylene carbonate (PC), γ-butyrolactone (GBL), N-methyl-2-pyrrolidone (NMP), and ethylene carbonate (EC). Strategies to improve the electrolyte's conducting properties include small amounts of N-methylbenzimidazole (NMBI) and guanidinium thiocyanate (GuSCN) as electric additives. In some cases, addition of co-absorbents improves the performance of an electrolyte by reducing charge recombination and may shift the CB edge of the TiO_2 to improve charge injection and regeneration of the oxidized dye, resulting in its long-term stability and ultimately higher open-circuit voltage.

6.4.2 Quasi-solid electrolytes

Characteristically, liquid electrolyte sealing, and long-term durability have limited the performance of DSSCs. As such, efforts have focussed on alternatives to liquid electrolytes, including developing quasi-solid-state and solid-state electrolytes. Quasi-solid-state electrolytes are non-volatile, viscous, and ionic with good charge transport characteristics. They include ionic liquid redox couples such as 1-propargyl-3-methylimidazolium iodide, bis(imidazolium) iodides and 1-ethyl-1-methylpyrrolidinium, as well as ionic/electronic polymer gels such as poly(ethylene oxide), poly(vinylidinefluoride) and poly(vinyl acetate) [41]. Although these exclude hermetic sealing of the compartment, their high viscosity lowers diffusion rates and limits efficient hole transport. To circumvent this, polymer electrolytes characterized as having ionic conductors embedded in a polymer host matrix have been investigated. These are particularly attractive because they retain the beneficial properties of liquid electrolytes such as, high ionic conductivity, diffusive transport and interfacial contact properties in combination with the mechanical benefits of the polymer, such as their durability and flexibility. These gel polymer hosts have a high capacity for the electrolyte, and their excellent interfacial properties between the electrodes result in fast dye regeneration, whilst their high conductivity ensures fast charge transport to the CE. Common examples include polyacrylonitrile (PAN), polyethylene oxide (PEO) derivatives, and conducting polymers that include polypyrrole (PPy), polyanaline, bisphenol-A-ethoxylate dimethacrylate (BEMA), poly(methyl methacrylate) (PMMA) and poly (ethylene glycol)methyl ether methacrylate (PEGMA). BEMA is characterized as having a 3D network and the addition of PEGMA as a copolymer influences the propagation reaction and changes the architecture of the polymeric matrix, thus affecting its properties. In an early reported example, a PCE of 8.03% was achieved by using a polymer electrolyte based on PMMA [43]. In this system, the ionic conductivity and diffusivity of the iodine/1-butyl-3-methylimidazolium iodide (BMII) redox system embedded were comparable to those of liquid electrolytes, and the resulting cells showed improved stability compared to traditional liquid DSSCs. Noteworthily, corresponding PCEs of DSSCs based on quasi-solid-state electrolytes are not very high and because of their thermodynamic instability under high temperature, they still suffer from solvent leakage.

6.4.3 Solid-state electrolytes

Much of the recent research and development in DSSCs is now focussed on solid-state DSSCs (ssDSSCs) comprised of solid-state electrolytes. Solid-state electrolytes are different from liquid and quasi-solid electrolytes since charge diffusion occurs within the material, and the mechanism of hole transport involves intermolecular hopping across the material layer rather than ion diffusion. These solid-state HTM electrolytes are usually deposited from solution onto the mesoporous semiconductor layer, where they tend to become embedded into their pores. However, they suffer from inadequate pore-filling of the solid phase after solvent removal. To circumvent this, the mesoporous layer of HTM-based ssDSSCs is made much thinner compared to their liquid counterparts and in some cases, an alternative approach is to use HTMs that have low melting or glass transition temperatures where they are directly embedded by melting onto the photoanode layer.

Most HTMs can be categorized as inorganic solids, organic polymers, or *p*-conducting molecules. However, their functionality is still unmatched relative to iodine/iodide redox electrolytes. This is because there is poor electronic contact and conduction between the HTM and the dye due to incomplete penetration from the solid HTM into the photoanode. This results in incomplete dye regeneration which amplifies charge recombination resulting in a DSSC of lower open-circuit voltage and PCE, as compared to the iodine/iodide redox counterpart. Currently, the HTM of choice is 2,2′,7,7′-*tetrakis*(*N*,*N*-di-p-methoxyphenyl-amine)9,9′-spirobifluorene (spiro-OMeTAD) and is the benchmark for the performance of newly developed

HTMs [4, 10]. It is much better than other types of organic HTMs and produced a high PCE of 15% in perovskite-based ssDSSCs [44]. However, it has some inherent limitations such as poor conductivity and hole mobility, poor long-term stability, difficulty to synthesize and high cost [41]. Importantly, the design strategy towards developing better HTMs encompasses materials whose VB maximum is slightly above the CB minimum of the oxidized dye, as well as no absorption in visible region, and photochemically stability.

6.4.3.1 Inorganic molecule HTMs

Inorganic semiconducting materials are particularly attractive as HTMs because of their good electronic properties, conductivity and high thermal stability. However, their use as HTMs in DSSCs is quite limited because of their insolubility in solvents required for solution processing onto the photoanode. Additionally, few have adequate bandgap alignment with dyes and the *n*-type photoanode, and bandgap tunability is also a limitation. However, the few materials that can be used in ssDSSCs have better performances than most organic HTMs, although charge recombination at grain boundary defects and insufficient photoanode pore-filling diminishes electronic contact and ultimately poor device efficiencies. Of the range of inorganic HTM materials studied, CuI/CuSCN possesses high hole mobility [41]. However, fast crystallization rates during solution fabrication result in poor filling into the photoanode films, and thus the ssDSSC shows a poor PCE of 3.8% [45]. On the other hand, Kanatzidis and co-workers [46] reported that solution processable *p*-type direct bandgap semiconductor $CsSnI_3$ can be used for hole conduction. The resulting solid-state dye-sensitized solar cells consisted of $CsSnI_{2.95}F_{0.05}$ doped with SnF_2, nanoporous TiO_2 and the dye N719, and showed conversion efficiencies of up to 10.2%. They effectively showed $CsSnI_3$, possessing high hole mobility, low cost, abundant elements, and low-cost processing, as another promising *p*-type semiconductor HTM.

6.4.3.2 Small organic molecule HTMs

Small organic molecules are the most widely employed as HTMs in DSSCs because they possess tunable electronic properties, easily form films with the photoanode, have wide-range high solubility allowing for easy processability, and are generally synthetically reproducible. However, though their relatively small size allows them to infiltrate the photoanode mesoporous structure of DSSCs, this partially hinders intermolecular charge transfer and therefore they generally suffer from poor charge conductivity, especially in their pristine form. Most small organic HTMs constitute the conjugated triphenylamine (TPA) moiety as a core component [10]. The lone pairs of the donor nitrogen atom render TPA a good hole acceptor along with the three phenyl rings that contribute to the delocalization of the resulting charge, stabilizing the cation. The electronic and hole-transporting properties of these HTMs are tuned by the various electron donating substituents including phenyl and methoxy groups, making use of the electron delocalization arising from the ring conjugation. Notably the previously discussed spiro-OMeTAD has been the benchmark for small organic molecular HTMS. However, a range of other types have been investigated. For example, X51 (9,9′-([1,1′-biphenyl]-4,4′-diyl)bis(N3,N3, N6,N6-*tetrakis*(4-methoxyphenyl)-9H-carbazole-3,6-diamine)) is a high-performance carbazole-based HTM small organic molecule suitable for use in DSSCs. In combination with a metal-free dye such as DN-F05, device PCE of 6.0% exceeding that of spiro-OMeTAD (5.5%) has been achieved [47]. Tris(4-methoxyphenyl)amine is a small and versatile redox-active molecule which can be employed in both liquid and solid-state DSSCs. When used with liquid electrolytes such as metal-based redox couple it enables exceptionally fast dye regeneration, allowing device PCEs over 10%. In solid-state solar cells it may be used as a low cost HTM, either by itself or combined with another material such as P3HT [48].

6.4.3.3 Polymer molecule HTMs

Conjugated organic polymers offer attractive prospects in ssDSSCs because unlike small-molecule solid-state HTMs, they are more stable and likely more efficient because they offer superior charge transport properties [10]. These materials, which are characterized by extended delocalization of their conjugated carbon backbone, are able to transport holes through their VB. Their typical long chains ensure very fast intramolecular charge transfer and their entangled morphology in the solid-state provides many points of contact for intermolecular charge hopping. Overall, they are more complex than small-molecule HTMs and are more difficult to penetrate the mesoporous photoanode. However, to circumvent this, most are prepared via *in situ* polymerization in the photoanode mesoporous structure. This allows for easy embedding of smaller monomer molecules into the pores and following polymerization, the generally higher conductivity of polymers can be exploited. However, the drawback is that polymers allow less control over electronic properties and the polymerization steps sometimes limit the chemical composition of the monomer. A less than facile control over the polymerization reaction limits reproducibility and material properties. Notwithstanding these limitations, many polymers have emerged, most notably polythiophenes such as poly(3,4-ethylenedioxythiophene) (PEDOT) [10]. Overall, the efficiencies of polymer HTMs are lower than those with small-molecule HTMs and liquid electrolytes. In some early studies, for example, the HTM—PEDOT, was prepared from the monomer adsorbed onto the photoanode through solution immersion and was subject to *in situ* photoelectrochemical polymerization [10]. The device with photoanode of three different dye molecules yielded different PCEs up to 4.34%, indicating that a good combination of dye and HTM is critical in enhancing charge generation and recombination suppression. Similarly, PEDOT was prepared via *in situ* electrochemical polymerization and the device with different dyes yielded PCEs up to 5.54%. However, HTM of PEDOT doped with PSS, (PSS-poly(styrene sulfonate)) yielded a device efficiency of 7.4%. Over time these HTMs have yielded devices with improved PCEs in various combinations of components. Also, a solid-state HTM of ProDOT, poly(propylene-dioxythiophene) was prepared by *in situ* polymerization of the monomer and drop casted onto the photoanode following solvent evaporation at 25 °C [49].

DBProDOT molecules deeply penetrated and polymerized to fill nanocrystalline TiO_2 pores with PProDOT, which functioned as an HTM for I_2-free ssDSSC with a platinum-coated FTO CE. With the introduction of an organized mesoporous TiO_2 (OM-TiO_2) layer, the energy conversion efficiency reached 3.5% at 100 mW cm^{-2}, which was quite stable up to at least 1500 h. The cell performance and stability was attributed to the high stability of PProDOT, with the high conductivity and improved interfacial contact of the electrode/HTM resulting in reduced interfacial resistance and enhanced electron lifetime. In recent years, many carbazole derivatives have been applied as HTMs because of their excellent hole-transporting capabilities and chemical stability. To that end, Kong and co-workers [50] reported on biscarbazole-based polymers with different molecular weights, (PBCzA-H and PBCzA-L) which were applied in combination with additives to produce ssDSSCs. A device with PBCzA-H showed a better J_{sc}, V_{oc}, *FF* than a device with PBCzA-L, resulting in 38% higher conversion efficiency. Compared to the PBCzA-L, the PBCzA-H with a higher molecular weight showed faster hole mobility and larger conductivity, leading to elevations in J_{sc} via rapid hole transport, V_{oc} via rapid hole extraction, and *FF* via lowered series and elevated shunt resistances. These results showed that carbazole derivatives such as PBCzA-H are useful HTMs for replacing liquid electrolytes in DSSCs.

6.5 Counter electrode development

The CE of DSSCs is an electrocatalyst cathode surface that collects electrons from the external circuit and reduces the redox electrolyte or HTMs in ssDSSCs. It also functions as a reflective surface, redirecting unabsorbed light into the cell, enhancing sunlight utilization. Accordingly, an optimal CE should possess the primary requirements of high catalytic activity, high conductivity, high reflectivity, as well as chemical, electrochemical and mechanical stability, and energy level overlap with the potential of the redox couple/HTM. Along with low cost, other important physical features include high surface area, high porosity, optimum thickness and good adhesivity with transparent conducting oxides. The most widely used CE materials include noble metals, such as platinum, gold and silver because of their high electrocatalytic activity, with platinum the most commonly used. However, because of their high cost and corrosion in liquid electrolytes, a range of alternative CE materials have emerged over the years. Several of these include carbon materials, transition metal compounds, conductive polymers, and composites. Accordingly, there are a range of solution-synthetic methods for preparing CE materials, including precursor thermal decomposition, electrochemical deposition, chemical reduction, chemical vapour deposition, hydrothermal reaction and *in situ* polymerization, as the primary ones. Importantly, these methods are utilized to influence the particle size, surface area, and morphology, as well as catalytic and electrochemical properties of these electrodes. Invariably, electrode nanoparticles with larger surface areas will produce more catalytic active sites, thereby improving their electrocatalytic activity, and therefore show great promise.

6.5.1 Metal CE materials

The original Grätzel cell featured platinum as the CE because of its high electrical conductivity, catalytic activity towards triiodide reduction, and high reflecting properties, and has since been the standard CE for DSSCs. Though platinum is typically sputter-coated onto a conducting substrate as the CE, there are several alternative solution-processed preparations including electrochemical deposition, thermal vapour deposition, spray pyrolysis, cyclic electrodeposition, electrochemical reduction and thermal decomposition [51]. One of the earliest approaches reported involved preparation of platinum nanoparticles onto a TCO-coated glass substrate via thermal decomposition of a drop casted precursor solution of H_2PtCl_2 in isopropanol at 385 °C for 10 min [52]. The resulting highly transparent electrode had very low platinum loadings (<3 mg cm^{-2}) and low charge transfer resistances (<1 Ω cm^2). Consequently, this economical approach at producing electrochemically/chemically stable electrodes with superior mechanical stability and good adherence to substrates has been a standard approach for producing platinum CEs. The development of platinum CEs has evolved to include 3D nanostructures with high surface areas and enhanced functionality. These structures include nanowires, nanoflowers, nanotubes, multipods, examples of which are shown in figure 6.9 [51].

The facile cyclic electrodeposition (CED) method has emerged as one of the most cost-effective solution-processed methods to synthesize 3D platinum nanostructures with controllable morphology and high purity. For example, platinum nanoclusters, nanosheets, nanograsses and nanoflowers have been produced via CED at room temperature in solution containing H_2PtCl_6 precursor and $NaNO_3$, with variation of CED scan rate and precursor concentration to control morphology [53]. Of these,

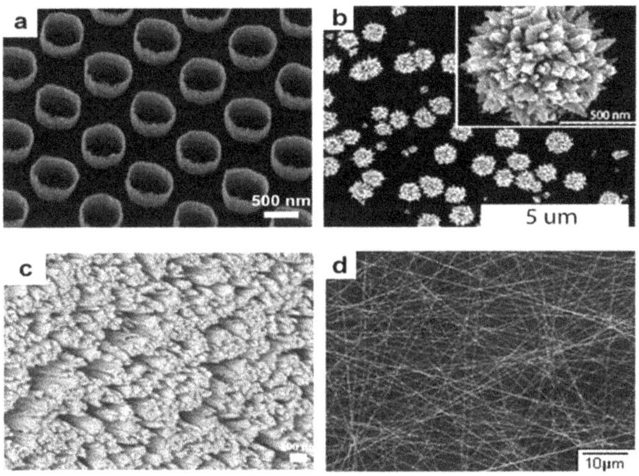

Figure 6.9. 3-D nanostructured CE materials. EM images of (a) Pt nanocups. (b) Pt nanoflowers. (c) Pt nanowires. (d) Pt nanofiber networks. Reproduced with permission from [51]. Copyright 2017 Royal Society of Chemistry.

films with uniform nanograss morphology were the most suitable as CEs because of their higher surface activity, giving rise to high electrocatalytic performance and intrinsic light scattering. In this study, the DSSCs which incorporated N719 dye attained a PCE of 9.61%–12% higher than platinum CEs fabricated via thermal decomposition under similar experimental conditions, with cells optimized to PCE of 10.62% consisting of a thicker TiO_2 layer with the Z907 dye.

Whilst platinum CEs are the benchmark for DSSCs they present some limitations because of their high cost and availability. Additionally, platinum is electrochemically unstable in cells consisting of liquid electrolytes of the I^-/I_3^- redox couple. In effect, platinum may become oxidized, redepositing on the photoanode layer along with reduction of iodides, which ultimately results in a short-circuiting of the cell and lower efficiency. Consequently, other noble metals have been investigated as its counterparts. These primarily include, ruthenium, gold, silver and titanium with desirable CE properties [51]. For example, ruthenium is attractive owing to its high electrical conductivity, high electrocatalytic activity, excellent electrochemical stability and lower cost compared to platinum. Catalytic gold has higher conductivity (4.3×10^5 S cm^{-1}) than platinum (0.9×10^5 S cm^{-1}) and excellent corrosion resistance. Silver exhibits the highest electrical conductivity, thermal conductivity and reflectivity of any metal as well as high corrosion resistance and inertness. Titanium is the most stable and resistant to corrosion owing to formation of a thin surface-passivating oxide layer. However, its resistivity of 42×10^{-8} Ω m is higher than that for platinum (10.6×10^{-8} Ω m), which diminishes its performance as an electrocatalytic material. Notwithstanding the desirable properties of these metals for use as CEs, their performance is not as good compared to platinum and their application is limited since they are also expensive. As such, a range of synthetic protocols are required to produce them as 3D nanoarchitectures to enhance their electrocatalytic properties. Despite that, one strategy to circumvent these limitations is to produce alloys with inexpensive, earth abundant metals that also have good electrical properties. Such alloys provide synergistic effects in the properties of the metals and are generally cheaper. In many of these, cost-effective and scalable solution-processed methods, including electrodeposition, mild hydrothermal synthesis and chemical plating, have yielded alloy CE thin films of various morphologies, with superior electrocatalytic activity, charge-transfer ability, good electrical conduction, high reflectance and better sample loading for DSSCs, which make them promising electrodes.

6.5.2 Carbon CE materials

Carbon materials as CEs are quite attractive alternatives to platinum CEs owing to their low cost, high surface area, high catalytic activity, high electrical conductivity, high thermal stability, corrosion resistance towards iodine, high reactivity for triiodide reduction, etc [51]. The primary congeners of graphene, carbon nanotubes, carbon nanofibers, activated carbon, graphite, and carbon black have demonstrated promise as effective CEs in DSSCs with notably good performances. The electrochemical and conductive properties of these materials is attributed to their graphitic nature due to the sp^3 hybridized molecular bonding. The first DSSC comprising a carbon-based CE was prepared by Grätzel using a graphitic-carbon black composite [54]. The functionalized graphite served to enhance electronic conductivity, whilst the high-surface area carbon black increased the catalytic activity towards triiodide reduction, yielding a cell with PCE of 6.67%. Typically, carbon black is characterized as spherical nanostructures usually consisting of graphitic layers, giving a high surface-area-to-volume ratio critical for high electrocatalytic activity. Its high electrical conductivity, variable nanoparticle morphology and surface characteristics such as particle size, surface area, porosity, crystallinity, thickness, shape, and purity are critical features for an effective CE. One of the limitations associated with the fabrication of carbon-based CEs is the use of expensive vacuum and vapour deposition methods. However, carbon black CE is fabricated in DSSCs using several methods such as the doctor-blade method followed by thermal treatment, and spin/spray coating. It has been found that preparation conditions have been effective in controlling the film thickness, particle size and morphology, which affect the electrochemical properties and ultimately the performance of the DSSCs. Despite that, more cost-effective methods are needed to prepare these CEs allowing for film uniformity and more control of properties.

CNTs also show good promise as CEs because they have extraordinary electrical conductivity, thermal conductivity and mechanical strength. However, their use is limited because production is very expensive compared to other carbon nanostructures. Similar to carbon black, they have been fabricated into CEs using the typical doctor-blade method as pastes and spin/spray coating. For example, single- and multi-walled CNTs (SWCNTs and MWCNTs) mixed with poly(ethylene glycol) (PEG) to form a paste were applied by the doctor-blade method onto FTO substrates to form CEs [55]. Following heat treatment to remove the PEG binder, the CEs resulted in DSSCs with PCEs up to 7.8% owing to low charge transfer resistance. Another approach included fabricating a CE of SWCNT/graphene composite by spin coating [56]. The CE architecture was characterized with a large surface area and good surface hydrophilicity, enabling enhanced electrolyte–electrode interface for good electrolyte diffusion. The resulting DSSC exhibited a good PCE of 8.31% with its counterpart of platinum-based CE (PCE of 7.56%). The better performance was attributed to the high electrical conductivity and good electrocatalytic activity of the SWCNTs, and excellent catalytic properties of graphene arising from the enhanced interfacial characteristics. Additionally, it has been demonstrated that DSSCs incorporating CEs fabricated from the screen printing of MWCNTs paste onto FTO glass exhibited relatively good PCE of 8.03% in comparison to platinum-based CE (PCE of 8.80%) [51]. These promising results were attributed to a larger surface area for electrocatalytic activity and charge conductivity of the CNTs. Recently, Hasan and Susan [57] reported the fabricating of a new CE using a binary composite of heteroatom-doped carbon dots (C-dots) and functionalized multi-walled carbon nanotubes (o-MWCNTs). They demonstrate that this binary composite exhibited superior performance to pristine o-MWCNTs. The PCE of the o-MWCNT/C-dots composite was measured at 4.28%, significantly outperforming the pristine o-MWCNT electrode, which yielded an efficiency of 2.24%. The enhanced performance of the o-MWCNT/C-dots composite was attributed to the synergistic effects of heteroatom-doped C-dots since their binding to the o-MWCNTs by activated oxygenic surface functional groups increases the surface area from 218 to 253 m^2 g^{-1}.

Despite the unusual and attractive properties of graphene, such as high carrier mobility (10 000 cm^2 V^{-1} S^{-1}), high specific area (2630 m^2 g^{-1}), excellent thermal conductivity (3000 W m^{-1} K^{-1}) and high optical transparency (97.7%) its application as CEs in comparison to platinum requires much more work [51]. Though there are several chemical approaches to prepare CEs of graphene, most have been prepared involving the spin coating of rGO.

6.5.3 Polymer CE materials

Conductive polymers are primarily used as HTMs, but their use as CEs in DSSCs makes them attractive alternatives to platinum CEs because of their facile synthesis, porous structure, electrical conductivity, low cost, abundance and favourable catalytic properties. A striking feature and advantage of organic conductive polymers is their tunable conductivity between metal and semiconductor by varying substituents. Their conductive properties useful as CEs are derived from their conjugated structure and primarily include derivatives of polyacetylene, polyaniline, polypyrrole or polythiophenes. An important attribute of these conductive polymers is their high threshold for dopant inclusion which considerably increases their photocatalytic and conductive properties as functional CEs. Critically, conductive polymer CEs are flexible, transparent and benefit from simple and scalable solution-processing methods of *in situ* polymerization including electro-polymerization and deposition, along with spin coating. Overall, conducting polymer CEs show reasonable performance in comparison to platinum CEs. Of these, PEDOT [poly (3,4-ethylene-dioxythiophene)] is the most popular and exhibits the best performance, however, its cost is comparable to that of platinum. PPy (polypyrrole)-based CEs are cheaper but their performance is slightly inferior to that of PEDOT, and PANI (polyaniline)-based CEs are low cost with comparable performance.

PEDOT exhibits excellent conductivity much higher than polyaniline, polypyrrole, and polythiophene. One of the challenges was its solubility but when doped with poly(styrene sulfonate) (PSS)—PEDOT-PSS improves considerably, which is widely applied in DSSCs. Importantly, the morphology of PEDOT and other conductive polymer CEs which can be controlled during preparation affects their electrochemical properties and PV performance of the devices. For example, PEDOT nanoporous layers have been deposited onto FTO through electro-oxidative polymerization in hydrophobic ionic liquids. In early studies, these CE films in DSSCs yielded a PCE of 7.93%, relative to 8.71% of the platinum counterpart. In another case, CEs of electropolymerized PEDOT nanotubes (PEDOT-NTs) deposited onto FTO resulted in a PCE of 8.3% which was comparable to the device with platinum CE (8.5%). Similarly, PEDOT-nanofibers (PEDOT-NFs) with diameters of 10–50 nm prepared using sodium dodecyl sulfate micellar solution were spin coated onto a substrate from a methanol-based colloidal dispersion. The PEDOT-NF CE incorporated into a DSSC achieved a relatively high PCE of 9.2% compared with bulk PEDOT (6.8%) and its platinum counterpart of PCE 8.6%. On the other hand, the PCE values of the DSSCs with PProDOT poly (3,4-propylene-dioxythiophene) CEs are notably higher than DSSCs using most other types of CEs (PProDOT), including platinum CEs. This is attributed to high conductivity arising from its ultrahigh surface area and its ability to avoid the formation of charge hindering passivation layers at the electrode/electrolyte interface.

Like the polythiophine conductive polymers, PANI is also very attractive as CE because of its high conductivity, high thermal and chemical stability, good redox properties, low cost and easy synthesis. PANI nanostructures are particularly useful for CE functionality because of their porosity, high surface area and well-connected microporous structure, which enhances charge mobility. The first reported use of PANI as CE involved DSSCs with a PCE of 7.15% in which PANI nanoparticles of 100 nm diameter were polymerized by aqueous oxidation using perchloric acid in the presence of ammonium persulfate. Furthermore, it has been shown that 1D morphology including nanofibers, nanobelts and nanotubes influences their CE functionality. For example, *in situ* polymerized oriented PANI nanowire array showed higher electrocatalytic activity for Co^{3+}/Co^{2+} redox couple and their DSSC (PCE of 8.24%) outperforms that made using random PANI nanostructures (5.97%) and its platinum-based counterpart at 6.78%. This improved performance was attributed to higher surface area which enhances catalytic reduction of the oxidized species in the electrolyte. PANI also has a high threshold for dopant ions such as SO_4^{2-}, F^-, Cl^-, PSS, etc, which affect the morphology and electrocatalytic properties of its films. For example, DSSC based on PANI–SO_4–F doped CEs displayed a comparably good PCE of 8.8% relative to its platinum counterpart at 9.0%. The electrode was prepared by spin coating a solution of the doped polymer in hexafluoro-isopropanol (HFIP) at room temperature. Though the good photoelectric properties, easy preparation and low cost of PANI CEs present as promising alternatives to platinum-based CEs, their use has been stymied because of instability, self-oxidation and carcinogenic effects.

Oxidized derivatives of PPy are good conductors and are well suited as CEs in DSSCs because of their good electrocatalytic activity, low cost, easy polymerization in high yield, along with reliable air, thermal and chemical stability. They are particularly attractive as CEs because their films usually consist of 1D nanostructures characterized by crosslinking and chain hopping which are important in enhancing conductivity and catalytic activity. Like other conductive polymers, they have good dopant capacity and their conductivity is controlled by the facile method of synthesis and deposition onto substrates. However, conductivity is generally lower than for other polymer CEs. The first reported use of PPy as CEs involved nanoparticles of diameter of 40–60 nm, synthesized by chemical polymerization using iodine as initiator and coated onto FTO. The microporous compacted CE film yielded a DSSC of PCE of 7.55% relative to a DSSC (6.9%) made with a platinum CE. Similar to other polymer CEs, the morphology is critical for their conducting and catalytic properties. In this respect, ultrathin PPy nanosheets (UPNSs) have

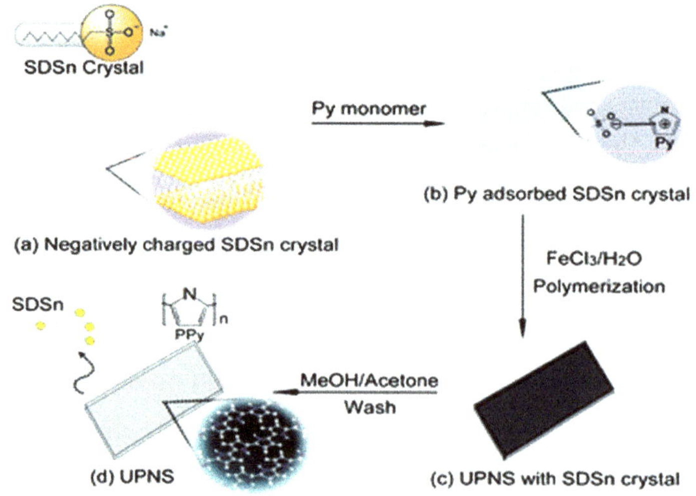

Figure 6.10. A diagram of the UPNS preparation. Reprinted with permission from [51]. Copyright 2017 Royal Society of Chemistry.

been synthesized by organic single-crystal surface-induced polymerization (OCSP) using sodium decylsulfonate (SDSn) as a template (shown in figure 6.10). These UPNSs with higher surface area and active sites were deposited as 2D structures with fewer grain boundaries onto FTO, to produce CEs of high transparency (94%) in DSSCs. It was shown that doping using HCl vapour increased conductivity and catalytic activity of the electrodes, which resulted in a PCE of 6.8%, 19.3% higher than the untreated CE, and comparable to that of platinum CE-based DSSCs (7.8%).

6.5.4 Transition metal compound CEs

A variety of transition metal compounds such as sulfides (e.g. CoS_2, $CuInS_2$, Cu_2ZnSnS_4, Co_9S_8, Sb_2S_3, Cu_2S and $CoMoS_4$), carbides (e.g. TiC, MoC, NbC, TiC), nitrides (e.g. Fe_2N, TiN, ZrN, Mo_2N, MoN), phosphides (e.g. Ni_2P and Ni_5P_4), tellurides (e.g. CoTe and $NiTi_2$), and metal oxides (e.g. WO_2 and V_2O_5) have also been studied as alternative effective CE materials [41]. This is because they have electronic structures similar to platinum, tunable bulk phase structures, nanostructures and morphology. One of the advantages of these materials is that they can be synthesized by a range of standard inorganic and electrochemical methods, and can be easily solution-processed onto conductive substrates. Among these methods, precursor decomposition synthesis, hydrothermal and solvothermal synthesis, cyclic voltametric/electrochemical deposition, composite mediated methods, electrophoretic deposition, ion exchange deposition, have been successful. It has been shown that the PV performance of the DSSCs based on CEs of transition metal compounds is relatively lower than that of the devices with other platinum-free CE materials, such as conductive polymers and carbon-based materials. However, there is opportunity to expand this field of CEs by exploring approaches to improve electrical conductivity, especially with respect to defect and interfacial chemistry.

In particular, transition metal carbides are attractive CE materials because of their low cost, high catalytic activity, good selectivity, high electrical and thermal conductivity, and good thermal and chemical stability. Moreover, N-doped carbides, such as N–TiC, N–VC, and N–NbC show better catalytic activity and PV performance than the corresponding undoped structures, where it seems that the synergetic effect of N and C atoms influence their electronic structures. Transition metal chalcogenides (TMCs) are particularly attractive for CE applications because they have various compositions, molecular and phase structures, and distinctive electronic and electrochemical properties. TMCs are more diverse than the carbides and nitrides since they offer much more opportunity for tunable structural and electronic properties by virtue of their synthesis methods. For example, highly crystalline and well-defined nanostructured NiS thin films were deposited onto flexible polyethylene terephthalate (PET) substrate by a simple and effective solution-based method [58]. The flexible DSSC with the NiS CE exhibited a high PCE of 9.50% relative to 8.97% for the device with platinum CE. In another case, CoS and NiS nanostructures were deposited onto FTO by a simple electrodeposition method with the morphology tuned by varying the reaction conditions [59]. DSSCs made from these CEs exhibited good PCEs of 9.23% (CoS) and 9.65% (NiS), respectively, in comparison to DSSCs made from corresponding platinum CE (8.12%).

Similar to 2D graphene-based materials, layered transition metal dichalcogenides, MX_2 (M = W, Mo, Hf, Nb, Re, Ta; X = S, Se, Te) have become of great interest due to their unique electronic and photonic properties and a very wide range of applications and are particularly useful for electrodes. For example, Al-Mamun and co-workers [60] deposited MoS_2 nanoscale thin films onto FTO substrates using a low-temperature one-pot hydrazine assisted hydrothermal method. DSSCs fabricated using these surface exposed layered nanosheets of MoS_2 CEs exhibited PCEs as high as of 7.41%, higher than the Pt electrode-based DSSCs using TiO_2 photoanodes sensitized by N719 dye. They showed that both the hydrothermal reaction temperature as well as the different molar ratio of reaction precursors was found to impact the structure and performance of MoS_2 films used as CEs for DSSCs.

6.5.5 Composite/hybrid CE materials

In order to improve the performance and adaptability of CEs, much attention has been devoted to developing a range of hybrids or composite CEs. These hybrids which are comprised of two or more components take advantage of the synergistic effects of individual components, thereby improving the performance of CEs. These hybrid CEs can be divided into platinum-loaded and platinum-free hybrids. The latter can be classified based on the components accordingly: TMCs/carbon, carbon/polymers and polymers/TMCs. Overall, the PV performance of most DSSCs based on hybrid CEs is better than that of the devices with their corresponding components. Such improved performance can be attributed to the synergetic effects of the different components in the hybrids opening up new routes for the development of high-performance and low-cost DSSCs.

In particular, 2D layered materials can be combined into a single hybrid structure, for example, MoS_2 and graphene which can synergistically enhance photovoltaic properties and therefore have been investigated as CEs for DSSCs [61]. For example, Sun and co-workers [62] reported an all electrochemical strategy to prepare MoS_2/graphene composite films that directly act as CEs of DSSCs without post-treatment. This involved electrodeposition and electroreduction of GO and subsequent electrodeposition of MoS_2 on reduced GO layers. The DSSCs based on optimized MoS_2/graphene CEs exhibited a high PCE of 8.01%, which was comparable to 8.21% of the platinum CE.

In order to make use of the light transparent and conductive features particularly useful for solar cell applications for BIPV that make use of the light from the interior of the building as well as the outside, there has been some focus on conductive polymer materials such as PEDOT, PProDOT, PANI, etc as CE materials. However, the catalytic performance of some of these CE alone is not as good as that of the platinum CE. In an effort to optimize the characteristics, Xu and co-workers [63] reported on transparent organic–inorganic hybrid composite films of molybdenum disulfide and poly(3,4-ethylenedioxythiophene) (MoS_2/PEDOT) to take full advantage of the conductivity and electrocatalytic ability of the two components. MoS_2 was synthesized by hydrothermal method and spin coated to form the MoS_2 layer, and then PEDOT films were electrochemically polymerized on top of the MoS_2 film to form the composite CEs. The DSSC with the optimized MoS_2/PEDOT composite CE showed a PCE of 7% under front illumination and 4.82% under back illumination. Compared with the DSSC made by the PEDOT CE and the platinum CE, the DSSC fabricated by the MoS_2/PEDOT composite CE improved the PCE by 10.6% and 6.4% for front illumination, respectively.

6.6 Transparent and flexible conducting substrates

The selection of suitable conducting substrates is a critical factor to determine the cost and the fabrication process of DSSCs, and therefore to influence the cell performance and long-term stability. Traditionally, TCO glasses were utilized as the substrate of DSSCs. These include, fluorine-doped tin oxide (SnO_2:F) (FTO), tin-doped indium oxide (In_2O_3:Sn) (ITO), and aluminium-doped zinc oxide (ZnO:Al) (AZO), as the traditional types. However, despite their good stability against oxygen and water, their practical applications are restricted due to high cost, fragility, rigidity, shape limitations, and the roll-to-roll production process cannot be used for fabrication. Therefore, flexible substrates with light weight and easy transportation properties have become alternatives.

Primary among these flexible substrates is conducting plastics including ITO-coated polyethylene terephthalate (ITO/PET) and ITO-coated polyethylene naphthalate (ITO/PEN) [51]. Notwithstanding the attractive flexible properties, the critical challenge for conducting plastics lies in the low-temperature preparation of the TiO_2 layer, since their maximum endurable temperature is around 150 °C. In fabricating DSSCs, the TiO_2 layer is normally sintered at around 400 °C–500 °C for optimal electrical contact among particles, and the organic binder and viscous solvents of the TiO_2 paste are removed during this high-temperature sintering. Ko and co-workers [64] reported a completely TCO-free and flexible DSSC fabricated on a plastic substrate using a unique transfer method and back-contact architecture. By adopting unique transfer techniques, the WEs and CEs were fabricated by transferring high-temperature-annealed TiO_2 and platinum/carbon films, respectively, onto flexible plastic substrates without any exfoliation. Importantly, the completely TCO-free and flexible DSSC exhibited a remarkable efficiency of 8.10%. Also, Lai and co-workers employed Al-doped ZnO (AZO) film as the TCO onto the glass substrate for DSSCs. To improve the electrical property of AZO, a film of indium gallium zinc oxide (IGZO) was introduced as the buffer layer between the AZO and the glass substrate. They reported that the resistivity of the AZO glass was reduced with the introduction of the IGZO buffer layer, and the PV efficiency was enhanced when the AZO/IGZO TCO was used in DSSC. In addition, it was found that the PV efficiency of the DSSC with AZO/IGZO glass was also higher than those of the DSSCs with conventional FTO and ITO glasses.

Other types of the flexible substrates includes metal foils such as stainless steel, zinc, and titanium, since they possess inherent high conductivity, light scattering, and high-temperature sintering abilities [51]. Among them, the best is the titanium foil-based DSSC because of its characteristics of higher conductivity and superior corrosion resistance due to the surface passivation by a thin layer of TiO_2. However, metal-based DSSCs are limited because these conducting substrates are non-transparent, reflecting a fraction of the incident radiation, and the formation of metal oxide passivation layers increases interfacial resistance affecting cell performance.

6.7 Cell architecture, device physics and performance

The concept of the TiO_2-based dye-sensitized solar cells was premised on the discovery of photocurrent generation based on a chlorophyll-sensitized zinc oxide electrode. However, despite numerous studies into the devices which had poor conversion efficiencies the breakthrough in device development was realized in 1991 by Grätzel and co-workers. The Grätzel cell was characterized with rough surfaced nanoporous TiO_2 electrodes that achieved a PCE of 7.1%–7.9% [1]. This development generated tremendous investigations resulting in a device of commercial promise with competitive record PCE of 12.3% (AM 1.5G) in 2011 composed of a zinc porphyrin dye (YD2-o-C8), co-sensitized with another organic dye (Y123), interfaced with a $Co^{2+/3+}$ tris(bipyridyl)-based redox electrolyte [42]. In 2015, a PCE of 14.3% was attained for a DSSC composed of metal-free organic co-sensitizers and the redox electrolyte based on $Co^{2+/3+}$ tris(phenanthroline) with many cells today featuring these types of electrolytes [65, 66]. Today, DSSC cell architectures have evolved from the Grätzel *n*-type cell to a range including, *p*-type DSSCs, solid and quasi-solid-state DSSCs, quantum dot DSSCs and tandem DSSCs with achieving record PCE of around 13.0% [67], these due to the development of various components and in tandem with methods for device modelling, as discussed in the following. In particular, for indoor photovoltaics, DSSCs are promising candidates for commercialization due to the outstanding PCE above 34% with great long-term device stability under ambient conditions [66].

6.7.1 Device modelling (*first principles*)

Parts of this subsection have been reproduced from [69], with permission from Springer Nature.

The complexity of DSSCs and the chemical and physical processes involved for efficient photocurrent generation requires a fundamental understanding and therefore herein is outlined generalized strategies involved in doing so. Accordingly, modern theoretical computational methods are useful to treat with the complex materials and processes that span several scales of space and time in these devices [3]. Processes of light harvesting, dye/electrode charge transfer, electron transport to the charge collector, oxidized dye regeneration, electrolyte diffusion, and reduction at the CE, occur at specific locations within the cell and with timescales ranging from femtoseconds to milliseconds. Therefore, the simulation approach must be multidimensional, starting from the elementary processes at the nanoscale and sequentially building to the larger (longer) space (time) scales [3]. Accordingly, a range of *ab initio* (*first principles*) computational methodologies have been developed with density functional theory (DFT) as the primary method for understanding the electronic structure of materials and interfaces, and time-dependent DFT (TD-DFT) which has enabled the effective description of dynamic excited state properties in these devices. These methodologies in modelling the electron transfer mechanisms require explicit quantum mechanical treatment. Importantly, in order to set up an approach to screen novel candidates, any method introducing empirical parameters not straightforwardly extendable to novel materials has to be excluded since this is a basis for the design of novel devices [68].

Whilst computational modelling of isolated device components such as sensitizers, semiconductor nanoparticles, redox shuttles, etc, is important, the interdependencies between these components require simulation methodologies to consider various pair interactions, including dye–semiconductor, redox shuttle–semiconductor, dye–redox shuttle, and dye–dye [69]. Intrinsic to these simulations are the dynamics involving kinetic and thermodynamic factors that govern these processes. Importantly, a computational approach to DSSC modelling must be stepwise, where individual components are accurately simulated independently before considering the dynamics of the more complex pairs. This can constitute the basis for an integrated multiscale computational description of the device functioning towards improving performance, along with providing the basic understanding of the device, necessary for further enhancing target characteristics, such as temporal stability and optimization of device components. In employing simulation methodologies, there are several items of basic information that are required. These include: (1) the dye molecular or crystal structure, ground and excited states oxidation potentials, and optical absorption properties; (2) the semiconductor electronic band structure properties; and (3) the electrolyte/hole conductor redox properties [69]. For the interacting dye/semiconductor pair, calculations should include: the modes of dye adsorption, the nature and localization of the dye semiconductor excited states, and the alignment of ground and excited state energy levels at the dye/semiconductor interface, which, along with an estimate of the electronic coupling, constitute the fundamental parameters controlling the electron injection and dye regeneration processes. Additionally, calculations involving the dye-sensitized semiconductor in the presence of the electrolyte solution and/or redox species, allow for an understanding of the dynamics of charge transfer, a critical element of device efficiency.

Overall, the information extracted from DFT and TD-DFT calculations constitute the basis for the explicit simulation of photo-induced electron transfer by means of quantum or non-adiabatic dynamics [69]. Here, various combinations of electronic structure/excited states and nuclear dynamics descriptions are applied to dye-sensitized interfaces. In most cases these approaches depend either on semi-empirical Hamiltonians or on the time-dependent propagation of single-particle DFT orbitals, with the nuclear dynamics being described within mixed quantum–classical or fully quantum mechanical approaches. Additionally, real-time propagation of the TD-DFT excited states has also emerged as a powerful tool to study photo-induced electron transfer events, with applications to dye-sensitized interfaces based on mixed quantum–classical dynamics. Important in these methodologies is to consider other complexities of DSSCs such as multiple dye-adsorption on TiO_2 relevant to aggregate properties, and the simulation of co-sensitized patterns involving both different dyes and dye co-adsorbent interactions.

6.7.2 Grätzel cell

As previously mentioned, the so-called Grätzel cell was the original DSSC firstly fabricated in 1991 by O'Regan and Grätzel [1]. The most distinguishing feature of the cell then, in relation to other solar cells is that it separates light absorption from charge carrier transport. The construction and mechanism of this DSSC photoelectrochemical system are illustrated in figure 6.1. The general structure of such cells comprises a photoanode as a WE, a CE and electrolyte interfacing the photoanode and CE. The photoanode is composed of an optically transparent conductive glass, coated with indium/fluorine-doped tin oxide (ITO/FTO), on which a dye-coated TiO_2 semiconductor of thickness, typically 10 μm was layered. The metal oxide semiconductor functions both as a support for the sensitizer and a carrier of photogenerated electrons from the dyes to the external circuit. In effect, current is generated when a photon is absorbed by the organometallic dye molecule giving rise to electron injection into the CB of the TiO_2 semiconductor. To complete the circuit, the dye must be regenerated by electron transfer from a redox species in an electrolyte solution which is then reduced at the CE. In preparing the original Grätzel cell, high surface area TiO_2 films of 10 μm thickness were deposited from colloidal solution onto the FTO [1]. Sintering at 450 °C resulted in a three-dimensional interconnected network of 15 nm sized nanoparticles which yielded a highly conductive photoanode layer. Onto this layer, a ruthenium-based dye,

$RuL_2(\mu\text{-}(CN)Ru(CN)L'_2$, where L is 2,2′-bipyridine-4,4′-dicarboxylic acid and L′ is 2,2′-bipyridine, was deposited from solution. The CE consisted of conducting glass coated with a few monolayers of platinum, placed directly on top of the WE and sealed at the edges to create a sandwich-type cell configuration. A thin layer of the I^-/I_3^- redox electrolyte which consisted of tetrapropylammonium iodide and iodine or potassium iodide–iodine in a mixture of ethylene carbonate and acetonitrile, interfaced the electrodes through capillary action. The high surface area of the semiconductor film and the ideal spectral characteristics of the dye enabled this original device to harvest a high proportion of the incident solar energy flux (46%) and showed exceptionally high efficiencies for the conversion of incident photons to electrical current (more than 80%) for these types of cells at that time. The overall conversion efficiency was 7.1%–7.9% in simulated solar light with AM 1.5 G spectral distribution. The notably high fill factors (comparatively speaking at that time) were between 0.69 and 0.76, and large current densities (greater than 12 mA cm^{-2}) were key factors for good cell efficiency. Importantly, performance under diffuse light conditions was because the spectral distribution overlaps favourably with the absorption of the dye-coated TiO_2 film. Overall, the performance of the cell was attributed to relatively low recombination losses, as described then in relation to other devices, and because the semiconductor was a conduit for charge injection and transfer, its surface roughness was not a factor in any recombination losses. The device also displayed exceptional long-term stability with the dye sustaining 5 million turnovers without decomposition. In effect, the high charge injection rate from the dye of 10^{12} s^{-1} was practically maintained over the long term. Overall, these factors and the low cost of the cell proved the potential practical application of these types of cells. However, these cells suffered from leakage or evaporation of the electrolyte, dye desorption and degradation of other components, which affected their stability. These issues alongside the need for competitive PCEs propelled the development of alternative device architectures.

6.7.3 Solid-state DSSCs

The primary difference in device structure between the solid-state-DSSC (ssDSSC) and the conventional Grätzel device is the use of a solid redox electrolyte or HTM layer instead of the liquid electrolyte. This architecture was developed to overcome the limitation associated with the stability issues of the Grätzel cell due to leakage or evaporation of the electrolyte, dye desorption and other component degradation due to the liquid media. These monolithic DSSC structures have advantages over the sandwich structure, since only one FTO glass substrate is used for the different layers. As such, ssDSSCs do not require extra sealing between the electrodes but have good encapsulation and series or parallel connection. A schematic of a typical ssDSSC device structure is shown in figure 6.11 [3]. Similar to the liquid DSSC, onto a conducting glass substrate, usually glass coated with FTO, a thin compact layer of TiO_2 is deposited, followed by a 1–3 μm mesoporous layer of TiO_2 [70]. The dense, compact TiO_2 layer prevents direct contact between the FTO and the HTM, which could lead to losses due to shunting arising from electron–hole recombination.

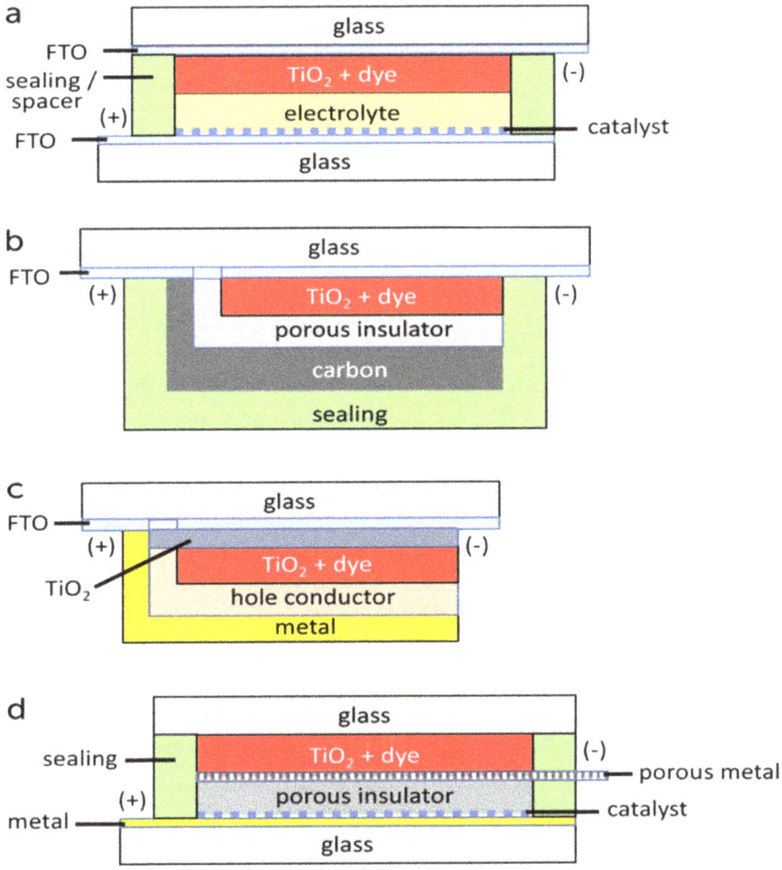

Figure 6.11. Device structures for DSSCs: (a) sandwich cell, (b) monolithic cell with carbon CE, (c) solid-state DSSC (monolithic), and (d) conducting glass-free DSSC design. Reprinted from [3] CC BY 3.0.

During fabrication, dye molecules are adsorbed from solution onto the mesoporous TiO_2 layer. Then, the HTM infiltrates the porous metal oxide structure, whilst also forming a thin overstanding layer. In effect, the mobility of the holes in the HTM is usually higher than the electron mobility in the mesoporous TiO_2. The CE usually consists of thin layer of metal (e.g. Au or Ag) or carbon-based materials deposited by drop casting or evaporation on top of the HTM layer.

The mechanism of light-to-power conversion in the ssDSSCs is similar in principle to that of the Grätzel cell, with some notable differences and improved efficiency. After the photogenerated electron from the excited dye enters the CB of the mesoporous semiconductor, the HTM reduces the oxidized dye and the generated hole travels from the sensitized electrode by means of charge hopping to a metal contact. In Grätzel cells, redox electrolyte ionic conductors travel via liquid diffusion but for solid-state devices, conductivity is via charge carrier migration controlled by the electric field, involving electronic conduction from a solid donor material, and holes transported to the CE [10]. In ssDSSCs, the oxidized dye returns to the ground state within a few hundred picoseconds, orders of magnitude faster than in the liquid-junction cells (microseconds). This fast regeneration is attributed to direct hole transfer via an energy gradient into the HOMO level of a solid-state hole transporter from the oxidized dye. These extremely fast regeneration dynamics lead to a rapid hole injection from the oxidized dye into the HTM, essentially on the same timescale as the electron injection or even higher. Importantly, the overstanding HTM layer is required to be very thin in order to avoid additional series resistance in the device [70]. However, too thin a layer could develop pinhole defects resulting in contact of the FTO with the mesoporous TiO_2 film, providing a loss mechanism of charges resulting in shunting losses. The HTM is finally regenerated by charge transfer at the CE.

Despite progress in the development of ssDSSCs, especially in relation to the Grätzel cell, even when one considers developments in sensitizer molecules, photoanodes, hole transport layer, etc, the highest performing ssDSSC is at 11.7% compared to around 14% for liquid cells [10]. The limitations in performance of ssDSSCs are largely due to energy level misalignment between the individual components, in particular the dye and the semiconductor oxide, and the CE and electrolyte for hole injection, as well as the kinetics of the charge separation and transfer processes which impact the photocurrent, cell potential and fill factor [10]. Primarily, photocurrent improvement relies on the absorption characteristics of the dye and its charge injection into the semiconductor. Strategies towards improvement include, use of dye absorbers of a wide spectral range and high absorptivity, improving semiconductor morphology and thickness to ensure sufficient pore-filling of the dye, and low series resistance by the HTM. Importantly, improvement in the V_{oc} can be achieved by lowering the potential for dye regeneration, which involves limiting charge carrier recombination between the excited dye and the HTM, to minimize back flow current and energy loss. Such strategies depend on improved charge transfer kinetics in the dye molecules and their very fast injection into the semiconductor layer. Increase of the fill factor can be achieved by improving the morphology of the photoanode, especially by incorporating a compact blocking layer between the TCO and the HTM. The blocking layer, typically a thin layer of TiO_2, is an electron-selective contact that impedes recombination of electrons in the TCO with holes in the HTM. It is important that there are no pinholes or cracks in this layer to avoid direct contact of the TCO with the HTM, but it remains as thin as possible to minimize resistance losses [70]. Additionally, improvement in the fill factor is dependent on the nature and morphology of the CE to ensure better contact with the HTM.

6.7.4 *p*-Type DSSCs and *p*-type solid-state DSSCs

Parts of this section have been reproduced from [74]. CC BY 3.0.

Whilst most of the research on DSSCs has been devoted to architectures based on *n*-type semiconductors, the typical Grätzel cell and ssDSSCs, there has been focus on *p*-type DSSC (*p*-DSSC) architectures. These utilize a *p*-type semiconductor as photocathode instead of an *n*-type photoanode with the anchoring group of the

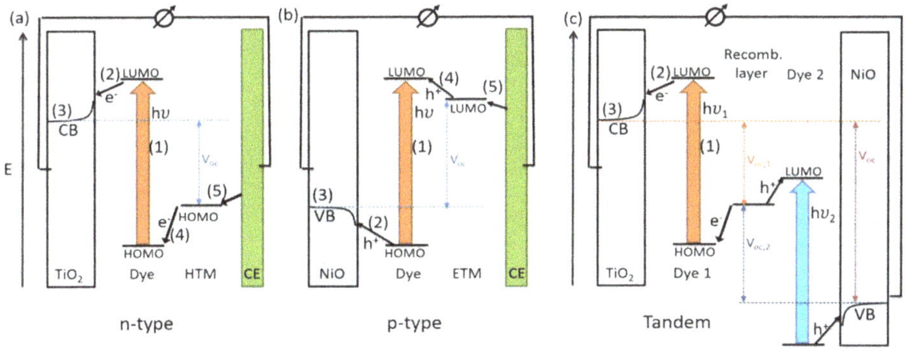

Figure 6.12. Charge transfer processes in (a) *n*-type, (b) *p*-type and (c) tandem ssDSSCs. Reprinted with permission from reference [10]. Copyright 2018 Royal Society of Chemistry.

dye bound to the electron-donor. The working principle is similar to that of the Grätzel *n*-type DSSC, but the mechanism of charge transfer is different in which the excited state sensitizer injects holes into the VB of the *p*-type semiconductor upon illumination, as shown in figure 6.12 [10, 71]. Here, the generated hole travels through the *p*-type semiconductor to the back contact and then through the external load to the CE, where it oxidizes the redox couple, thereby regenerating the reduced sensitizer. The *p*-DSSC was first reported in 1999 by Lindquist and co-workers [72], in which a porphyrin dye was used as a photosensitizer, nanoporous NiO acted as a wide bandgap *p*-type semiconductor photocathode for dye loading, and a liquid I^-/I_3^- redox electrolyte used for electron transport between the CE and the photocathode [73, 74]. However, these devices typically suffered from low photovoltages due to a small energy gap (ΔE) between the VB of the NiO photocathode and the redox potential of the I^-/I_3^- redox couple. Accordingly, several studies have been focussed on the development of alternative *p*-type semiconductors, however NiO maintains the highest conductivity and is more processable relative to the alternatives. Additionally, other strategies to improve photovoltage have involved use of other redox couples, such as tris(1,2-diaminoethane)cobalt(III)/(II) and tris(acetylacetonato)iron(III)/(II) with more negative reduction potentials than I^-/I_3^- to widen the ΔE [74]. However, such redox couples are sensitive to oxygen, which makes device fabrication challenging. Additionally, the issue with electrolyte leakage persists, since organic solvents are used.

As an alternative to *p*-DSSCs, Tian and co-workers [73] pioneered the solid-state *p*-type cell (*p*-ssDSSC) in 2016, an example of which is shown in figure 6.13. The cell was composed of mesoporous NiO with the organic P1 dye and solid-state ETM, phenyl-C_{61}-butyric acid methylester ($PC_{61}BM$) instead of the I^-/I_3^- redox electrolyte for dye regeneration with electron transport between the dye-sensitized photocathode and a 100 nm Al back contact. The compact NiO layer was placed between the FTO and nanoporous NiO layers in order to prevent physical contact between the FTO and the penetrated $PC_{61}BM$ for suppressing electron and hole recombination. The device based on NiO/P1/$PC_{61}BM$ rendered a J_{sc} of 50 mA cm^{-2} and a V_{oc} of 620 mV. Stipulations for an effective *p*-ssDSSC are that the ETM should be

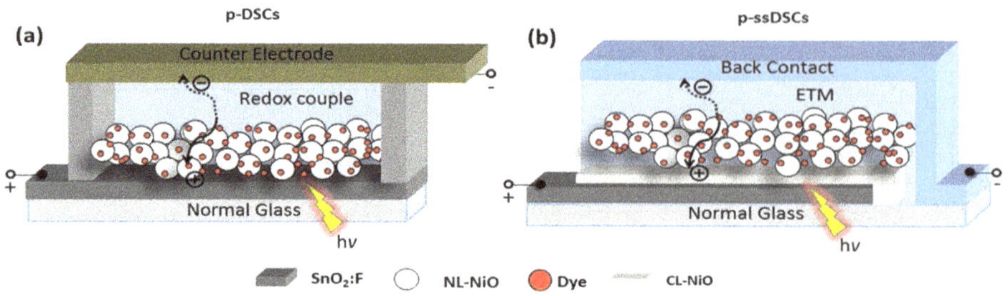

Figure 6.13. Schematic of the configurations of a conventional liquid *p*-DSSC (a) and a solid-state device *p*-ssDSSC (b). Reprinted from [74] CC BY 3.0.

able to infiltrate the nanoporous *p*-type photocathode layer with good contact with the dye [74]. This requires the ETM to have good solubility for solution processing or be prepared *in situ*. In order to obtain a satisfactory photovoltage, the ΔE between the VB of the *p*-type semiconductor and the reduction potential of the ETM should be maximum, but the reduction potential of the ETM has to enable dye regeneration. The ETM should also enable good electron mobility to efficiently transport electrons to the back contact. In the pursuit of further development of *p*-ssDSSCs, in 2017, Pham and co-workers [75] sensitized NiO with a diketopyrrolopyrrole (DPP) dye and achieved a device with J_{sc} of 0.45 mA cm^{-2}. Thereafter, Tian [74] was able to ascertain that charge recombination was probably caused by direct contact between NiO and TiO_2, and introduced a 1 nm Al_2O_3 blocking layer, whilst simultaneously protecting the dye. The resulting *p*-ssDSSCs based on NiO–PB6–Al_2O_3–TiO_2 significantly improved the J_{sc} and V_{oc} relative to NiO–PB6–TiO_2-based device attributed to prolonged charge separation lifetime.

6.7.5 Quantum dot-sensitized solar cells

Quantum dot-sensitized solar cells (QDSSCs) are one of the promising solar cell technologies analogous to the conventional DSSC. Overall, quantum dot-based solar cells can be classified into four kinds, including Schottky junction solar cells, *p*–*n* junction solar cells, hybrid QD–polymer solar cells, and QDSSCs, having received much attention in recent years because of the tremendous improvement in PCE beyond 18% [11, 67, 76]. The basic QDSSC architecture mimics that of the DSSC but differs in terms of the sensitizer and electrolyte. However, it is attractive due to the simple architecture derived from low-cost, simple fabrication methods including solution processing. The common feature of QDSSCs is quantum confinement of the exciton in the quantum dot (QD) absorber, leading to particle size-dependent spectral absorption. In these cells, the microscopic surface of a typical wide bandgap semiconductor such as TiO_2 is sensitized with a layer of QDs, whilst a redox electrolyte fills the free space around the nanostructures. The QD sensitizers become excited on absorption of light energy equivalent to the bandgap and inject electrons from their CB into the CB of the photoanode, whilst the oxidized QDs are recharged by the redox electrolyte. Charge transport to the front electrode, typically TCO, as well as transport of oxidized redox species to the CE are diffusion-driven

[77]. However, like other DSSCs, unwanted charge recombination circumvents the desirable charge transport properties, thereby limiting device performance [11].

QDs are attractive relative to organic or metal–organic dyes because they have several salient and versatile optical and electrical properties including, aforementioned quantum confined tunable bandgaps, larger extinction coefficient, higher stability towards water and oxygen, and multiple exciton generation (MEG) with single-photon absorption [78]. There are many reports [78] examining a range of QDs for DSSCs and include narrow bandgap semiconductor QDs, such as CdS, CdSe, PbS, InAs, $CuInS_2$ [79], $AgInSe_2$, PbSeS, and Ag_2Se as conventional examples. Overall, smaller QDs are preferred in order to possibly achieve higher loading and more compact photoelectrode thin films. Smaller QDs with wider bandgaps have also demonstrated a higher electron injection rate than larger ones because of better band alignment of the CB edges of the wide bandgap photoanode [78]. Therefore, controlling quantum size confinement in monodisperse QD samples is the most effective means of aligning the energy levels with respect to the wide bandgap nanostructure, since a decrease in size widens the bandgaps, thereby increasing the CB minimum, which results in better band alignment with the oxide photoanode. Additionally, surface passivation by dipoles provides a simple and efficient way to shift the QD energy levels, and this has been achieved with exchanging capping ligands with molecular dipoles [77].

One of the challenges with QDSSCs is the range of reliable and efficient redox electrolytes required for efficient hole extraction from the QD to prevent recombination and degradation, with the most common examples being I^-/I_3^- and polysulfide [77]. The polysulfide is more reliable for sulfide QDs, but its redox chemistry is complex, and it is challenging to prepare CEs without significant losses due to high charge-transfer resistances. Additionally, the redox potential of these electrolytes is relatively high, leading to energy loss and consequently a low V_{oc}. This results in slow QDs regeneration, leading to an accumulation of holes in them, thereby decreasing their stability and increasing the recombination rate [11]. Overall, the fill factors for QDSSCs with a polysulfide electrolyte are generally low but are much improved for I^-/I_3^- couple used with a platinum CE. However, the presence of iodine often causes photo-corrosion of the QDs. An important strategy in utilizing polysulfide electrolytes has been to incorporate them with additives, which improves performance by either inhibiting charge recombination at the photoanode/electrolyte interface or by shifting the CB edge of TiO_2. For example, Zhong and co-workers utilized water-soluble polymers, including poly(ethylene glycol) (PEG) and poly(vinyl pyrrolidone) (PVP) as additives in the polysulfide electrolyte and observed a remarkable enhancement in the PCE [11, 80]. They further employed tetraethyl orthosilicate (TEOS) as an additive and reported an improvement in PCE from 11.75% to 12.34%, attributing this to reduction of recombination losses at the photoanode/electrolyte interface due to the presence of the TEOS [11, 81]. Additionally, systems with quasi-solid-state and solid-state HTMs have been focussed on for use in QDSSCs. For example, in the former case, Meng and co-workers [11, 82] applied the natural polysaccharide konjac glucomannan (KGM) as the polymer matrix to prepare a gel electrolyte with a Cu_2S CE. The QDSSCs exhibited a comparable PCE of 4.06% to its liquid electrolyte-based counterpart (4.22%) with improved stability. In general, although quasi-solid/all-solid-state QDSSCs possess advantages with respect to improving long-term stability of a device, their PCEs are still low compared to the liquid-junction QDSSCs [11]. To date, the highest PCE for an all-solid-state QDSSC is about 8% [46].

Fabrication of QD-photoanodes for QDSSCs can be achieved via two typical approaches: (1) *in situ* growth of QDs directly from precursor solutions, and (2) *ex situ* approaches, in which pre-synthesized colloidal QDs are deposited onto the metal oxide substrate through direct adsorption, electrophoretic deposition, or linker-assisted deposition methods, as typical examples [11, 74, 78]. However, the latter usually yields QDSSCs with lower conversion efficiency, largely due to the difficulty in achieving sufficient coverage of QDs on the photoanode. The former typically includes chemical bath deposition and successive ionic layer absorption and reaction (SILAR), and typically produces cells of better performance due to the easy processability, good reproducibility, and high QD-loading. Of the *ex situ* methods utilized, the main challenge for the use of pre-synthesized QDs in QDSSCs, is how to achieve high QD-loading [11] with much of the early development in these cells not achieving good performance. The first notable outcome was reported by Toyoda and co-workers in 2007 [11, 83], who achieved QDSSCs of PCEs over 2%. An important development was made by Lee and co-workers in 2009 with their CdS/CdSe co-sensitization structure through SILAR deposition, improving the PCEs to 4.22% [84]. This issue of achieving uniform and high QD-loading on the photoanode was then possible using the effective capping-ligand-induced self-assembly (CLIS) method developed by Zhang and co-workers [85]. Here, they deposited high-quality pre-synthesized colloidal QDs onto TiO_2 by changing the capping ligand, linker molecule concentration and pH of the QD solution. This provided a way to introduce high-quality QDs as sensitizers in QDSSCs with cell performances improving thereafter [11]. Through the exploration of superior colloidal QD sensitizers and interfacial engineering, the highest PCE of QDSSCs has been improved from 5% to around 16% in recent years, resulting in QDSSCs becoming more attractive and competitive of the emerging technologies [85].

6.7.6 Perovskite-sensitized solar cells

The precursor to current types of perovskite solar cells was a perovskite-sensitized solar cell. This liquid-based sensitized solar cell structure was constructed by adsorption of methylammonium lead halide ($CH_3NH_3PbX_3$; X = Br, I) perovskite sensitizer from solution onto a nanocrystalline TiO_2 surface, yielding a photocurrent with PCE up to 3.8% [86, 87]. The poor efficiency was probably due to non-optimized conditions such as a diluted precursor solution for perovskite sensitizer coating. In 2011, the PCE was optimized to 6.5% with the use of $CH_3NH_3PbI_3$ and an I^-/I_3^- electrolyte [86]. Unfortunately, the liquid-based perovskite solar cell construction made little progress due to stability issues involving dissolution of the perovskite in the liquid electrolyte. However, a more long-term, stable, and higher efficiency cell of 9.7% was developed in 2012 by substituting the liquid electrolyte with the solid HTM, spiro-MeOTAD through efforts in developing solid-state devices. Thereafter, higher efficiency of 12.3% of a solid-state $CH_3NH_3PbI_3$-sensitized mesoscopic solar cell was demonstrated using polytriarylamine as HTM. However, it was later shown that replacement of TiO_2 with Al_2O_3 to yield a device of structure FTO/TiO_2 blocking layer/Al_2O_3/perovskite/spiro-MeOTAD, where $CH_3NH_3PbI_2Cl$-adsorbed mesoporous Al_2O_3, demonstrated highest efficiency of 10.9%. In this cell, the Al_2O_3 acts as a scaffold without any electron injection from the perovskite, since it acts as the electron transport layer and absorber material. Today, progress in perovskite solar cells has been tremendous with PCEs close to 30% [67, 88]. As such, these developments are covered in its dedicated chapter.

6.7.7 Tandem DSSCs

In order for DSSCs to be considered commercially viable, their efficiencies need to be comparably competitive on the order of 15%–20% or even higher. Considering the advantages and limitations of *n*-type and *p*-type DSSCs, one strategy in doing so is with tandem structured DSSCs (*t*-DSSCs), since the theoretical efficiency limit is around 42% [89]. For a range to cell types, tandem (multijunction) solar cells have been shown to be an effective way to harvest light from a large part of the solar spectrum. In these, two or more sub-cells with complementary absorption characteristics are stacked and connected either in series or in parallel. They have demonstrated that increasing the number of junctions in a device is an efficient way to overcome the Shockley–Queisser limit of 33% PCE for a single-junction device. As such, *t*-DSSCs can be categorized according to: (1) *n*-type DSSC + *n*-type DSSC; (2) *n*-type DSSC + *p*-type DSSC; and (3) *n*-type DSSC + other solar cell, also called a hybrid-DSSC [90].

With the capabilities of *n*-type DSSCs, the so-called *n-n t*-DSSCs involving *n*-type DSSC plus *n*-type DSSC architecture would be expected to offer better performance over the single-junction type. The *n-n t*-DSSC architecture can be divided into three categories: (1) with two separate cells connected in series or in parallel; (2) with multilayered photoanode of different dyes; and (3) two electrodes sensitized with two dyes, with one floating porous electrode in the middle of the cell [90]. For the first type, depending on different combinations of dyes, especially more efficient near-infrared absorption dyes, the tandem design based on two completed cells achieved improved performances. The advantage of this structure is that the performance of both cells could be optimized separately. However, the primary drawback is the light absorption by the platinum layer of the top cell which affects the amount of light available for the bottom cell. The first of this type of cell was reported by Shozo and co-workers [90, 91] involving a cell with N719 sensitization on top of a black dye-sensitized cell. The *n-n t*-DSSC was improved relative to the individual cells of PCE of 7.6% with series connection. Here, the electrode of the front cell with N719 embedded TiO_2 nanoparticles was highly transparent, allowing for most light absorption by the black dye of the bottom cell. Following this development, Nelles and co-workers [92] reported a PCE of 10.5% for the cell connected in parallel, and Masatoshi and co-workers [93] optimized the thickness of

TiO_2 films in the separated cells with different connection modes, achieving a PCE of 10.6%. As an alternative to the double cell configuration, the *t*-DSSC with multilayered photoanode was developed by Park and co-workers [94] which involved a controlled method of desorption of one dye over another. However, the cell performance was poor attributed to insufficient dye loading. Consequently, Ma and co-workers [95] optimized the dye loading using a simple and low-cost film-transfer method achieving an efficiency of 11.05%. The structure of the DSSC consisted of a TiO_2/dye 1/transferred TiO_2/dye 2 multilayered photoanode configuration. The bottom layers were prepared by coating the TiO_2 paste onto FTO glass using the doctor-blade method, followed by drying, coating with a second layer of TiO_2 paste, then sintering and then dye loading by immersion of a N719 dye solution. To prepare the layer for transfer, the pastes of P25 and ST41 were coated onto the smooth surface of a ceramic tile and then sintered. Subsequently, the two films were placed face-to-face and the porous P25 layer was easily transferred from the ceramic tile to the surface of the bottom layer under the friction acting on the two layers, and then compression. After transferring the scattering layer, the obtained photoanode was sensitized with dye 2. The third type of configuration was developed by Murayama and co-workers [96, 97] in which two dye-sensitized nanocrystalline TiO_2 films were placed face-to-face as WEs, and a platinum mesh sheet with suitably high optical transmittance was inserted between them as a CE. The cell connected in series or in parallel exhibited a PCE of 3.9%, but with an increase in thickness of the front electrode, the PCE improved to 4.7%.

In a typical *n-p t*-DSSC, the *n*-type photoanode (e.g. TiO_2) and *p*-type photocathode (e.g. NiO) can be sensitized with dyes of different spectral absorption to maximize the light harvesting in the solar cell [10]. In these, the V_{oc} of the cell is the energetic offset between the Fermi level of electrons in the CB of the TiO_2 and the Fermi level of holes in the VB of the NiO. The first *n-p t*-DSSC was reported with an efficiency of 0.39% [98] with further development over time achieving modest improvements in performance. For example, Gibson and co-workers [3, 99] reported on *n-p t*-DSSCs of PCE of 0.55% based on a N719-sensitized TiO_2 photoanode and a perylene-sensitized NiO photocathode with a cobalt redox mediator. Following that, Nattestad and co-workers [100] reported a device with PCE of 1.91%. However, the first all-solid-state tandem DSSC was described in 2009 by Bruder and co-workers [101]. It consisted of a charge recombination layer of Ag sandwiched by a top cell with indoline-sensitized TiO_2/spiro-OMeTAD and a bottom cell of zinc phthalocyanine/C60, and demonstrated an V_{oc} of 1.36 V and PCE of 6.0%.

However, the architectures of *t*-DSSCs can extend beyond the typical heterojunctions hitherto described. These include various forms of inorganic/organic solar cells, known as hybrid tandem solar cells, involving dye-sensitized solar cells, DSSC/GIGS, DSSC/GaAs, and DSSC/perovskite. For example, Chae and co-workers [102] developed a highly stable monolithic tandem solar cell by combining a DSSC and solution-processed copper indium gallium (CIGS) thin film solar cells from a nanoparticle ink. They showed that the durability of the tandem cell, which consisted of the effective CE (PEDOT:PSS) and sensitizer (Y123) was dramatically enhanced by replacing the redox couple from I^-/I_3^- to $[Co(bpy)_3]^{2+}/[Co(bpy)_3]^{3+}$, and a 1000 h durability test in ambient conditions resulted in only a 5% decrease in record PCE of 6.11%. Moon and co-workers [103] reported a similar DSSC/CIGS hybrid tandem cell fabricated via a simple solution-based process by preparing the CIGS absorber film for the bottom cell from a nanoparticle ink. The cell exhibited a high PCE of 13.0% compared to the corresponding single-junction DSSC (7.25%) and CIGS (6.2%), respectively. However, one of the major issues in the fabrication of efficient tandem DSSC/CIGS cells is to control the electrical/optical properties of the interface between the DSSC and the CIGS. Typically, a platinum-based catalytic layer is deposited on the TCO layer of the CIGS thin film solar cell such as Al-doped ZnO (AZO) and Sn-doped In_2O_3 (ITO). However, the platinum layer in a DSSC usually prepared via thermal decomposition using a platinum precursor (e.g. H_2PtCl_6), followed by sintering at 380 °C–450 °C is unsuitable for fabrication of a DSSC/CIGS tandem cell because the bottom CIGS cell, fabricated prior to the platinum layer, is vulnerable to high-temperature treatments. Accordingly, in their study, Moon and co-workers employed arc-plasma deposition (APD) as a low-temperature method to minimize damage to the AZO and CIGS absorber films. In one study, Yuan and co-workers [104] fabricated a series of four-terminal solution-processed hybrid tandem solar cells with highest PCE of 12% using DSSCs as top-cells and lead sulfide colloidal quantum dot solar cells (CQDSSCs) as bottom-cells. The fabricated device comprised the structure of FTO/c-TiO_2/mp-TiO_2/dye with a redox electrolyte and PEDOT/FTO back contact as the DSSC, and the PbS CQDSSC comprised ITO/AZO/PbS–PbI_2/PbS-EDT/Au structure. This device demonstrated enhanced performance due to the excellent infrared photon absorption by the PbS quantum dots. In another example, tandem solar cells were developed for improved efficiency up to 10.54% from perovskite and dye-sensitized solar cells. The organic dye based on thioindigo and N719 was applied as sensitizers in fabricating bottom-cells with the top perovskite cell. The V_{oc} of the tandem solar cells was about 1.1 V, but the J_{sc} was limited by the dye-sensitized solar cells to 12.8 mA cm^{-2} [105].

6.8 Cell degradation and failure

6.8.1 Cell stability

In order for DSSCs to be considered commercially viable, they must exhibit long-term stability in addition to good efficiency. For DSSCs, the complex nature of the various cell architectures renders them prone to several degradation mechanisms. As such, the approach to evaluating their degradation and failure must involve analyzing the degradation mechanisms specific to the individual components and their interfaces, and to develop protocols for long-term and accelerated testing. Unlike silicon solar cells, no standardized stability testing protocol has been established for traditional DSSCs or ssDSSCs. The first report exclusively dedicated to long-term stability tests was by Kay and Grätzel in 1996 [54] in which they monitored the I_{sc} and V_{oc} as a function of time over 110 days for cells incorporating a ruthenium dye embedded nanostructured TiO_2 WE, a carbon CE and tetra-butylammoniumiodide/I_2 electrolyte. Though the cells were stable with respect to the overall efficiency at a light intensity of 800 W m^{-2}, they exhibited a decline in V_{oc} with an an increase in I_{sc}.

Overall, the degradation mechanisms of DSSCs can be summarized based on their respective cell components accordingly [3]:

1. *Dye degradation.* Many organometallic and organic dyes undergo photo-oxidation when bound to the TiO_2 surface [106]. This degradation is mainly triggered by the reactivity of the radical and the oxidized species produced due to light excitation and injection into the TiO_2. They can also desorb from the electrode, often accelerated at higher temperatures. In order to stabilize the sensitizer dye, it is important that the electron injection and the recovery of the oxidized form by the redox couple are fast enough to suppress side reactions. In some cases, additives that are not photochemically active have been incorporated to stabilize the sensitizers.
2. *Electron collection.* In this respect, the properties of the nanostructured TiO_2 are crucial. The electrode can hinder electron collection primarily in three ways: (1) shifting of the Fermi level relative to the dye and the redox electrolyte; (2) leakage of charge from the CB by electron trapping associated with surface defect states; and (3) loss of electrical contact between neighbouring particles or with the FTO substrate. In the case of the Fermi level, a lowering in respect of the redox couple will decrease V_{oc}, however, an increase may result in less efficient injection of electrons from the excited state of the dye, resulting in a decreased I_{sc}. Additionally, surface states involving Ti^{4+} ions at the surface of the TiO_2 particles has been a concern, since electron transfer from the CB of the TiO_2 to the electrolyte takes place mainly through these states; this could impact V_{oc} and I_{sc}.
3. *Redox electrolyte.* Overall, for liquid electrolytes, their chemical change has been a primary contributor to cell degradation. For example, photobleaching of the redox electrolyte consisting of I^-/I_3^- at high light intensities is characterized by decomposition of I_2 which decreases reduction of the oxidized dye, leading to a degradation of the dye over time. This degradation which also accompanies formation of other degradation products on the TiO_2 electrode surface increases internal resistance which lowers fill factor. Additionally, loss of solvent from ineffective sealing will affect the stability and efficiency of the dissolved redox couple.
4. *Counter electrode.* This can be unstable due to the corrosive nature of the redox mediator, or it can be poisoned by other degradation species. The stability of the platinum has been of focus since it remains the most effective CE. In this respect, several processes have been identified as factors of performance loss. These involve loss of platinum atoms or clusters through the mechanism: (1) diffusion to the WE, then adsorbing onto the FTO substrate resulting in an increase in back-electron transfer, thereby lowering I_{sc}; (2) adsorption onto the TiO_2 surface and changing its photoelectrochemical properties; and (3) loss of the catalytic properties of the CE, decreasing the recovery of the oxidized redox couple, resulting in a lowering of fill factor and I_{sc}. Additionally, degradation products from the electrolyte might

adsorb onto the platinum surface, hindering catalytic activity and lowering fill factor and I_{sc}.

5. *Sealing*. The encapsulation of the cells should prevent the exchange of material between the cell and the environment. Therefore, imperfect sealing can lead to loss of electrolyte and its solvent, and/or introduction of water and oxygen into the system, which can lead to detrimental chemical changes.

Solid-state alternatives to liquid electrolyte-based DSSCs were developed to circumvent the inherent thermal instability of volatile electrolyte solvents and sealing issues in traditional DSSCs. Consequently, ssDCSCs don't exhibit such degradation mechanisms of the electrolyte and are much more stable relative to their liquid counterparts, especially since they do not have sealing issues [10]. Since semiconductor metal oxides generally exhibit excellent photo and thermal stability, degradation pathways in ssDSSCs largely involve the dye molecules and the electrolyte/hole-transporting material. This largely stems from the absorption of UV photons since the bandgap of the widely used TiO_2 is 3.2 eV. Consequently, the sensitizers and electrolyte/HTM components can suffer from destructive, irreversible reactions. Accordingly, typical methods to evaluate long-term stability have involved strategies to eliminate higher energy UV photons with use of cut-off filters to prevent prolonged exposure which can degrade the sensitizer dye and damage to the TiO_2 layer by highly oxidizing holes.

In addition to employing solid HTMs in solid-state devices, other strategies have included gelation of the redox electrolyte couple by embedding in a conducting polymer matrix to improve efficiency and/or long-term stability. Another approach has involved including organic/polymer additives which do not participate in the primary photoelectrochemical processes [106]. Additives, can offer improvement in several ways such as with the redox couple potential, the semiconductor surface state, the semiconductor CB edge, recombination kinetics, and photovoltaic parameters [3]. In terms of stability, they can facilitate the removal or minimization of unwanted chemical and physical interactions with other components of the solar cell.

6.8.2 Failure diagnostics

Overall, the primary approach involved in predicting and diagnosing cell degradation and failure especially in the early stages of operation must consider fundamental and intrinsic parameters of kinetics of electron transfer processes or charge transfer resistances [106]. This can be done with techniques such as impedance spectroscopy which analyses photocurrent loss with time, as well as photo-induced absorption spectroscopy which evaluates back-electron transfer rate. Importantly, such methods should include various interactions under normal solar light conditions to anticipate failure.

In order to reliably test the long-term stability of DSSCs, accelerated testing is desirable. These simulated long-term working conditions of cells usually involve monitoring various photoconversion characteristics along with thermal, mechanical and UV stability for degradation of up to several thousand hours of continuous light exposure. However, accelerated testing highlights several challenges because of the need to establish equilibrium of kinetic and thermodynamic parameters [106], and therefore careful extrapolation of device performance is required. In particular, the I_{sc} saturates at high light intensities depending on the kinetics and the diffusion within the cell, and extended illumination leads to insufficient dye recovery resulting in long-lived unstable oxidized dyes. Overall, these processes are not expected to occur under normal sun illumination and therefore many years of operation cannot be compared with testing for a short period of time at artificially high light intensities. As it relates to DSSC module stability, several challenges exist. For example, in systems with serial connections, undesired mass transport of ions between adjacent cells are factors that limit long-term stability and performance [3]. Additionally, serial-connected cells face the possibility of reverse bias degradation effects, i.e. one or several cells in a module that are electrically mismatched, from e.g. partial shade, are exposed to high currents which reduces module performance and device degradation.

6.9 Conclusion, challenges and future prospects

DSSCs have emerged as a less expensive and promising alternatives to the traditional silicon-based solar cells for harnessing and converting solar energy to electrical energy. Since the development of the Grätzel cell in 1972, DSSCs have evolved into various other architectures including, solid-state and tandem devices with tremendous progress in terms of conversion efficiency and long-term durability. This has been facilitated by developments in various device components such as a range of effective photosensitizer absorber molecules, photoanode materials and morphology, redox electrolyte and hole transport materials, CEs and substrates, and overall improvements in the charge separation, transport and interfacial dynamics that can improve cell performance as well as stability. Notwithstanding these improvements, the cell performance of DSSCs still lags behind that of traditional cells and some emerging cells such as the perovskite devices. Therefore, to make DSSC technology more competitive, it is crucial to improve the device efficiency and stability along with reducing materials and manufacturing costs.

Developing efficient and novel sensitizers with broader spectral absorption especially in the NIR region into multiple electrons, constructing well-defined structures of semiconductors with suitable morphology for improved dye-loading capacity, enhancing charge transfer within the photoanode, and designing suitable CE with improved catalytic properties for accelerating the redox reactions in the electrolyte are necessary to achieve more highly efficient DSSCs. To achieve this, quantitative methods should be developed to identify, rationally design and model these components and their assembly, with an essential understanding of the mechanisms to maximize light harvesting and minimize the electron losses. In particular, detailed understanding of charge recombination is currently the major cause of efficiency loss in DSSCs. For example, when one of the components (dye, redox shuttle, or semiconductor) is modified, many processes are impacted, thereby influencing the overall performance. Therefore, quantitative efforts with optimized and new theoretical/computational tools harnessing increased computer power coupled with artificial intelligence and machine learning will be essential to model these systems. *In silico* design and optimization of materials will need to shift from single components to ensembles such as dye/electrode or, ideally, electrode/dye/electrolyte.

Of the various types of DSSC devices, tandem and solid-state DSSCs represent two attractive device architectures for application due to their high efficiency and excellent stability, and form the basis for a new generation of meso-superstructured solar cells with the massive developments in perovskite solar cells. Overall, attributes that make DSSCs attractive for the future are: (1) they are easy to fabricate; (2) they are manufactured from low-cost materials; (3) they are environmentally friendly; (4) they have relatively comparable conversion efficiencies; and (5) they perform well in diffused light and at high temperatures. At the basis of this however, is the ability for facile and cost-effective solution-processed fabrication of these devices. Overall, it is anticipated that the development work in DSSCs can lead to devices with PCEs of up to 20% under normal sunlight and 45% for ambient light. Furthermore, the high sensitivity and efficiency of DSSCs in low and ambient light conditions is a major benefit since they can be used where diffused solar light prevails over direct solar illumination. Accordingly, the essential use of DSSCs in building windows is particularly appealing and practical and their commercial viability is promising.

References

[1] O'Regan B and Grätzel M 1991 A low-cost, high-efficiency solar cell based on dye-sensitized colloidal TiO_2 films *Nature* **353** 737–40

[2] Carella A, Borbone F and Centore R 2018 Research progress on photosensitizers for DSSC *Front. Chem.* **6** 481

[3] Muñoz-García A B *et al* 2021 Dye-sensitized solar cells strike back *Chem. Soc. Rev.* **50** 12450–550

[4] Sharma K, Sharma V and Sharma S S 2018 Dye-sensitized solar cells: fundamentals and current status *Nanoscale Res. Lett.* **13** 381

[5] Wang P, Zakeeruddin S M, Comte P, Charvet R, Humphry-Baker R and Grätzel M 2003 Enhance the performance of dye-sensitized solar cells by co-grafting amphiphilic sensitizer and hexadecylmalonic acid on TiO_2 nanocrystals *J. Phys. Chem.* B **107** 14336–41

[6] Wang P, Klein C, Moser J-E, Humphry-Baker R, Cevey-Ha N-L, Charvet R, Comte P, Zakeeruddin S M and Grätzel M 2004 Amphiphilic ruthenium sensitizer with 4,4′-diphosphonic Acid-2,2′-bipyridine as anchoring ligand for nanocrystalline dye sensitized solar cells *J. Phys. Chem.* B **108** 17553–9

[7] Chen W-C, Kong F-T, Li Z-Q, Pan J-H, Liu X-P, Guo F-L, Zhou L, Huang Y, Yu T and Dai S-Y 2016 Superior light-harvesting heteroleptic ruthenium(II) complexes with electron-donating antennas for high performance dye-sensitized solar cells *ACS Appl. Mater. Interfaces* **8** 19410–7

[8] Tachibana Y, Haque S A, Mercer I P, Durrant J R and Klug D R 2000 Electron injection and recombination in dye sensitized nanocrystalline titanium dioxide films: a comparison of ruthenium bipyridyl and porphyrin sensitizer dyes *J. Phys. Chem.* B **104** 1198–205

[9] Xie Y, Tang Y, Wu W, Wang Y, Liu J, Li X, Tian H and Zhu W-H 2015 Porphyrin cosensitization for a photovoltaic efficiency of 11.5%: a record for non-ruthenium solar cells based on iodine electrolyte *J. Am. Chem. Soc.* **137** 14055–8

[10] Benesperi I, Michaels H and Freitag M 2018 The researcher's guide to solid-state dye-sensitized solar cells *J. Mater. Chem.* C **6** 11903–42

[11] Pan Z, Rao H, Mora-Seró I, Bisquert J and Zhong X 2018 Quantum dot-sensitized solar cells *Chem. Soc. Rev.* **47** 7659–702

[12] Taylor R A and Ramasamy K 2017 Colloidal quantum dots solar cells *Nanoscience* **vol 4** ed P J Thomas and N Revaprasadu (London: The Royal Society of Chemistry) p 0

[13] Ghosh Chaudhuri R and Paria S 2012 Core/shell nanoparticles: classes, properties, synthesis mechanisms, characterization, and applications *Chem. Rev.* **112** 2373–433

[14] Yang J, Wang J, Zhao K, Izuishi T, Li Y, Shen Q and Zhong X 2015 CdSeTe/CdS Type-I core/shell quantum dot sensitized solar cells with efficiency over 9 *J. Phys. Chem.* C **119** 28800–8

[15] Jiao S, Shen Q, Mora-Seró I, Wang J, Pan Z, Zhao K, Kuga Y, Zhong X and Bisquert J 2015 Band engineering in core/shell ZnTe/CdSe for photovoltage and efficiency enhancement in exciplex quantum dot sensitized solar cells *ACS Nano* **9** 908–15

[16] Yue L, Rao H, Du J, Pan Z, Yu J and Zhong X 2018 Comparative advantages of Zn–Cu–In–S alloy QDs in the construction of quantum dot-sensitized solar cells *RSC Adv.* **8** 3637–45

[17] Du J, Du Z, Hu J-S, Pan Z, Shen Q, Sun J, Long D, Dong H, Sun L, Zhong X *et al* 2016 Zn–Cu–In–Se quantum dot solar cells with a certified power conversion efficiency of 11.6% *J. Am. Chem. Soc.* **138** 4201–9

[18] Zhang L, Pan Z, Wang W, Du J, Ren Z, Shen Q and Zhong X 2017 Copper deficient Zn–Cu–In–Se quantum dot sensitized solar cells for high efficiency *J. Mater. Chem.* A **5** 21442–51

[19] Ming S-K, Taylor R A, McNaughter P D, Lewis D J, Leontiadou M A and O'Brien P 2021 Tunable structural and optical properties of $CuInS_2$ colloidal quantum dots as photovoltaic absorbers *RSC Adv.* **11** 21351–8

[20] Wang J, Li Y, Shen Q, Izuishi T, Pan Z, Zhao K and Zhong X 2016 Mn doped quantum dot sensitized solar cells with power conversion efficiency exceeding 9 *J. Mater. Chem.* A **4** 877–86

[21] Alex P, Jacob S-B, Erika F and Malkiat J 2012 Investigating new materials and architectures for Grätzel cells *Third Generation Photovoltaics* ed F Vasilis (London: IntechOpen) ch 5

[22] Haggerty J E S *et al* 2017 High-fraction brookite films from amorphous precursors *Sci. Rep.* **7** 15232

[23] Adachi M, Murata Y, Okada I and Yoshikawa S 2003 Formation of titania nanotubes and applications for dye-sensitized solar cells *J. Electrochem. Soc.* **150** G488

[24] Mor G K, Shankar K, Paulose M, Varghese O K and Grimes C A 2006 Use of highly-ordered TiO_2 nanotube arrays in dye-sensitized solar cells *Nano Lett.* **6** 215–8

[25] Mali S S, Kim H, Shim C S, Patil P S, Kim J H and Hong C K 2013 Surfactant free most probable TiO_2 nanostructures via hydrothermal and its dye sensitized solar cell properties *Sci. Rep.* **3** 3004

[26] Pauporté T 2018 Synthesis of ZnO nanostructures for solar cells—a focus on dye-sensitized and perovskite solar cells *The Future of Semiconductor Oxides in Next-Generation Solar Cells* ed M Lira-Cantu (Amsterdam: Elsevier) ch 1 pp 3–43

[27] Ko S H, Lee D, Kang H W, Nam K H, Yeo J Y, Hong S J, Grigoropoulos C P and Sung H J 2011 Nanoforest of hydrothermally grown hierarchical ZnO nanowires for a high efficiency dye-sensitized solar cell *Nano Lett.* **11** 666–71

[28] Wali Q, Fakharuddin A and Jose R 2015 Tin oxide as a photoanode for dye-sensitised solar cells: current progress and future challenges *J. Power Sources* **293** 1039–52

[29] Sengupta D, Das P, Mondal B and Mukherjee K 2016 Effects of doping, morphology and film-thickness of photo-anode materials for dye sensitized solar cell application—a review *Renew. Sustain. Energy Rev.* **60** 356–76

[30] Guo X, Lu G and Chen J 2015 Graphene-based materials for photoanodes in dye-sensitized solar cells *Front. Energy Res.* **3** 50

[31] Makal P and Das D 2021 Reduced graphene oxide-laminated one-dimensional TiO_2–bronze nanowire composite: an efficient photoanode material for dye-sensitized solar cells *ACS Omega* **6** 4362–73

[32] Tong Z, Peng T, Sun W, Liu W, Guo S and Zhao X-Z 2014 Introducing an intermediate band into dye-sensitized solar cells by W^{6+} doping into TiO_2 nanocrystalline photoanodes *J. Phys. Chem.* C **118** 16892–5

[33] Latini A, Cavallo C, Aldibaja F K, Gozzi D, Carta D, Corrias A, Lazzarini L and Salviati G 2013 Efficiency improvement of DSSC photoanode by scandium doping of mesoporous titania beads *J. Phys. Chem.* C **117** 25276–89

[34] Kim C, Kim K-S, Kim H Y and Han Y S 2008 Modification of a TiO_2 photoanode by using Cr-doped TiO_2 with an influence on the photovoltaic efficiency of a dye-sensitized solar cell *J. Mater. Chem.* **18** 5809–14

[35] Yan L-T, Wu F-L, Peng L, Zhang L-J, Li P-J, Dou S-Y and Li T-X 2012 Photoanode of dye-sensitized solar cells based on a ZnO/TiO_2 composite film *Int. J. Photoenergy* **2012** 613969

[36] Song W, Gong Y, Tian J, Cao G, Zhao H and Sun C 2016 Novel photoanode for dye-sensitized solar cells with enhanced light-harvesting and electron-collection efficiency *ACS Appl. Mater. Interfaces* **8** 13418–25

[37] Yu H, Bai Y, Zong X, Tang F, Lu G Q M and Wang L 2012 Cubic CeO_2 nanoparticles as mirror-like scattering layers for efficient light harvesting in dye-sensitized solar cells *Chem. Commun.* **48** 7386–8

[38] Yang N, Zhai J, Wang D, Chen Y and Jiang L 2010 Two-dimensional graphene bridges enhanced photoinduced charge transport in dye-sensitized solar cells *ACS Nano* **4** 887–94

[39] Anish Madhavan A, Kalluri S, K Chacko D, Arun T A, Nagarajan S, Subramanian K R V, Sreekumaran Nair A, Nair S V and Balakrishnan A 2012 Electrical and optical properties of electrospun TiO_2–graphene composite nanofibers and its application as DSSC photo-anodes *RSC Adv.* **2** 13032–7

[40] Singh A, Saini Y K, Kumar A, Gautam S, Kumar D, Dutta V, Lee H-k, Lee J and Swami S K 2022 Property modulation of graphene oxide incorporated with TiO_2 for dye-sensitized solar cells *ACS Omega* **7** 44170–9

[41] Ye M, Wen X, Wang M, Iocozzia J, Zhang N, Lin C and Lin Z 2015 Recent advances in dye-sensitized solar cells: from photoanodes, sensitizers and electrolytes to counter electrodes *Mater. Today* **18** 155–62

[42] Yella A, Lee H-W, Tsao H N, Yi C, Chandiran A K, Nazeeruddin M K, Diau E W-G, Yeh C-Y, Zakeeruddin S M and Grätzel M 2011 Porphyrin-sensitized solar cells with cobalt (II/III)-based redox electrolyte exceed 12 percent efficiency *Science* **334** 629–34

[43] Tsai C-H, Lu C-Y, Chen M-C, Huang T-W, Wu C-C and Chung Y-W 2013 Efficient gel-state dye-sensitized solar cells adopting polymer gel electrolyte based on poly(methyl methacrylate) *Org. Electron.* **14** 3131–7
[44] Burschka J, Pellet N, Moon S-J, Humphry-Baker R, Gao P, Nazeeruddin M K and Grätzel M 2013 Sequential deposition as a route to high-performance perovskite-sensitized solar cells *Nature* **499** 316–9
[45] Hodes G and Cahen D 2012 All-solid-state, semiconductor-sensitized nanoporous solar cells *Acc. Chem. Res.* **45** 705–13
[46] Chung I, Lee B, He J, Chang R P H and Kanatzidis M G 2012 All-solid-state dye-sensitized solar cells with high efficiency *Nature* **485** 486–9
[47] Xu B, Sheibani E, Liu P, Zhang J, Tian H, Vlachopoulos N, Boschloo G, Kloo L, Hagfeldt A and Sun L 2014 Carbazole-based hole-transport materials for efficient solid-state dye-sensitized solar cells and perovskite solar cells *Adv. Mater.* **26** 6629–34
[48] Hao Y, Yang W, Zhang L, Jiang R, Mijangos E, Saygili Y, Hammarström L, Hagfeldt A and Boschloo G 2016 A small electron donor in cobalt complex electrolyte significantly improves efficiency in dye-sensitized solar cells *Nat. Commun.* **7** 13934
[49] Kim B, Koh J K, Kim J, Chi W S, Kim J H and Kim E 2012 Room temperature solid-state synthesis of a conductive polymer for applications in stable I2-free dye-sensitized solar cells *Chem. Sus. Chem.* **5** 2173–80
[50] Kong M, Kim K S, Nga N V, Lee Y, Jeon Y S, Cho Y, Kwon Y and Han Y S 2020 Molecular weight effects of biscarbazole-based hole transport polymers on the performance of solid-state dye-sensitized solar cells *Nanomaterials* **10** 2516
[51] Wu J, Lan Z, Lin J, Huang M, Huang Y, Fan L, Luo G, Lin Y, Xie Y and Wei Y 2017 Counter electrodes in dye-sensitized solar cells *Chem. Soc. Rev.* **46** 5975–6023
[52] Papageorgiou N, Maier W F and Grätzel M 1997 An iodine/triiodide reduction electrocatalyst for aqueous and organic media *J. Electrochem. Soc.* **144** 876
[53] Li L-L, Chang C-W, Wu H-H, Shiu J-W, Wu P-T and Wei-Guang Diau E 2012 Morphological control of platinum nanostructures for highly efficient dye-sensitized solar cells *J. Mater. Chem.* **22** 6267–73
[54] Kay A and Grätzel M 1996 Low cost photovoltaic modules based on dye sensitized nanocrystalline titanium dioxide and carbon powder *Sol. Energy Mater. Sol. Cells* **44** 99–117
[55] Mei X, Cho S J, Fan B and Ouyang J 2010 High-performance dye-sensitized solar cells with gel-coated binder-free carbon nanotube films as counter electrode *Nanotechnology* **21** 395202
[56] Wu M, Lin X, Wang T, Qiu J and Ma T 2011 Low-cost dye-sensitized solar cell based on nine kinds of carbon counter electrodes *Energy Environ. Sci.* **4** 2308–15
[57] Hasan A M M and Susan M A B H 2024 Synergism in carbon nanotubes and carbon-dots: counter electrode of a high-performance dye-sensitized solar cell *RSC Adv.* **14** 7616–30
[58] Shinde D V, Patil S A, Cho K, Ahn D Y, Shrestha N K, Mane R S, Lee J K and Han S-H 2015 Revisiting metal sulfide semiconductors: a solution-based general protocol for thin film formation, hall effect measurement, and application prospects *Adv. Funct. Mater.* **25** 5739–47
[59] Huo J, Wu J, Zheng M, Tu Y and Lan Z 2015 Effect of ammonia on electrodeposition of cobalt sulfide and nickel sulfide counter electrodes for dye-sensitized solar cells *Electrochim. Acta* **180** 574–80

[60] Al-Mamun M, Zhang H, Liu P, Wang Y, Cao J and Zhao H 2014 Directly hydrothermal growth of ultrathin MoS_2 nanostructured films as high performance counter electrodes for dye-sensitised solar cells *RSC Adv.* **4** 21277–83

[61] Singh E, Kim K S, Yeom G Y and Nalwa H S 2017 Two-dimensional transition metal dichalcogenide-based counter electrodes for dye-sensitized solar cells *RSC Adv.* **7** 28234–90

[62] Li S, Min H, Xu F, Tong L, Chen J, Zhu C and Sun L 2016 All electrochemical fabrication of MoS_2/graphene counter electrodes for efficient dye-sensitized solar cells *RSC Adv.* **6** 34546–52

[63] Xu T, Kong D, Tang H, Qin X, Li X, Gurung A, Kou K, Chen L, Qiao Q and Huang W 2020 Transparent MoS_2/PEDOT composite counter electrodes for bifacial dye-sensitized solar cells *ACS Omega* **5** 8687–96

[64] Yoo K *et al* 2015 Completely transparent conducting oxide-free and flexible dye-sensitized solar cells fabricated on plastic substrates *ACS Nano* **9** 3760–71

[65] Wu J, Lan Z, Lin J, Huang M, Huang Y, Fan L and Luo G 2015 Electrolytes in dye-sensitized solar cells *Chem. Rev.* **115** 2136–73

[66] Masud and Kim H K 2023 Redox shuttle-based electrolytes for dye-sensitized solar cells: comprehensive guidance, recent progress, and future perspective *ACS Omega.* **8** 6139–63

[67] NREL 2024 Best Research-Cell Efficiency Chart. National Renewable Energy Laboratory: https://nrel.gov/pv/interactive-cell-efficiency.html

[68] Labat F, Le Bahers T, Ciofini I and Adamo C 2012 First-principles modeling of dye-sensitized solar cells: challenges and perspectives *Acc. Chem. Res.* **45** 1268–77

[69] Pastore M and De Angelis F 2014 Modeling materials and processes in dye-sensitized solar cells: understanding the mechanism, improving the efficiency *Multiscale Modelling of Organic and Hybrid Photovoltaics* ed D Beljonne and J Cornil (Berlin: Springer) pp 151–236

[70] Jinbao Zhang M F, Hagfeldt A and S G B 2018 Olid-state dye-sensitized solar cells *Molecular Devices for Solar Energy Conversion and Storage* ed G B Haining Tian and H Anders (Singapore: Spinger)

[71] Li M-H, Yum J-H, Moon S-J and Chen P 2016 Inorganic p-type semiconductors: their applications and progress in dye-sensitized solar cells and perovskite solar cells *Energies* **9** 331

[72] He J, Lindström H, Hagfeldt A and Lindquist S-E 1999 Dye-sensitized nanostructured p-type nickel oxide film as a photocathode for a solar cell *J. Phys. Chem.* B **103** 8940–3

[73] Zhang L, Boschloo G, Hammarström L and Tian H 2016 Solid state p-type dye-sensitized solar cells: concept, experiment and mechanism *Phys. Chem. Chem. Phys.* **18** 5080–5

[74] Tian H 2019 Solid-state p-type dye-sensitized solar cells: progress, potential applications and challenges *Sustain. Energy Fuels* **3** 888–98

[75] Pham T T T *et al* 2017 Toward efficient solid-state p-type dye-sensitized solar cells: the dye matters *J. Phys. Chem.* C **121** 129–39

[76] Shilpa G, Kumar P M, Kumar D K, Deepthi P R, Sadhu V, Sukhdev A and Kakarla R R 2023 Recent advances in the development of high efficiency quantum dot sensitized solar cells (QDSSCs): a review *Mater. Sci. Energy Technol.* **6** 533–46

[77] Rühle S, Shalom M and Zaban A 2010 Quantum-dot-sensitized solar cells *Chem. Phys. Chem.* **11** 2290–304

[78] Tian J and Cao G 2013 Semiconductor quantum dot-sensitized solar cells *Nano Rev.* **4** 22578

[79] Chung N T K, Nguyen P T, Tung H T and Phuc D H 2021 Quantum dot sensitized solar cell: photoanodes, counter electrodes, and electrolytes *Molecules* **26** 2638
[80] Yang J and Zhong X 2016 CdTe based quantum dot sensitized solar cells with efficiency exceeding 7% fabricated from quantum dots prepared in aqueous media *J. Mater. Chem.* A **4** 16553–61
[81] Yu J, Wang W, Pan Z, Du J, Ren Z, Xue W and Zhong X 2017 Quantum dot sensitized solar cells with efficiency over 12% based on tetraethyl orthosilicate additive in polysulfide electrolyte *J. Mater. Chem.* A **5** 14124–33
[82] Wang S , Zhang Q-X, Xu Y-Z, Li D-M, Luo Y-H and Meng Q-B 2013 Single-step *in situ* preparation of thin film electrolyte for quasi-solid state quantum dot-sensitized solar cells *J. Power Sources* **224** 152–7
[83] Diguna L J, Shen Q, Kobayashi J and Toyoda T 2007 High efficiency of CdSe quantum-dot-sensitized TiO_2 inverse opal solar cells *Appl. Phys. Lett.* **91** 023116
[84] Lee Y-L and Lo Y-S 2009 Highly efficient quantum-dot-sensitized solar cell based on Co-sensitization of CdS/CdSe *Adv. Funct. Mater.* **19** 604–9
[85] Zhang Z, Wang W, Rao H, Pan Z and Zhong X 2024 Improving the efficiency of quantum dot-sensitized solar cells by increasing the QD loading amount *Chem. Sci.* **15** 5482–95
[86] Park N-G 2013 Organometal perovskite light absorbers toward a 20% efficiency low-cost solid-state mesoscopic solar cell *J. Phys. Chem. Lett.* **4** 2423–9
[87] Park N-G 2015 Perovskite solar cells: an emerging photovoltaic technology *Mater. Today* **18** 65–72
[88] Liu S, Biju V P, Qi Y, Chen W and Liu Z 2023 Recent progress in the development of high-efficiency inverted perovskite solar cells *NPG Asia Mater.* **15** 27
[89] Wei J, Shao Z, Pan B, Chen S, Hu L and Dai S 2020 Toward current matching in tandem dye-sensitized solar cells *Materials* **13** 2936
[90] Xiong D and Chen W 2012 Recent progress on tandem structured dye-sensitized solar cells *Front. Optoelectron.* **5** 371–89
[91] Kubo W, Sakamoto A, Kitamura T, Wada Y and Yanagida S 2004 Dye-sensitized solar cells: improvement of spectral response by tandem structure *J. Photochem. Photobiol.,* A **164** 33–9
[92] Dürr M, Bamedi A, Yasuda A and Nelles G 2004 Tandem dye-sensitized solar cell for improved power conversion efficiencies *Appl. Phys. Lett.* **84** 3397–9
[93] Yanagida M, Onozawa-Komatsuzaki N, Kurashige M, Sayama K and Sugihara H 2010 Optimization of tandem-structured dye-sensitized solar cell *Sol. Energy Mater. Sol. Cells* **94** 297–302
[94] Lee K, Park S W, Ko M J, Kim K and Park N-G 2009 Selective positioning of organic dyes in a mesoporous inorganic oxide film *Nat. Mater.* **8** 665–71
[95] Miao Q, Wu L, Cui J, Huang M and Ma T 2011 A new type of dye-sensitized solar cell with a multilayered photoanode prepared by a film-transfer technique *Adv. Mater.* **23** 2764–8
[96] Murayama M and Mori T 2007 Dye-sensitized solar cell using novel tandem cell structure *J. Phys.* D **40** 1664
[97] Murayama M and Mori T 2008 Novel tandem cell structure of dye-sensitized solar cell for improvement in photocurrent *Thin Solid Films* **516** 2716–22
[98] He J, Lindström H, Hagfeldt A and Lindquist S-E 2000 Dye-sensitized nanostructured tandem cell-first demonstrated cell with a dye-sensitized photocathode *Sol. Energy Mater. Sol. Cells* **62** 265–73

[99] Gibson E A, Smeigh A L, Le Pleux L, Fortage J, Boschloo G, Blart E, Pellegrin Y, Odobel F, Hagfeldt A and Hammarström L 2009 A p-type NiO-based dye-sensitized solar cell with an open-circuit voltage of 0.35 V *Angew. Chem. Int. Ed.* **48** 4402–5

[100] Nattestad A, Mozer A J, Fischer M K, Cheng Y B, Mishra A, Bäuerle P and Bach U 2010 Highly efficient photocathodes for dye-sensitized tandem solar cells *Nat. Mater.* **9** 31–5

[101] Bruder I, Karlsson M, Eickemeyer F, Hwang J, Erk P, Hagfeldt A, Weis J and Pschirer N 2009 Efficient organic tandem cell combining a solid state dye-sensitized and a vacuum deposited bulk heterojunction solar cell *Sol. Energy Mater. Sol. Cells* **93** 1896–9

[102] Chae S Y, Park S J, Joo O-S, Jun Y, Min B K and Hwang Y J 2016 Highly stable tandem solar cell monolithically integrating dye-sensitized and CIGS solar cells *Sci. Rep.* **6** 30868

[103] Moon S H, Park S J, Kim S H, Lee M W, Han J, Kim J Y, Kim H, Hwang Y J, Lee D-K and Min B K 2015 Monolithic DSSC/CIGS tandem solar cell fabricated by a solution process *Sci. Rep.* **5** 8970

[104] Yuan L, Michaels H, Roy R, Johansson M, Öberg V, Andruszkiewicz A, Zhang X, Freitag M and Johansson E M J 2020 Four-terminal tandem solar cell with dye-sensitized and PbS colloidal quantum-dot-based subcells *ACS Appl. Energy Mater.* **3** 3157–61

[105] Hosseinnezhad M 2019 Enhanced performance of dye-sensitized solar cells using perovskite/DSSCs tandem design *J. Electron. Mater.* **48** 5403–8

[106] Figgemeier E and Hagfeldt A 2004 Are dye-sensitized nano-structured solar cells stable? An overview of device testing and component analyses *Int. J. Photoenergy* **6** 121739

Chapter 7

Perovskite-based solar cells

Perovskite solar cells (PSCs) are a leading third-generation solar cell. They have shown remarkable growth in power conversion efficiency (PCE) of over 25% in the short period from their introduction in 2009. High-efficiency, tunable bandgap from 1.3 to 2.2 eV, high absorption cross-section ($>10^4$ cm^{-1}) and low-cost manufacturing process are attractive factors for investigating PSCs for commercial applications. Thanks to perovskites' good compatibility with first- and second-generation solar cell materials, it is possible to achieve even higher PCEs by creating tandem solar cells with them. Another interesting aspect of perovskite solar cells is their aesthetically pleasing appearance due to their tunable coloration, which is a desire for accepting building integrated photovoltaics (BIPV). Nevertheless, PSCs are not free of challenges. The champion performance is achieved from a device that uses lead (Pb). Lead is a toxic element that is listed in the Restriction of Hazardous Substances (RoHS) regulation. Further, PSCs are found to be unstable in the presence of moisture and oxygen. There have been tremendous efforts dedicated towards addressing these challenges. For instance, perovskite devices partially or fully replacing lead with germanium and/or tin have been constructed to reduce lead toxicity and moisture barrier encapsulation as approaches to protect devices from moisture, have been investigated. The chapter is focussed towards providing a comprehensive analysis of perovskite material structure, different solar cells device architectures, various fabrication processes, defects and degradation mechanisms.

7.1 Structure and properties of perovskites

The prototypical perovskite for solar cells, methylammonium lead iodide ($MAPbI_3$) crystallizes in cubic crystal structure with a general formula of ABX_3 with a $Pm\bar{3}m$ space group. In the general formula, A is a monovalent cation, sits in the cuboctahedral positions in a cubic site. B, a bivalent cation, occupies the octahedral position. X is an anion, in this case halides [1]. The crystal structure of $MAPbI_3$ is given in figure 7.1. Theoretically, any set of compatible ions can form the perovskite

doi:10.1088/978-0-7503-3255-2ch7

© IOP Publishing Ltd 2025. All rights, including for text and data mining (TDM), artificial intelligence (AI) training, and similar technologies, are reserved.

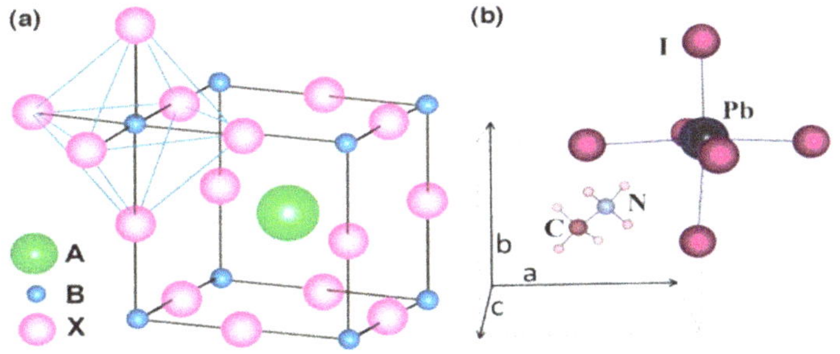

Figure 7.1. (a) Crystal structure of perovskite with ABX_3 construction (b) unit cell of cubic $CH_3NH_3PbI_3$ perovskite. Adapted with permission from [1]. Copyright 2015 Elsevier Ltd. CC BY-NC-ND 4.0.

structure if they satisfy the tolerance (t) and octahedral factor (μ). The tolerance factor can be calculated by taking the ratio of A–X and B–X lengths in a perfect solid-sphere model using the following formula.

$$t = (R_A + R_X)/\sqrt{2}(R_B + R_X)$$

where, R_A, R_B, R_x are ionic radii of the A, B and X ions.

Likewise, the octahedral factor (μ) can be derived from the ratio of the ionic radius of the divalent cation (R_B) and the radius of the anion (R_X). Lower tolerance factor ($t < 1$) yields tetragonal or orthorhombic structures, whereas higher tolerance factor ($t > 1$) stabilizes hexagonal structures with face-sharing octahedral.

At temperatures below 315–330 K, $MAPbI_3$ undergoes phase transition from cubic to tetragonal and at temperatures even below ~160 K, transforms into orthorhombic phase. The cubic-to-tetragonal transition temperature is decreased from ~330 to 177 K when the iodide is replaced with chloride in the perovskite structure. Similarly, when the A cation is replaced with a smaller formamidinium, the transition temperature decreases to 238 K, whereas with caesium (Cs), the temperature increases to ~403 K. It has been identified that the reorientation of the A cation is the reason for the phase transition.

7.1.1 Defect structure

Defects in the solar cell structures play a crucial role in their performance. Generally, deep-level defects provide non-radiative recombination channels for photogenerated carriers, which affect open-circuit voltage (V_{oc}) and short-circuit current (J_{sc}). Perovskites for solar cells are typically produced by solution state methods. The solution state methods often yield perovskites with point defects, surface defects, grain boundaries (GBs) and interfaces. Understanding the defects and controlling their formation is important for improving solar cells' performance. Many different calculations have been performed towards identifying the defects in $MAPbI_3$ [2–4]. Three different vacancy defects (V_{MA}, V_{Pb}, V_I), three different interstitial defects (Ma_i, Pb_i, I_i), and four different antisites substitution defects (MA_I, Pb_I, I_{MA} and I_{Pb}) are possible in $MAPbI_3$. All three vacancy defects and MA_i and I_i interstitial and MA_{Pb} substitutional defects show shallow transition energy

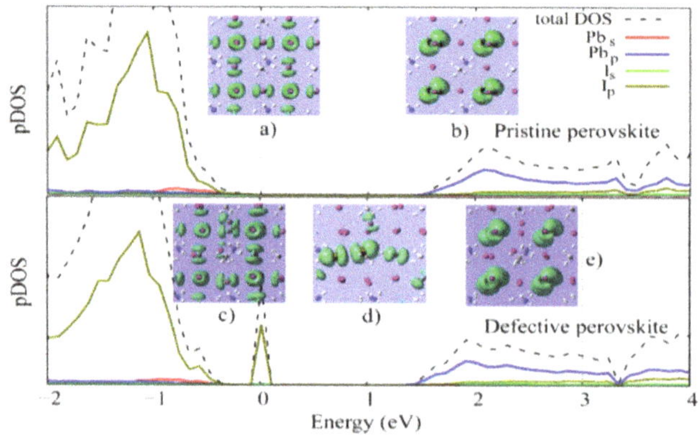

Figure 7.2. Total and projected DOS of pristine $CH_3NH_3PbI_3$ perovskite (top panel) and with the iodine interstitial defect (bottom panel). Charge densities of pristine $CH_3NH_3PbI_3$ (a) VBM and (b) conduction band minimum and $CH_3NH_3PbI_3$ with iodine interstitial defect (c) VBM and (d) conduction band minimum and (e) trap state. Adapted with permission from [3]. Copyright 2017 American Chemical Society.

levels. It has been estimated that the shallow level defects of both donors and acceptors have comparably low formation energies resulting in bipolar conductivities [5, 6]. Covalent coupling of Pb lone pair *s* orbital and I *p* orbitals results in an acceptor state, whereas shallow donors originate from the high ionicity. I_{MA}, I_{Pb}, Pb_i, MA_i and Pb_i defects are generated in deep levels with the most stable being MA_i^+ and halide interstitial. MA_i is not involved in the Pb–I framework, therefore it is stable only in the positive charged states. I_i^+ has a low formation energy with relatively high stability and density. Density of states (DOS) of $MAPbI_3$ with and without interstitial I defect are shown in figure 7.2. The figure shows the presence of an interstitial defect near valence band maximum (VBM). GBs are one of the common defects that affect electrical, mechanical and chemical properties of solar cell materials. Dangling bonds at the GB sites act as recombination centers for charge carriers. There have been a few studies on GBs in PSCs [7]. First principle calculations predict that the GBs in $MAPbI_3$ are benign since they do not create deep trap states making polycrystalline perovskite material behave like a single-crystalline material [8]. Another study showed that the PbI_2-terminated surface does not introduce sub-bandgap states [9]. While the full effect of GBs in PSCs is still being evaluated, some studies have shown that they could dissociate charge carriers, thereby increasing photocurrent density resulting in improved solar cell performance. Meanwhile, GB passivation investigations to prevent or suppress some of the negative effects are being evaluated.

7.2 Solution preparation of perovskites

The performance of solar cell devices is dictated by the microstructure of the perovskite film. The film quality, crystallinity, morphology and microstructures are manipulated by the perovskite preparation methods. Perovskites are generally prepared by low-temperature solution state methods. In the solution state synthesis, the perovskite films crystallize in a very short time, limiting control over the film formation. The kinetics of the film formation is controlled by the solvent, precursors interactions and by the protocol.

7.2.1 One-step and step-by-step processes

The growth of perovskite films is generally defined by either one single-step process where all constituent precursors are in single solution or two-step process where the perovskite precursors are coated step-by-step and followed by a thermal annealing.

7.2.1.1 One-step process

There have been a number of studies to deposit uniform, well-crystallized, and void-free perovskite films to obtain high-efficiency solar cells. One of the earliest ones to deposit perovskite was following a single-step process. In the process, precursors such as lead halide and methylammonium halide were dissolved in solvents like dimethyl sulfoxide (DMSO) or dimethylformamide (DMF) [10–12]. The precursor solution was directly spin-coated onto the substrate. During the spin-coating process, the concentration of the precursor reached supersaturation, inducing nucleation of perovskite film growth. The as-deposited film was then annealed in the temperature range of 100 °C–150 °C to dry off the residual solvent and also to improve the crystallinity. In this one-step process, the growth of perovskite was quick, which led to uncontrolled and irregular grain growth [13]. The lack of uniform coverage over the substrate impacted the efficiency. Slow evaporation of high boiling DMF solvent is regarded as the potential reason for the irregular grain growth. Therefore, there have been many efforts including introducing anti-solvents, heating the substrate, gas-blowing, and applying vacuum to remove the solvents quickly.

7.2.1.1.1 Anti-solvent approach

In the anti-solvent approach, a solvent that is miscible with the solvent that is responsible for dissolving the precursors is required, but it should not dissolve the precursors. This approach is borrowed from the colloidal nanocrystals isolation process. In the initial attempt, chlorobenzene was added to induce the nucleation over the substrate [14]. Films grown by this method were found to consist of large grains up to micron in grain size. A planar heterojunction device constructed with this film showed an average efficiency of 13.9% ± 0.7%. Following this effort, other solvents have also been investigated to improve the efficiency. In one of the studies, γ-butyrolactone and DMSO were used for dissolving the perovskite precursors and then toluene was drop casted as anti-solvent [15]. This led to uniform and dense perovskite layers, which reflected on the PCE of 16.2%. The PCE was further improved to 17.08% by using ethyl ether and n-hexane mixed anti-solvent [16].

Recently, to generalize the anti-solvent type and their effect on perovskite growth, 14 different solvents were investigated [17]. Based on their ability to dissolve organic and inorganic perovskite precursors, they were categorized into three different types. Type I solvents contained alcohols from ethyl to butyl. Type II solvents include ethyl acetate, trichloromethane, chlorobenzene, butyl acetate, dichlorobenzene, anisole and trifluorotoluene. Type III solvents consist of diethylether, xylene, toluene and mesitylene. The duration of solvent mixing with DMF/DMSO had an effect on the perovskite films' uniformity and their power conversion performance. In the fast addition, solvents were dispensed in $\sim$0.18 s, whereas in the slow addition, it took 1.3 s for dispensing. It was observed that the Type I solvents performed better when they were added fast with high V_{oc}, J_{sc}, fill factor (FF) and PCE. Interestingly, slow addition of Type I solvents affected the solar cells' performance negatively. Notably, ethanol yielded <5% efficient devices. Likewise, Type III solvents exhibited the opposite trend. Slow addition yielded good performing devices, but poor performing devices were obtained from the fast addition. Particularly, mesitylene exhibited a large variation. Slow addition of mesitylene resulted in competitively performing devices. A non-functioning device was obtained when mesitylene was added fast. The duration of Type II solvent addition did not affect the performance of the devices. Nevertheless, all tested solvents exhibited average efficiency of 18% PCE and over 21% in comparison with the champion devices. Figure 7.3 shows the summary of various solvent interactions with respect to addition duration.

The diffusivity of ions and interaction between precursors and solvent molecules are different for different compositions and solvents. Thus, a change in perovskite composition yielded variation in solubility, film formation and performance. Mixed perovskites films have a narrow processing window by this approach. Devices with PCE close to 20% were obtained using anisole as an anti-solvent [18]. Thanks to its low evaporation rate and interaction with DMF/DMSO, it provided a wide processing window.

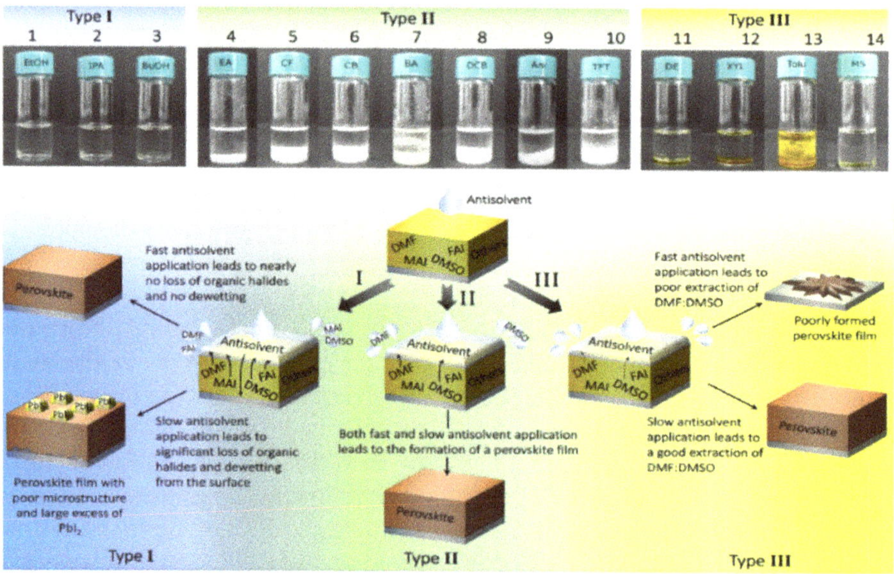

Figure 7.3. Schematic of perovskite film formation using three anti-solvent types. Adapted with permission from [17] CC BY 4.0.

Extending the anti-solvent approach, the use of additives has also been investigated. A solar cell device with over 18% efficiency was reported using the mixture of methylamine and acetonitrile. The mixture yielded defect-free and compact microstructure films. It was hypothesized that the methylamine formed an adduct with methylammonium lead iodide which resulted in direct crystallization of perovskite during the spin-coating process [19]. Another smooth film with an average crystallite size of 1 μm was obtained when the cyclic urea compound, 1,3-diemthyl-2-imidazolidinone (DMI) was used in the $MAPbI_3$ precursor solution [20]. This smooth film exhibited a PCE of 17.6%. The use of 2,2′-bipyridine (Bpy) and 2,2′:6′,2′-terpyridine (Tpy) as additives has been investigated to improve the crystallinity and to increase the grain size of the perovskite films [21]. The perovskite films with these additives showed PCEs of 19.02% for Bpy and 18.68% for Tpy. Perovskite precursors such as lead halides, and organic halides are Lewis acids. On reacting with Lewis bases these compounds undergo redox reaction or adduct formation through a dative bond. The formation of an adduct intermediate structure could facilitate homogeneous crystal growth. To test the hypothesis, ethylene carbonate (EC), propylene carbonate (PC) and poly(propylene carbonate) (PPC) were introduced with the perovskite precursors [22]. The devices prepared using these additives showed 18.65% for EC, 19.09% for PC and 19.35% for PPC, compared to 16.97% from the device that was prepared without the use of additives. PPC is a polymeric compound that stayed along with the perovskites during the annealing step, which was found to be passivating the defects at the GBs. Further, the hydrophobic nature of PPC passivated the perovskites from moisture that resulted in improved moisture stability.

7.2.1.1.2 Gas-blowing approach

One of the approaches to remove solvents from the perovskite layer is to blow a non-reactive gas (nitrogen or argon). Unlike the anti-solvent approach, this method does not involve the use of toxic chemicals such as toluene or chlorobenzene. Because of that the gas-blowing method is regarded as environmentally friendly. Further, the gas-blowing approach promoted a high degree of supersaturation and rapid homogeneous nucleation. Argon gas was blown over the spin-coated perovskite layer to promote solvent evaporation [23]. The gas-blown approach deposited ~300 nm thick film with densely packed crystallites covering the entire substrate. The planar device deposited by this approach showed the best efficiency of 17% with good reproducibility. Recently, expanding this approach, a photoconversion efficiency of over 20% was achieved when nitrogen gas-knife assisted meniscus coating was carried out for the perovskite layer [24]. A schematic of the air-knife (nitrogen) solvent evaporation method is shown in figure 7.4.

A cryogenic cooling process was investigated to separate nucleation and crystallization [25]. In this process, as-casted films were cooled by immersing in liquid nitrogen and then dried by passing nitrogen gas over the frozen films. This promoted uniform precipitation of precursors while evaporating residual solvents. Unlike the anti-solvent approach, this process is widely applicable for various compositions, which was demonstrated on three different types of mixed halide perovskites. The best performance device exhibited 21.4% efficiency with 1.14 V

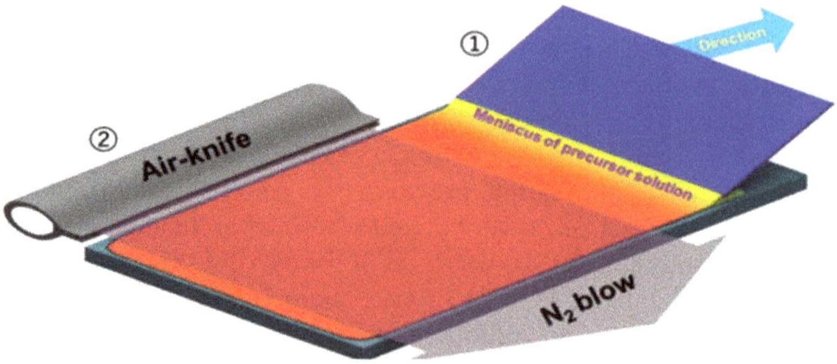

Figure 7.4. Schematic of perovskite film formation using air-knife assisted meniscus coating method. Adapted with permission from [24]. Copyright © 2019 Wiley-VCH Verlag GmbH & Co. KGaA, Weinheim.

open-circuit voltage, 23.5 mA cm^{-2} short-circuit current and with 80% FF. Three different mixed halide devices studied in this work included $MA_xFA_{1-x}PbI_3$, $Cs_{0.05}(MA_{0.17}FA_{0.83})_{0.95}$ $Pb(I_{0.84}Br_{0.16})_3$ and with chloride in the second composition. All three compositions showed improved performance when cryogenic cooling was introduced.

7.2.1.1.3 Heating approach

One of the traditional approaches to remove solvent is applying heat. Millimetre-sized perovskite crystals were obtained by dripping a hot solution (70 °C) containing a mixture of PbI_2 and MACl onto a spinning substrate at 180 °C [26]. A systematic increase in grain size was observed by increasing the substrate temperature from as-casted to 190 °C. The increase in grain size was reflected in the solar cell performance with ~1% efficiency for ~1 μm grain size films to ~18% efficiency for ~180 μm grain size films. This was also observed in other performance parameters such as short-circuit current from 3.5 to 22.4 mA cm^{-2}, open-circuit voltage from 0.4 to 0.92 V and FF from 45% to 82%. The increased grain size through the effect of hot-casting helped suppress charge trapping, thereby eliminating hysteresis and also allowing photo-generated carriers to propagate longer distance without encountering defects. Figure 7.5 shows a schematic of hot-casting film growth compared to conventional room temperature-casting method [27]. Continuing this heating approach, hot-blade coating was explored, in which a solution containing perovskite precursors along with a small amount of I-α-phosphatidylcholine surfactant was blade-coated over the pre-heated substrate (~70 °C–145 °C) [28]. The addition of surfactant helped alter the flow dynamics and also improved the adhesion between the perovskite solution and the hole transport layer (HTL). Further, the small amount of surfactant did not affect the opto-electronic properties of the solar cell devices. Solar cell devices fabricated by this hot-blade approach showed over 20% efficiency for 0.075 cm^2 scale devices and over 15% efficiency for 30 cm^2 sized devices.

7.2.1.1.4 Vacuum flash-assisted solution processing approach

Another classical method to remove solvents is by applying vacuum. Vacuum flash-assisted solution processing (VASP) was investigated for crystallizing the perovskite

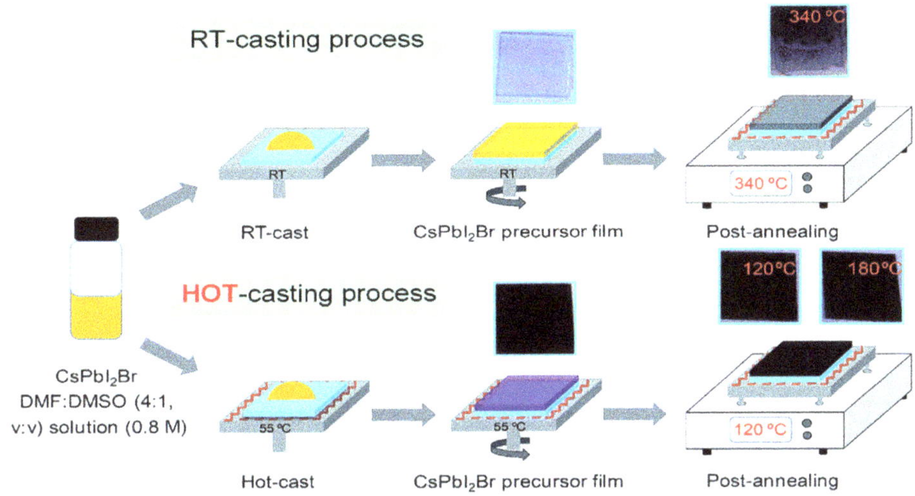

Figure 7.5. (a) Schematic of perovskite film formation by the hot-casting method in comparison to the conventional room temperature (RT)-casting process. Images of precursor and post annealed $CsPbI_2Br$ films prepared using the two different methods are also presented. Reproduced from [27] with permission from Royal Society of Chemistry.

layer and for removing DMSO [29]. In the VASP process, perovskite precursors with composition $FA_{0.81}MA_{0.15}PbI_{2.51}Br_{0.45}$ in DMSO was spin-coated on top of mesoporous TiO_2 and the film was placed under vacuum for a few seconds to remove most of the residual solvents. It was found that 20 Pa pressure was necessary for a 1 cm^2 sized film to obtain the best performance. The schematic of the VASP method is given in figure 7.6. The solar cell stack was completed by spin-coating [2, 7, 29]-tetrakis(N, N-di-p-methoxyphenylamine)-9,9-spirobifluorene (spiro-OMeTAD) in tert-butyl-pyridine and lithium bis(trifluoromethylsulfonyl)imide (Li-TFSI) HTL and 80 nm gold top layer. The VASP processed solar cell device performed at 20.38% efficiency with 1.14 eV open-circuit voltage, 23.19 mA cm^{-2} short-circuit current and 75.7% FF. In comparison, the conventionally processed device showed only 10.79% efficiency at this device scale. Later, the VASP method was improvised by introducing gas through the chamber to further enhance the solvent drying process [30, 31]. This process enabled fabrication of large-area devices up to 144 cm^2. However, at 0.1 cm^2 scale, devices fabricated with this approach showed highest efficiency of 20.44% and an average of 15.63% efficiency was obtained from the devices at 1.1275 cm^2 scale. Recently, this method was optimized to achieve efficiency over 25% at 0.06 cm^2 and over 20% efficiency at 1160 cm^2 scale [32]. Figure 7.6 illustrates an example of the perovskite film formation by VASP along with film morphology based on the precursor mixture [33].

7.2.1.2 Step-by-step process

In the step-by-step process, perovskite precursors are coated onto the substrate one precursor layer at a time. For instance, for fabrication of PSCs, lead halide precursor

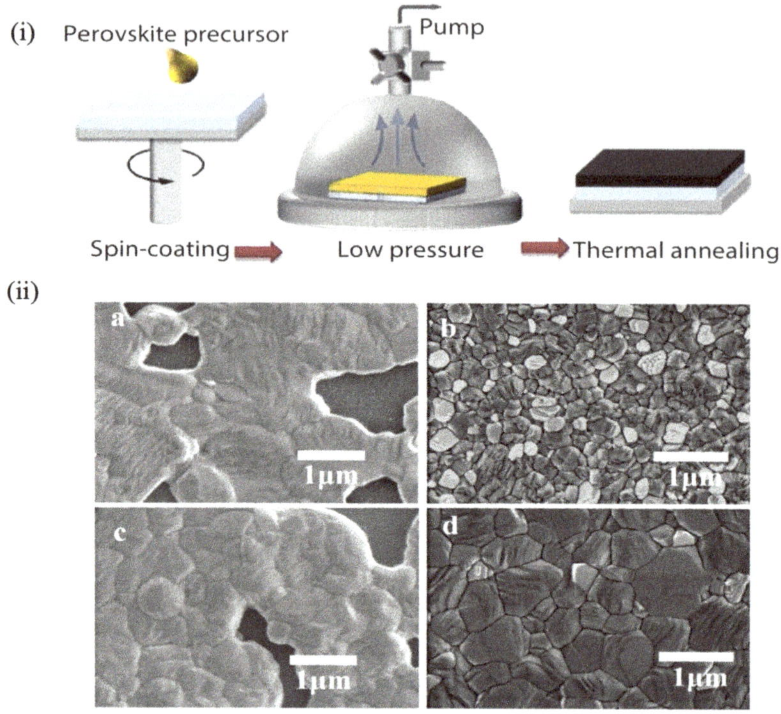

Figure 7.6. (i) Schematic of perovskite film formation by VASP. (ii) SEM images of perovskite films prepared from different precursor solutions. Adapted with permission from [33]. Copyright 2019 Wiley-VCH Verlag GmbH & Co. KGaA, Weinheim.

is deposited in the first step and then it is exposed to organic precursors in the second step. Generally, after the second step the film is taken for heat treatment which induces diffusion of ions that initiates perovskite nucleation and crystallization. Unlike the single-step process, this step-by-step process provides better control over perovskite crystal growth. This sequential perovskite growth was demonstrated by growing the layers on the glass substrates even before realizing their applicability in solar cells [34]. Lead or tin iodide was vacuum-deposited or spin-coated onto the glass substrate first and then this inorganic film was dipped into an organic ammonium iodide in toluene/2-propanol at room temperature for a short period. A single-phase perovskite layer obtained by this approach exhibited a photoluminescence peak indicating semiconducting behaviour of the perovskite films. However, the use of this approach for fabrication of solar cells was demonstrated only recently [35]. A slight modification from the original method was carried out, in which PbI_2 was spin-coated onto a TiO_2 layer from the DMF solution and the film was dried at 70 °C. After drying, the TiO_2/PbI_2 layer was dipped into the methylammonium iodide (MAI) solution for 20 s. The solar cell stack was completed by spin-coating the HTL and thermally evaporating the top contact gold layer. The device exhibited 15.0% photoconversion efficiency with 993 mV open-circuit voltage, 20.0 mA cm^{-2} short-circuit current and 73% FF. Further

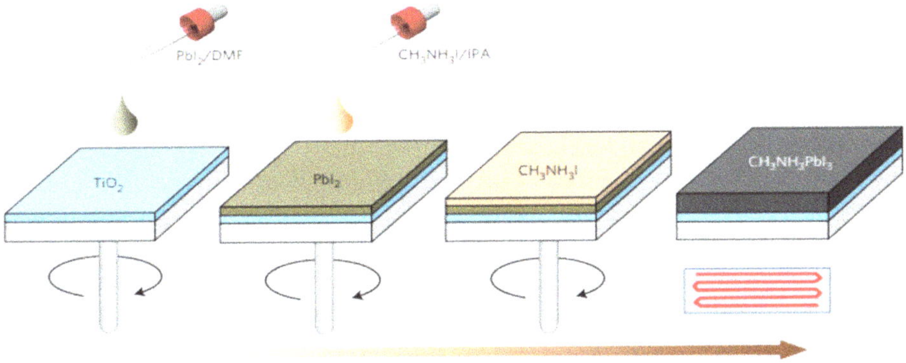

Figure 7.7. Schematic of two-step–step spin-coating for perovskite cuboid formation. Adapted with permission from [37]. Copyright 2014 Springer Nature.

optimizing this two-step crystal growth method, the efficiency of the device was improved to ~17%. The dip-coating of the organic layer was replaced with the spin-coating method and the resultant film was annealed at 100 °C for 5 min. A schematic of step-by step coating is given in figure 7.7. Interestingly, this spin-coating and annealing process deposited cuboid-shaped crystals with an average size ranging from 90 to 720 nm depending on the concentration of MAI solution. Higher concentration yielded smaller crystals, but resulted in lower light conversion efficiency. Following these works, pre-heating the fluorine doped tin oxide (FTO) substrate was investigated to increase PbI_2 filling over the TiO_2 pores [36]. Notably, the optimum temperature was found to be 50 °C. The devices fabricated without pre-heating showed only 11.16% efficiency, whereas the device fabricated by pre-heating the substrate at 50 °C performed at 15.31% efficiency. Upon increasing the heating temperature further to 60 °C, the device performance dropped to 10.49%, due to the larger PbI_2 crystals with a limited reactivity to MAI. In addition to mono-cation perovskites, mixed cation perovskites provide added advantage in terms of extending the light absorption region further into the redder region. Mixing formamide ammonium iodide along with MAI at different proportions during the spin-coating process, mixed cation devices were fabricated. Photoluminescence spectra of mixed cation perovskites showed a systematic increase in peak position on increasing the FA:MA ratio. Further, the mixed cation device exhibited 14.9% PCE with an improved short-circuit current of 21.2 mA cm^{-2}.

Although formamidinium lead iodide ($FAPbI_3$) absorbs longer wavelengths of light than the methylammonium derivative, that in principle should result in high performing devices. To fabricate $FAPbI_3$ perovskite devices, an intramolecular exchange process was developed [38]. PbI_2–DMSO complex was pre-formed by using toluene as an anti-solvent in the solution containing PbI_2 and DMSO. As-isolated PbI_2–DMSO complex was spin-coated onto the substrate by dissolving the PbI_2–DMSO complex in DMF and then formamidinium iodide (FAI) was spin-coated. A follow-up annealing of the films initiated DMSO and FAI exchange. A solar cell device fabricated after coating poly-triarylamine hole transport and gold top contact layers exhibited certified efficiency of 20.2% without hysteresis. The device showed 24.7 mA cm^{-2} short-circuit current, 1.06 V open-circuit voltage and 77.5% FF. The sequential growth of perovskites provided good, reproducible and different composition devices. To understand the reaction pathways involved in the sequential growth process, recently a systematic analysis at every step of the layer growth was performed using x-ray diffraction and electron microscope imaging. In the first step, nucleation and growth of PbI_2 was involved. Intercalation and reorganization of MAI was observed in the second step. In the third step, Ostwald ripening occurred where perovskite from TiO_2 was transported to the capping layer. Further, Ostwald ripening during the dipping steps ensured the growth of larger crystals [39].

7.2.2 Perovskites through intermediate complexes

When perovskite precursors were dissolved in DMSO for the spin-coating process either in the one-step process [15] or in the step-by-step process [38], DMSO was found to be forming a complex with the precursors. This DMSO complex structure apparently helped obtain better reproducible devices due to the slow release of DMSO during the annealing step, leading to controllable high-quality crystals. One of the investigations explored the effect of a DMSO complex on the solar cells' performance [40]. The DMSO complex of a perovskite possessing the chemical formula, $MA_2Pb_3I_8(DMSO)_2$ was prepared by reacting PbI_2, DMSO and MAI in DMF. Varying the ratio of PbI_2 to DMSO while fixing the MAI and PbI_2 concentration at 1.0 M, the optimum ratio for the formation of $MA_2Pb_3I_8(DMSO)_2$ intermediate was determined. When DMSO was not used, the $MAPbI_3$ perovskite was obtained. At a 1:1 ratio of PbI_2 and DMSO, $MAPbI_3$ along with a small amount of $MA_2Pb_3I_8(DMSO)_2$ was identified. Upon increasing the DMSO content, pure $MA_2Pb_3I_8(DMSO)_2$ intermediate complex was isolated at a 1:10 ratio of PbI_2: DMSO. On further increasing the ratio, again a mixture of $MAPbI_3$ and the $MA_2Pb_3I_8(DMSO)_2$ intermediate complex was detected. Planar solar cell devices fabricated using these different intermediate mixtures on top of the NiO layer showed a drastic improvement in performance. $MAPbI_3$ composition prepared without using DMSO complex showed only 4.64% efficiency, whereas the devices fabricated using the $MA_2Pb_3I_8(DMSO)_2$ intermediate complex performed at 18.47% efficiency. Further, the high-efficiency device showed a low hysteresis behaviour and performed longer than the devices that were fabricated without using the intermediate complex.

In a separate study, a similar observation was obtained in the TiO_2 mesoporous device structure [41]. The use of excess MAI yielded iodine rich phases of Pb, such as PbI_5 and PbI_6. Notably, the presence of PbI_6 aggravated the hysteresis issue in PSCs. Since PbI_2 is a Lewis acid, it tends to form an adduct complex with polar aprotic Lewis base solvents such DMSO, thiourea, pyridine, N-methylpyrolidone, thioacetamide and aniline. Likewise, ammonium iodide also acts as a Lewis base. The reactivity difference between the solvents and ammonium iodide often leads to different intermediate adduct species. When FAI was used to form perovskite, thiourea instead of DMSO was found to be a suitable agent for the formation of an intermediate adduct complex [42]. Similar to the methylammonium–DMSO case, formamidinium–thiourea also yielded large grains of perovskite films with long carrier lifetimes and better performing devices. In the follow-up report, N-methyl-2-pyrrolidone (NMP) was used for forming an intermediate adduct with $FAPbI_3$. This $FAPbI_3$-NMP adduct through perovskites enabled pin-hole free, uniform films that exhibited the best conversion efficiency of over 20% (average performance was 18.83% ± 0.73%) [43]. Density functional theory (DFT) calculation determined a high interaction energy between FA and NMP, suggesting NMP acting as a Lewis base solvent for adduct formation. Further, the calculations estimated some Lewis base selection criteria including hydrogen-bonding ability, steric hinderance and matching hardness and softness of the Lewis acid and base. Although NMP-assisted adduct yielded one of the best performing $FAPbI_3$ devices, it has been observed that $FAPbI_3$ exists in two different phases, the photoactive α-phase and photo-inactive δ phase. There have been a couple of investigations for δ to α phase transformation including annealing at 150 °C for a long time and replacing certain FA^+ with Cs^+ or MA^+ or DMSO solvent vapour annealing [44–47]. However recently, residual DMSO adduct in $MAPbI_3$ was exploited for phase transformation of $FAPbI_3$ [48]. This was demonstrated in the bifacial stamping process, where $MAPbI_3$ was spin-coated onto one substrate and onto another substrate, δ-phase $FAPbI_3$ was coated—subsequently both films were combined and annealed at 100 °C. Transport of DMSO from $MAPbI_3$ to $FAPbI_3$ during the annealing step helped the phase transition. This approach led to devices which achieved 18.34% efficiency in $FAPbI_3$ and ~13% efficiency in $EAPbI_3$ (EA = ethylammonium).

7.3 Cell architecture, device physics and performance

The PSC architecture is a derivative of either dye-sensitized solar cells (DSSCs) or organic solar cells. They are broadly classified as mesoscopic if the structure is similar to a DSSC cell configuration utilizing porous material for charge transport or planar, in which a number of thin layers are stacked together. Irrespective of the architecture, they generally consist of a bottom contact layer, ETL, absorber layer, HTL and top contact. Based on how the charge is extracted, the cell structure has been further classified as *n–i–p* or *p–i–n* configuration. In the *n–i–p* configuration, the ETL resides on top of the bottom contact layer where photo-separated electrons are collected, whereas in the *p–i–n* configuration, the HTL is deposited to extract holes. Figure 7.8 shows a schematic of mesoscopic and planar cell structures with *n–i–p* and *p–i–n* configurations. The planar structure is advantageous over the mesoscopic structure for fabrication of flexible devices or tandem cells since they are generally processed at low temperatures. Nevertheless, some of the best performing devices are fabricated utilizing the mesoscopic structure. Functions and different types of materials that are used in different layers are detailed in the following section.

7.3.1 Electron transport layer

The purpose of the ETL is to collect photogenerated electrons from the perovskite layer and to transfer the charge to the electrode layer. The ETL is sandwiched

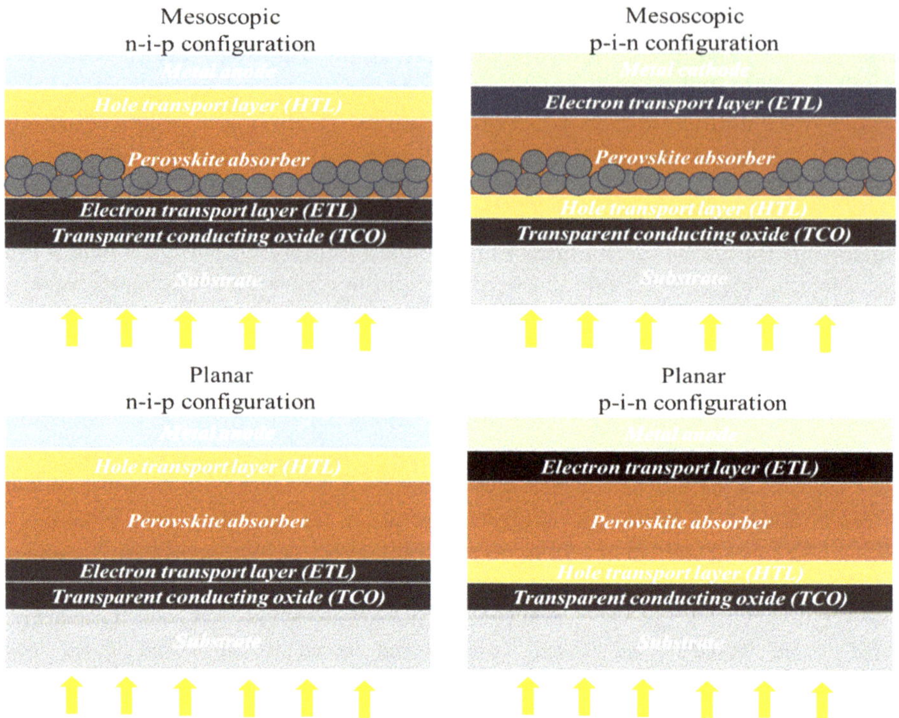

Figure 7.8. Schematic diagram of different PSC configurations. Arrows indicate light illumination direction.

between the perovskite and the FTO electrode layer. Materials should possess certain characteristics in order to be used as an ETL. The material should be a wide bandgap material with the conduction band energy level lower than the perovskite conduction band energy level, so that the photogenerated electrons from the perovskite layer can be injected into the ETL. The ETL material should be chemically inert with the perovskite layer and thermally stable for high-temperature processing.

7.3.1.1 Inorganic electron transport layer

TiO_2 has been a major choice of material for the ETL in PSCs because they have been improvised from the DSSC structure. TiO_2 is a wide bandgap material with good electron conductivity and suitable band alignment with the perovskite layer, which makes it a material of choice for ETL. However, TiO_2 generally requires high-temperature processing, it has a capacitance issue that causes current–voltage (I–V) hysteresis and is amenable for UV induced photocatalytic activities that can affect long-term reliability. Because of these drawbacks, there have been concerted efforts in addressing these issues or for identifying alternative materials for ETL. For instance, a Sb_2S_3 layer was introduced between TiO_2 and the perovskite layer to block off the UV light to prevent UV induced catalytic degradation [49]. Similarly, europium doped strontium cerium oxide down conversion phosphor was deposited on top of TiO_2 [50]. The phosphor-coated device retained 80% of its original photoconversion efficiency under UV light irradiation for 70 h by converting UV light to visible and maintained 78% of efficiency for 75 days under ambient condition with 20%–25% relative humidity. Most of the earlier investigations adopted the mesoscopic architecture because the planar structure constructed using TiO_2 thin films showed a high I–V hysteresis. The rutile phase (~100) of TiO_2 has a high relative capacitance in comparison to anatase phase TiO_2 (~40). The capacitance in TiO_2 has been identified as a major contributor for I–V hysteresis in PSCs [51, 52]. Therefore, a significant amount of effort was directed towards optimizing TiO_2 phase, morphology, surface engineering and doping [53].

Nanoparticles of rutile and anatase phase TiO_2 were synthesized with 60.6% and 49.1% porosity, respectively [54]. Mesoscopic perovskite devices fabricated by a one-step process showed photoconversion efficiency of 8.19% from the rutile phase and 7.23% from the anatase phase. Likewise, the two-step fabrication process resulted in higher efficiency devices with PCE of 13.75% from rutile and 13.99% anatase. Recently, a planar perovskite device was fabricated by sputter-coating rutile and anatase phases of TiO_2 onto FTO substrate. The devices exhibited a comparable performance with an average efficiency of 16% ± 0.67% from anatase and 15.6% ± 0.66% from rutile [55]. Nevertheless, none of the investigations explicitly detailed if the anatase phase reduces I–V hysteresis in perovskite films. Besides TiO_2 polymorphs, interfacial defects and charge accumulations are other potential causes for the I–V hysteresis. Treating TiO_2 with chlorine or annealing for a longer time was found to improve the defects related to hysteresis [44, 56]. Further, it has been found that extracting the electron from the perovskite layer using [6,6]-phenyl C61 butyric acid methyl ester ($PC_{61}BM$) helped reduce the charge accumulation, thereby helping to obtain a low hysteresis in the conventional *n*–*i*–*p* PSCs [57].

Another approach to stable, hysteresis-free PSCs is to use light-stable metal oxides with an appropriate band alignment. Tin oxide (SnO_2) has been investigated as one of the alternative ETLs for PSCs because of its photostability and low-temperature processability (<200 °C) [58]. The planar solar cells fabricated using SnO_2 exhibited reduced hysteresis with efficiency over 20% [59, 60]. Although the exact reason for the reduced hysteresis is still debatable, the consensus is that fast charge extraction and low interfacial charge accumulation contributes towards reducing the hysteresis. Defect passivation and effectively linking perovskite and tin oxide with a chemical linker has been demonstrated as one of the ways to improve efficiency in the SnO_2-based PSCs. 4-imidazoleacetic acid hydrochloride (ImAcHCl) was used for linking perovskite and SnO_2, which helped to realign conduction and valence band positions and to reduce non-radiative recombination and to improve carrier lifetime [61]. The solar cells fabricated with ImAcHCl showed an improved efficiency of 20.2% from an 18.6% efficiency device that was fabricated without the chemical linker. Recently, photoconversion efficiency over 25% was achieved by utilizing potassium 4-methoxysalicylate (MSAK) as a defect passivating compound at the ETL, perovskite layer and ETL/perovskite interface [62]. MSAK was found to be coordinating with under-coordinated Pb^{2+} ions in the perovskite and carboxylic and hydroxyl groups attaching to SnO_2. The device fabricated using MSAK was exhibiting 25.47% efficiency with reduced hysteresis. Notably, the defect passivated cells were stable in ambient condition for 60 days retaining 90% of initial PCE, whereas the unpassivated device dropped efficiency to ~70% for the same duration. Ethylene diamine tetraacetic acid (EDTA) was employed to coordinate with SnO_2 due to its strong chelating ability with metal oxides. A planar device fabricated with EDTA complexed with SnO_2 showed photoconversion efficiency of 21.5%, whereas the device with only SnO_2 or EDTA showed 18.67% and 16.34% efficiency, respectively. It was reported that the Fermi level of EDTA-complexed SnO_2 is better matched with the conduction band of the perovskite that helped attain a high open-circuit voltage, and faster electron mobility contributed to reduced hysteresis. One of the advantages of SnO_2 ETL layer is the low-temperature processability, which was utilized for fabricating PSCs on flexible poly (ethylene terephthalate) (PET)/ITO. The flexible device showed 18.28% efficiency, somewhat lower than the rigid glass/ITO substrate. Higher sheet resistance of PET/ITO led to reduced V_{oc} and *FF* values that are responsible for lower efficiency in flexible devices. In addition, the EDTA–SnO_2 device showed improved photostability under continuous irradiation at 100 mW cm^{-2} maintaining 86% of initial efficiency after 120 h, whereas the device using only SnO_2 retained only 38%. Likewise, the same devices in the dark, under ambient atmosphere for 2880 h retained 92% and 74% of efficiency. Figure 7.9 shows photostability and *J–V* curves of perovskite devices fabricated using SnO_2 and EDTA–SnO_2.

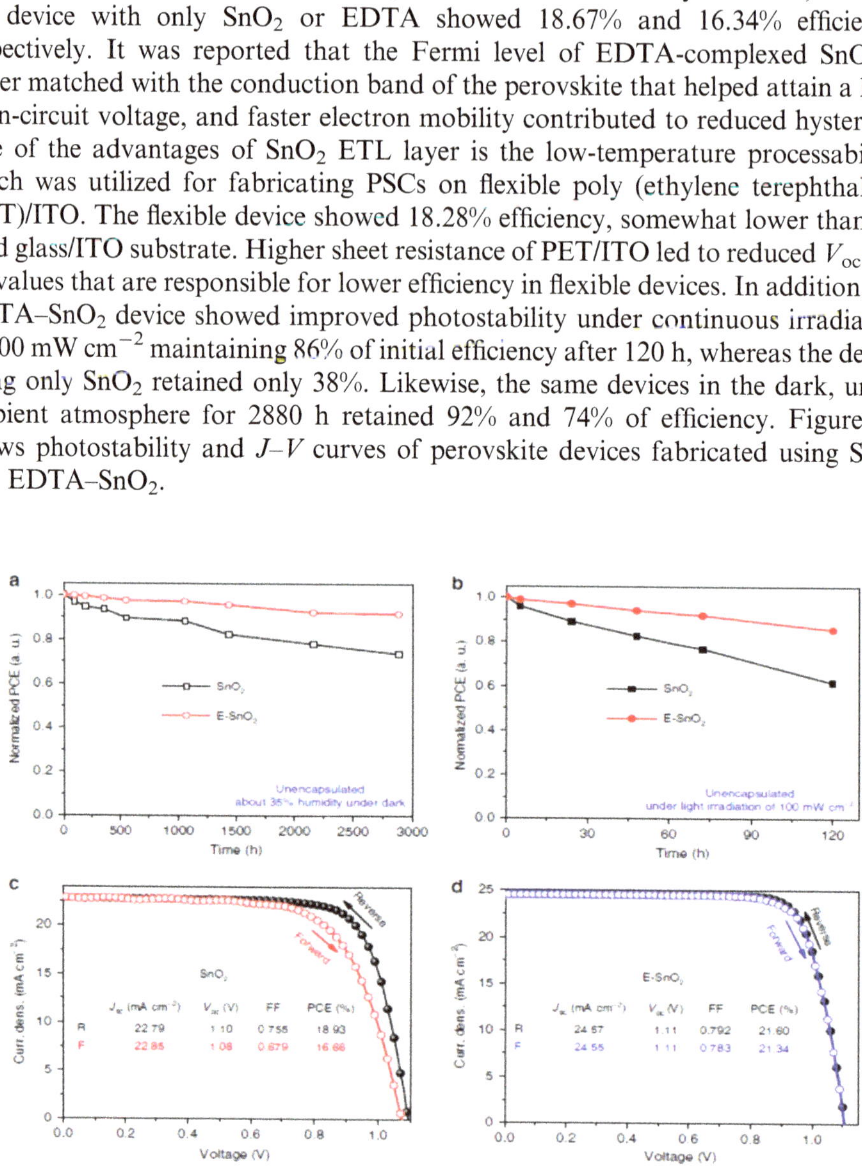

Figure 7.9. Stability data of perovskite with SnO_2 (a) under ambient condition (b) under illumination of 100 mW cm^{-2}, (c) *J–V* curves of the device with (c) SnO_2, (d) EDTA–SnO_2. Adapted from [63] CC BY 4.0.

Owing to high conductivity, low-temperature processability and favourable band alignment, ZnO has been studied as an ETL [64–66]. For instance, a planar perovskite device was fabricated by spin-coating ZnO nanoparticles onto ITO substrate. ZnO nanoparticles were synthesized by reacting zinc acetate dihydrate and KOH in methanol at 65 °C. Solar cell performance as a function of ZnO thickness was studied by varying the thickness from 0 to 70 nm. A device fabricated using 25 nm ZnO showed the best efficiency of 14.4% with 20.5 mA cm^{-2} short-circuit current and 1.01 V open-circuit voltage and 69.6% FF. Following the initial ZnO-based devices, indium zinc oxide (IZO) was investigated for ETL [67]. Devices fabricated using an IZO ETL layer showed 16.25% efficiency in comparison to 13.83% from the TiO_2-based devices. Surface passivation using chemical chelators has shown a significant performance improvement in SnO_2-based PSCs. A similar effort was carried forward to a ZnO ETL layer [68, 69]. ZnO surface was passivated using a thin layer of MgO and protonated ethanolamine (EA). It was noted that MgO inhibited the interfacial charge recombination and EA promoted electron transport from the perovskite to ZnO, helping to eliminate hysteresis. The combination of MgO and EA passivating layers pushed the photoconversion efficiency over 21.0% from a hysteresis-free device. Likewise, sulfidation of ZnO was investigated for reducing interfacial charge recombination [69]. Several other metal oxides and metal sulfides with suitable electrical and optical properties such as zinc tin oxide (ZTO), barium tin oxide and molybdenum sulfide (MoS_2) were explored as an ETL [70–73].

7.3.1.2 Organic electron transport layer

The organic ETL provides certain advantages over inorganic ones such as low-temperature solution processing that enables *p–i–n* type devices with high electron mobility and good charge extraction properties. Use of an organic ETL in PSCs is adapted from organic solar cells; therefore, the fabrication process is transferrable. Fullerene and fullerene derivatives, perylene diimides, naphthalene diimides and azaacenes are some of the widely studied organic compounds for the ETL [74]. For instance, an investigation compared pristine C_{60} and C_{70} as an ETL instead of the frequently studied $PC_{61}BM$ due to their higher electron mobility [75]. PCE of 14.04% was obtained using a mixture of C_{60} and C_{70} as an ETL in comparison to 13.74% from $PC_{61}BM$. Perylene diimides are non-fullerene molecules exhibiting suitable properties for ETL with very good stability. Recently, tetrachloroperylene diimide (TCl-PDI) was used as an ETL for fabricating PSCs along with poly(3-hexylthiophene) (P3HT) as a hole transport agent [76]. The device structure of ITO/TCl-PDI/perovskite/P3HT/MoO_3/Ag performed at 14.73% efficiency, whereas the same device structure with TiO_2 as an ETL had a PCE of 12.78%. Although PDI showed better performance relative to TiO_2, it was still lower than phenyl-C61-butyric acid methyl ester (PCBM)-based devices. Another aspect of PDI that limited its use in PSCs is π–π stacking, which inhibited solubility in a majority of solvents. Structurally similar non-fullerene molecule naphthalene diimide (NDI) was explored for ETL. The solubility of the molecule was modified using different alkyl chains. For example, using asymmetric-shaped chiral (R)-1-phenylethyl group in naphthalene-1,4,5,8-tetracarboxylic diimide, very good solubility was achieved in various organic solvents which provided good film forming ability [77, 78]. This good film forming ability was exploited for construction of PSCs with NDI as an ETL. The cells performed up to 20.5% efficiency with an average of 18.74% ± 0.95%, which is higher than the PCBM-based devices that are fabricated using the same process. Despite organic ETL materials performing comparable to inorganic ones, their chemical stability and cost are prohibiting their widespread use in PSCs.

7.3.2 Hole transport layer

Similar to ETL, the HTL's function is to collect photogenerated holes from the perovskite layer. The HTL sits between the perovskite layer and top contact layer. Since PSCs are derived from dye-sensitized solar cells, the HTL still predominantly utilizes state of art Spiro-OMeTAD especially in the *n–i–p* configuration. Due to its poor conductivity and low hole mobility, Spiro-OMeTAD requires a dopant to function as an effective HTL. Lithium bis(trifluoro methane)sulfonimide (Li-TFSI) and 4-tert-butylpyridine (tBP) are some routinely used dopants for Spiro-OMeTAD. These molecules are hygroscopic which creates stability issues in the devices. There have been investigations to find suitable alternative organic and inorganic HTLs in recent years towards addressing stability and cost drawbacks of Spiro-OMeTAD.

7.3.2.1 Organic hole transport layer

For a material to be a good HTL in PSCs, it should have a good hole mobility, suitable band alignment with perovskite and chemical and thermal stability. Organic molecules are advantageous for HTL over inorganic constituents since their properties can be tuned by modifying their substituents. Since the dopants are the main cause of instability in Spiro-OMeTAD-based devices, there has been significant effort vested in developing dopant-free hole transport materials (HTMs). For instance, tetrathiafulvalene derivative was introduced as a dopant-free HTM in perovskite materials that performed comparably to traditional spiro-OMeTAD-based devices with PCE of over 11% [79]. Following this initial result, a similar performance (11.8% efficiency) was obtained using 6,13-bis(triisopropylsilylethynyl) pentacene (TIPS-pentacene) as an HTL [80]. Continuing this effort, it was identified that introducing extended π-conjugation improves hole conductivity. Two π-conjugated molecules, namely $N^{2\prime}$,$N^{2\prime}$,$N^{7\prime}$,$N^{7\prime}$-tetrakis(4-methoxyphenyl)-spiro[dibenzo [c,h]xanthene-7,9′-fluorene]-2′,7′-diamine (X61) and $N^{2\prime}$,$N^{2\prime}$,N^{5},N^{5},$N^{7\prime}$,$N^{7\prime}$,N^{9},N^{9}-octakis(4-methoxyphenyl)spiro[dibenzo-[c,h]xanthene-7,9′-fluorene]-2′,5,7′,9-tetra-amine (X62), with spiro-[dibenzo[c,h] xanthene-7,90 -fluorene] (SDBXF) as the skeleton were synthesized and used as HTL for PSCs [81]. X62 showed higher conductivity and hole mobility combined with better film forming ability than X61 and spiro-OMeTAD, which was reflected in the solar cell device performance. The device fabricated using X62 as HTL exhibited 15.9% efficiency with 70.4% FF value, whereas devices with X61 and spiro-OMeTAD showed 8.0% and 10.8% efficiencies, respectively. Figure 7.10 shows the energy level diagram of PSC structure with X61, X62 and spiro-OMeTAD HTLs, schematic of device structure, *J–V* curve and

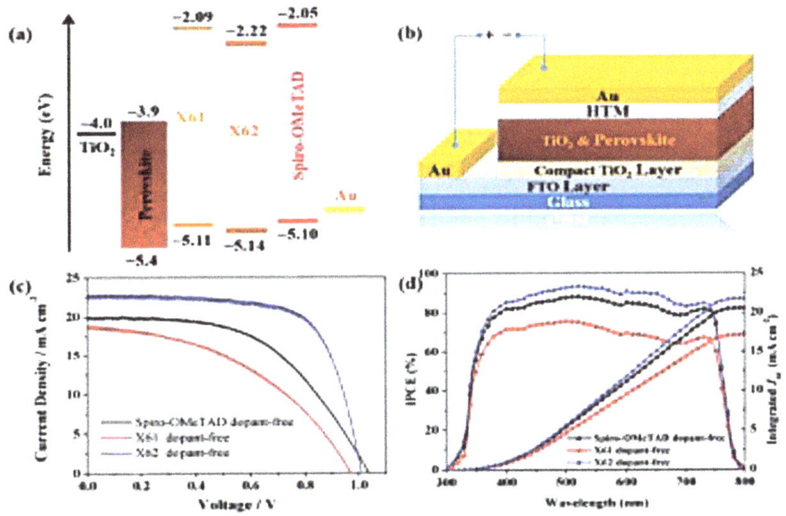

Figure 7.10. (a) Energy level diagram, (b) schematic of perovskite device configuration, (c) *J–V* curves of the devices, (d) IPCE spectra along with current density values. Adapted with permission from [81]. Copyright 2018 Royal Society of Chemistry, CC BY-NC 3.0.

incident-photon-to-current efficiency (IPCE) spectra. The polymeric compound poly (3-hexylthiophene) (P3HT) was found to be a low-cost alternate possessing requisite properties for HTL in PSCs [82]. A record efficiency device of 22.7%, very low hysteresis and good stability in humid conditions was fabricated using P3HT as HTL. The device maintained 95% of the efficiency under 1 Sun condition at 85% relative humidity after 1370 h, emphasizing better interfacial contact between the perovskite and the HTL in reducing the hysteresis and with good stability.

7.3.2.2 Inorganic hole transport layer

Instability is a major concern in organic HTMs. Therefore, *p*-type inorganic semiconductor materials have been explored as alternative stable materials for the HTL. Nickel oxide (NiO) is one of the widely investigated *p*-type semiconductors due to its wide bandgap, suitable band alignment, high optical transparency and amenability for various processing methods [83]. A planar device fabricated using NiO_x as HTL and $PC_{61}BM$ as ETL showed a decent efficiency of 7.8% [83]. Nevertheless, NiO_x HTL suffers from poor hole conductivity, which seems to be influenced by the synthesis methods and the morphology of NiO_x [84]. Recently, chemical bath deposition (CBD) was successfully implemented to grow conformal, uniform NiO_x films. A device constructed using this NiO_x film showed efficiency over 22% comparable to a device with organic spiro-OMeTAD HTL material [85]. In addition, NiO_x HTL requires high-temperature processing that has limited its use only in *p–i–n* type devices. In order to overcome issues with NiO_x HTL, copper thiocyanate (CuSCN) was proposed as an alternative, owing to its high carrier mobility, optical transparency and good thermal stability. Importantly, it is soluble in a variety of organic solvents that enabled CuSCN to be used in both *n–i–p* and *p–i–n* type devices. An over 20% efficient *n–i–p* device was fabricated using CuSCN that was deposited by a rapid solvent removal process. The device showed good thermal stability under long-term heating, however, operational stability was poor. It was found that the reactivity of CuSCN with Au top contact was responsible for the poor operation stability [86]. Introducing a thin layer of conductive reduced graphene oxide between CuSCN and the Au contact improved the operational stability—maintaining over 95% of its initial efficiency after 1000 h under full sun condition at 60 °C. Although good solubility of CuSCN enabled both types of devices, CuSCN was found to react with perovskite during the annealing process, which is generally required for obtaining good crystallinity in perovskite. This chemical reactivity with perovskite degrades the solar cell performance causing poor stability [87]. Cu_2O was explored as a HTL for mixed halide perovskite *n–i–p* and *p–i–n* type devices [88]. The initial device showed a decent PCE of 8.23% with $PC_{61}BM$ as an ETL. Although Cu_2O can be used for both types of devices, optical loss was observed when Cu_2O was used in a *p–i–n* type device due to the narrow bandgap. Therefore, a significant number of efforts have focussed on widening the bandgap of Cu_2O by making alloys with Cr or Ga to form $CuCrO_2$ (3.0 eV) and $CuGaO_2$ (3.6 eV) [89, 90]. Relative stability was achieved using $CuCrO_2$ as HTL in mixed halide and triple cation PSCs. $CuCrO_2$ was synthesized as nanoparticles formed by a hydrothermal synthesis process. The device exhibited a PCE of 16.7% with very low hysteresis and retaining ~83% of initial PCE after 60 days in an ambient condition, whereas the device fabricated using spiro-OMeTAD as HTL lost ~76% of its original efficiency [91]. Solar cell efficiency of over 20% was achieved by indium doping into $CuCrO_2$ and using it as HTL [92]. Figure 7.11 shows the energy level diagram of PSCs with In doped $CuCrO_2$ and $PC_{61}BM$ as a HTL and ETL, and cross-sectional SEM image of the cell along with *J–V* curves. Utilizing solution-processed $CuGaO_2$ nanoplates, an *n–i–p* configuration device with efficiency over 18.0% was reported [93]. Similar to other inorganic HTMs devices, use of $CuGaO_2$ also exhibited relatively stable performance over a month in an ambient atmosphere with 30%–55% humidity, but the device with spiro-OMeTAD lost almost all of its performance in 10 days. $CuGaO_2$ nanoplates used in this study were synthesized by reacting $Cu(NO_3)_2 \cdot 3H_2O$ and $Ga(NO_3)_3 \cdot xH_2O$ in ethylene glycol and KOH at 230 °C for 2 h in a Teflon-coated autoclave. Due to low-cost synthesis, amenable for solution processing, stable performance and competing efficiencies using inorganic HTMs is encouraging over organic ones. In particular, $CuCrO_2$ and $CuGaO_2$ are promising materials and likely their potential will be exploited for fabricating large-scale perovskite modules.

7.3.3 Lead (Pb)-free perovskite solar cells

PSCs have shown remarkable growth in a short period of time with efficiency of over 25%. One of the bottlenecks in bringing perovskites solar cells to large-scale deployment is the use of lead. Lead is a toxic and RoHS restricted element even though PSCs benefit a lot from it. Ever since the rise of PSCs, there has been some amount of effort focused on developing lead-free PSCs with the aim of comparable

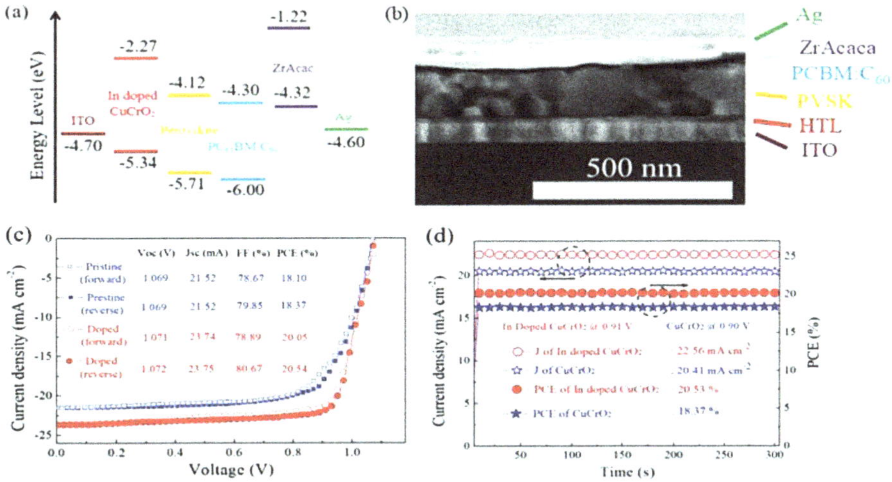

Figure 7.11. (a) Energy level diagram, (b) cross-sectional SEM image of perovskite device, (c) *J–V* curves of the devices, (d) photocurrent and efficiency of the best PSCs based on $CuCrO_2$ and In doped $CuCrO_2$ nanoparticles. Adapted with permission from [92]. Copyright 2019 Wiley-VCH Verlag GmbH & Co. KGaA, Weinheim.

or exceeding performance. Tin has a similar electronic configuration and coordination geometry as Pb and tin-based perovskites show low exciton binding energies and very good carrier mobilities. Theoretically, it has been predicted that tin-based perovskites can perform at 33% efficiency. A complete tin-based perovskite mesoscopic solar cell fabricated using $CH_3NH_3SnI_3$ on mesoporous TiO_2 exhibited over 6% efficiency [94]. Like Pb-based perovskites, the bandgap of tin-based ones was also tuned by substituting halides in the perovskite structure [95]. The efficiency was improved to over 9% by incorporating formamidinium in the perovskite structure. The cells also showed impressive stability retaining 92% of initial efficiency after 1440 h and without significant hysteresis [96]. Despite tin's promising prediction, their performance is still lagging behind Pb-based perovskites. One of the reasons for the lower performance for tin-based perovskites is rapid oxidation of Sn^{2+}. Recently, utilizing pyridyl-substituted fulleropyrrolidines (PPF) with *cis* and *trans* isomers as a precursor additive, better performing devices with efficiency of over 15% have been obtained [97]. It was suggested that the presence of PPF molecules slowed perovskite crystallization and also helped suppress Sn^{2+} oxidation. Further, the tin perovskite devices nearly retained all of their performance for 3000 h of storage and lost only about 7% under continuous illumination for 500 h. A number of theoretical studies in order to find alternative Pb-free materials predicted germanium-based perovskites as promising candidates [98–100]. Earlier studies synthesized $CsGeI_3$ and $MAGeI_3$ perovskites and fabricated solar cell devices using them. However, the cells performed rather poorly with efficiencies of only 0.11% and 0.20% due to sub-optimum bandgaps. By substituting part of iodine with bromine, the bandgap was tuned that improved the device performance slightly [101].

Nevertheless, the tendency of Ge^{2+} to oxidize easily, poor surface configuration and formation of imperfect crystals continue to hinder the performance of Ge-based perovskite cells. One of the strategies to improve air-stability and to stabilize Ge-based perovskites was to alloy with tin. A couple of studies reported the preparation of tin-germanium mixed perovskites $FA_{0.75}MA_{0.25}Sn_{1-x}Ge_xI_3$ and fabricated solar cells using the mixed structures to evaluate the effect of mixed cations on their performance. The devices showed somewhat better performances than pure germanium-based devices [102, 103]. Besides tin- and germanium-based perovskites, recently bismuth ($CsBiSCl_2$) and antimony ($MA_{1.5}Cs_{1.5}Sb_2I_3Cl_6$) containing perovskites have been investigated for solar energy conversion applications by fabricating solar cell devices [104–106]. Devices using $CsBiSCl_2$ showed an impressive efficiency of 10.38% with relatively stable performance for over 150 days in air. But antimony perovskites are still in their infancy.

7.3.4 Tandem solar cells

Combining two or more different solar active materials with variable bandgaps in a parallel or series configuration is generally referred to as tandem structure, the possible way to achieve PCEs over the Shockley–Queisser (S–Q) limit. In the tandem structure, a wide range of the solar spectrum is absorbed that minimizes thermal loss of the photon's energy and thereby provides a higher efficiency than single-junction solar cells. Tandem solar cells using perovskites are mostly fabricated on Si cells. The bandgap of both materials has to be matched optimally so that photocurrent can be maximized. The ideal bandgaps for Si and perovskites are 1.12 eV and 1.58 eV in perovskite/Si tandem cells. One of the earlier tandem devices used two-terminal multijunction architecture on an *n*-type silicon solar cell [107]. The tandem structure was fabricated by depositing a TiO_2 heterojunction layer on the planar n^{++} c-Si front surface. Perovskite was deposited on mesoporous TiO_2 and *p*-type spiro-OMeTAD, a HTL was spin-coated on top of the perovskite. Ag nanowire was deposited to complete the stack. The tandem structure exhibited 13.7% efficiency, 1.58 V open-circuit voltage, 11.5 mA cm^{-2} short-circuit current and 75% FF, suffering from low efficiency mainly because of parasitic absorption of light by the HTL. Further, reducing the layer thickness and switching to a heterojunction silicon cell, the efficiency was improved to 21.2%, higher than individual cells [108]. These reports highlighted that minimizing the parasitic absorption of window layers is one of the ways to reduce the optical loss and to improve efficiency. Switching to tin oxide (SnO_2)/zinc tin oxide (ZTO) window layer, high-efficiency perovskite/Si tandem cell with efficiency over 23% was achieved [109]. Atomic layer deposition or pulsed-chemical vapour deposition of window layers instead of sputtering-deposited conformal, produced uniform and transparent window layers. Figure 7.12 shows a schematic of perovskite/silicon tandem cell, optical microscope images, *J*–*V* curve and total absorbance spectrum.

Encouraged by these results, several studies were conducted by texturing the silicon surface and solution processing a perovskite layer, and overcoming phase segregation by introducing chloride to form a triple halide composition—the

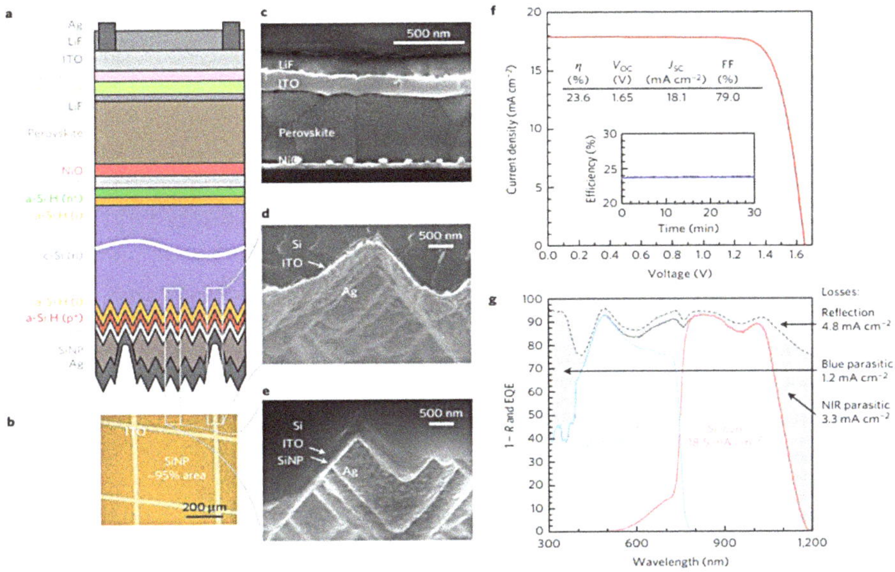

Figure 7.12. (a) Schematic of perovskite/silicon tandem solar cell. (b) Optical microscope image of the silicon cell. (c) Cross-sectional SEM image of perovskite top device. Cross-sectional SEM image of rear side of the silicon cell (d) without silicon nanoparticles (e) with silicon nanoparticles, (f) *J–V* curves of the devices, (g) total absorbance, EQE of top perovskite and bottom silicon cells. Adapted with permission from [109]. Copyright 2017 Springer Nature.

efficiency of perovskite/silicon tandem cells achieved over 25% routinely [110–114]. The tandem cells promise that S–Q efficiency limit can be surpassed by double or multijunction architecture in perovskite cells. However, this has not been achieved until recently. A record efficiency perovskite/silicon tandem cell was constructed by heavily textured rear surface and mildly textured front surface. The device exhibited 33.89% efficiency (slightly over S–Q limit of 33.7%) with 83% FF and 1.97 V open-circuit voltage [115]. Charge recombination at the perovskite and ETL was one of the hinderances affecting the efficiency in tandem cells. In the record device, utilizing an ultrathin layer of lithium fluoride (LiF) and additional deposition of ethylene diammonium diiodide (EDAI), the recombination loss was suppressed while maintaining efficient electron extraction. In addition, the LiF and EDAI passivation helped retain 90% of efficiency storage stability over 50 days, in comparison to only 82% from the control device. Likewise, under maximum power tracking condition, at room temperature in a nitrogen atmosphere, the bilayer-passivated cells had 80% of their original efficiency after 1200 h, while only LiF treated cells lost about 40% of their efficiency.

Combining perovskite with thin films solar cell technologies such as CIGS or CdTe offers additional advantages that are generally difficult to achieve from perovskite/silicon tandem cells. Recently, combining perovskites with CIGS and CIS thin-film cells, an absolute efficiency gain of 2.9% and 4.8% have been demonstrated by using semi-transparent PSCs [116]. Moreover, the devices exhibited good temperature coefficient of −0.18% $°C^{-1}$ in the temperature range 25 °C–65 °C. Theoretically, over 44% efficiency can be achieved by pairing 1.1 eV bandgap thin films CIGS cells with 1.6–1.75 eV bandgap perovskite cells [117]. However, achieving this theoretical goal has been challenging mainly because of difficulties in obtaining good efficient wide bandgap perovskite cells. Bimolecular additive engineering has been proposed as a solution for improving efficiency in perovskite/CIGS tandem cells. Utilizing complimentary properties of phenyl ammonium iodide (PEAI) and lead thiocyanate ($Pb(SCN)_2$) a ~20% efficient wide bandgap $(FA_{0.65}MA_{0.20}Cs_{0.15})Pb(I_{0.8}Br_{0.2})_3$ cell was reported [118]. Mechanically stacking this device with CIGS thin films, a four-terminal polycrystalline perovskite/CIGS thin-film tandem cell was constructed, exhibiting PCE of 25.9%. Although mechanically stacking CIGS with perovskite cells does not present any current matching issues, optical and electrical loss in wide bandgap semi-transparent perovskite cells is still persistent. Introducing methyldiammonium diiodide and adjusting the optical interference spectrum, the losses have been suppressed, and an over 20% efficient perovskite device at 81.5% average near infrared transmittance was achieved [119]. By pairing this semi-transparent device with a CIGS bottom device, a four-terminal tandem device was constructed that exhibited record efficiency close to 30%.

Compositional variability available in perovskites offers bandgap tunability. It has been proposed that when combining different bandgap perovskite compositions, tandem cells efficiency exceeding the S–Q limit can be realized. In one of the demonstrations, infrared absorbing composition of $FA_{0.75}Cs_{0.25}Sn_{0.5}Pb_{0.5}I_3$ with bandgap around 1.2 eV was combined with wider bandgap composition, $FA_{0.83}Cs_{0.17}Pb(I_{0.5}Br_{0.5})_3$. A two-terminal tandem device showed efficiency of 17% and mechanically stacked four-terminal tandem devices exhibited a little higher efficiency of 20.3% [120]. It was identified that, low-bandgap cells was the reason for the muted efficiency in the all-perovskite tandem cells. To improve the efficiency of low-bandgap cells, guanidinium thiocyanate was introduced that reduced the defect densities by a factor of >10, increased carrier lifetime to >1 μs, diffusion length to 2.5 μm and improved film morphology. These improvements reflected in the PCE of >20% from narrow bandgap single-junction cells, 25% from four-terminal and 23.1% from two-terminal tandem cells [121]. Recently, record efficiency among all-perovskite tandem cells was reported with efficiency of 28.5% and 23.8% from a Pb–Sn narrow bandgap device. The tandem device was fabricated by depositing wide bandgap lead halide perovskite on top of mixed Pb–Sn narrow bandgap perovskite through the evaporation solution method [122]. Figure 7.13 shows an SEM image of an all-perovskite tandem cell, performance metrics and stability data. The performance of all-perovskite tandem cells is rapidly improving, however, the underlying challenges affecting their efficiencies still exist. For instance, narrow bandgap composition always utilizes tin, which suffers from oxidation. Similarly, wide bandgap compositions exhibit open-circuit voltage deficit and halide segregation issues resulting in lower photostability.

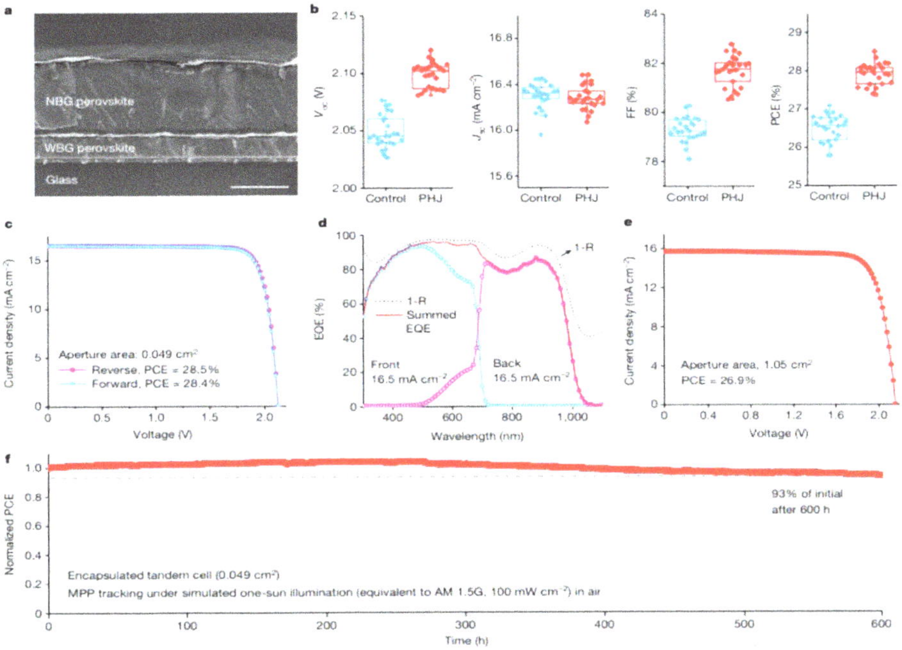

Figure 7.13. (a) Cross-sectional SEM image of all-perovskite tandem solar cell, (b) performance comparison of control and tandem cells replicates, (c) *J–V* curves of the devices, (d) total absorbance curves of the champion tandem device, (e) *J–V* curve of large-area tandem device, and (f) continuous MPP tracking of an encapsulated tandem cell. Adapted with permission from [122]. Copyright 2023 Springer Nature.

7.4 Cell degradation and failure diagnostics

For perovskite cells to see commercial success, they need to go through and pass a series of reliability scrutinization measures that are mandated for all type of solar cells. One of the challenges of bringing PSCs for mass production are their reliability. PSCs often benefit from materials that are sensitive to environmental factors such as moisture, oxygen, heat and sunlight. For instance, Spiro-OMeTAD, HTL utilizes lithium bis(trifluoro methane)sulfonimide (Li-TFSI) dopant, which is moisture sensitive. A recent review article emphasized the standards set by the International Electrotechnical Commission (IEC) for the minimum stability assessment for any photovoltaic cells [123]. Figure 7.14 shows a flow chart of tests that need to be performed on perovskites to make them qualify for commercial use. In the IEC standard, temperature and humidity tests are crucial factors. Methylammonium lead halides ($MAPbX_3$) is known for reacting with moisture and oxygen, weakening the original bond between MA^+ and Pb–X. Further, CH_3NH_2 and HX could also be produced in the presence of moisture and halide could be oxidized in an oxygen environment [124]. Also, $CH_3NH_3PbI_3$ is found to decompose to PbI_2 at temperatures as low as 100 °C [125]. In general, the cells are encapsulated using moisture barrier films such as polyolefin (POE) or ethylene vinyl acetate (EVA). Recently, encapsulating perovskite cells using glass/POE or EVA/

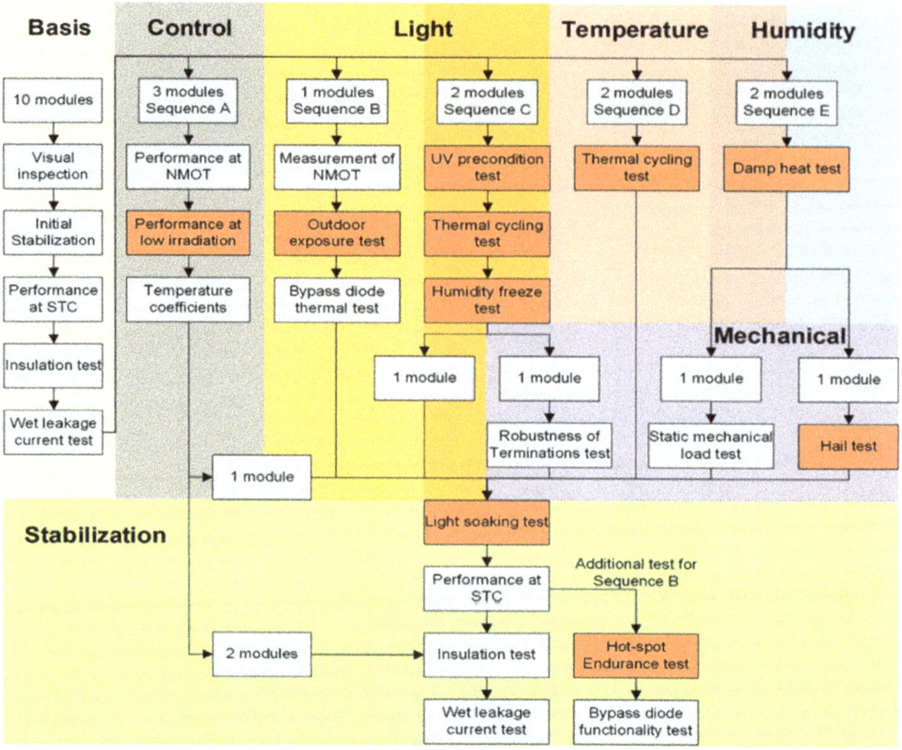

Figure 7.14. Test flow for design qualification and type approval of PV modules. Orange highlighted ones show tests that are already reported for perovskite. STC: standard test condition, NMOT, nominal module operating temperature. Adapted from [123]. CC BY 4.0.

glass, and sealing the edge with butyl rubber ICE 61 646, damp heat and thermal cycling tests were carried out. The cells performed at ~108% of initial PCE after 1000 h at 85 °C/85% RH and 200 thermal cycles [126]. Unlike moisture and humidity studies, where the cells can be encapsulated, in photostability studies the cells cannot be protected from the light. It has been reported that $MAPbI_3$ does not degrade upon illumination for several hundreds of hours, but the ionic conductivity is increased by an order of magnitude during the light soaking [127]. However, lead halides such as PbI_2, $PbBr_2$ and $PbCl_2$ are light sensitive, which are generally found in excess during the perovskite synthesis. These lead halides undergo photodissociation and form neutral halogen and neutral lead [128, 129].

Besides perovskites, other functional materials such as ETL are affected by light soaking, particularly for TiO_2 ETL [130]. TiO_2 is known for promoting photocatalytic reactions in the presence of UV. Further, UV soaking desorbs the oxygen passivation layer on TiO_2 surface leading to exposed oxygen vacancy sites that can cause non-radiative recombination [131]. The use of UV blockers or down-conversion materials or completely replacing photosensitive TiO_2 with SnO_2 or other electron transport materials are some of the mitigations explored to address this issue [132, 133]. In addition to light-induced degradation, in the presence of oxygen and light, various components undergo photooxidation process. Organic molecules such as fullerene-based ones undergo photooxidation, affecting their charge transport properties and thereby the efficiency. Likewise, photooxidation is observed in perovskite as well. Oxygen is found to be diffusing through iodide vacancies in perovskite crystals almost immediately under illumination [134]. Another degradation is found to be occurring at the metal contacts. Perovskite itself does not react with the metal contact, but the perovskite decomposition products such as iodide or methylamine can corrode metal contacts. Further, Pb^{2+} is found to be forming a redox couple with metal electrodes facilitating formation of metallic Pb. Metals from electrodes could also diffuse into the perovskite layer and replace lead to form a metal halide [135, 136]. When silver was used as a top contact in $MAPbI_3$ solar cells, discoloration was observed after several days, indicating Ag becoming corroded along with a drop in efficiency values [137]. There have been efforts dedicated to addressing this metal contact degradation including the use of carbon-based materials, transparent conductive oxide or use of barrier layer in between the metal electrode and the charge transporting layer to prevent corrosive material reaching the electrode layer [138–140]. There are several accelerated studies reported to demonstrate the effect of individual components on the overall cell stability. However, the degradation mechanism of the cells in real-world conditions may be different, which needs to be thoroughly evaluated and understood [141].

7.5 Conclusion, challenges and future prospects

PSCs have seen a remarkable growth in the last decade with single-junction cells efficiencies exceeding 25% and tandem cells efficiencies over 33%. The uniqueness of perovskite cells is solution processing. Unlike, other solar cell technologies, the perovskite cell architecture can be completely constructed by solution processing. This enables rapid and low-cost scale-up and manufacturing. Cost analysis performed for 12% efficiency cells with 15 years expected lifetime estimated the levelized cost of electricity (LCOE) for PSCs is 3.5–4.9 US cents/kWh lower than other PV technologies and energy sources [141]. In addition, the compositional tunability available in perovskites provides multiple options for fully harnessing their potential. For instance, just by tuning the composition alone, perovskite/perovskite tandem structures with efficiencies more than single-junction cells have been achieved. However, there are a few challenges and numerous opportunities to bring perovskite cells to commercial deployment. Thus far, high-efficiency cells have been focussed on methylammonium lead iodide composition, but other compositions may exhibit certain advantages, like perovskites using caesium showing better thermal stability. One of the pressing issues that has been drawing much attention in recent years is reliability. Due to the ionic structure of perovskite crystals, they are prone to oxygen- and moisture-related issues. This is one area of research that continues to present challenges and opportunities. Similarly, perovskite cells are often fabricated using environmentally sensitive chemicals and materials such as Spiro-OMeTAD and TiO_2. Substitute materials have been in development; however, their efficiencies still have room for improvement. Another major scope of perovskite cells is their lead-free composition. Potential replacement elements for lead such as tin, germanium, bismuth and antimony are being investigated, but they need significant uplift to bring their performance on a par with lead-based perovskites. Despite challenges, perovskite cells will soon see sunshine, outperforming conventional Si cells in the field.

References

[1] Park N-G 2015 Perovskite solar cells: an emerging photovoltaic technology *Mater. Today* **18** 65–72

[2] Meggiolaro D and De Angelis F 2018 First-principles modeling of defects in lead halide perovskites: best practices and open issues *ACS Energy Lett.* **3** 2206–22

[3] Li W, Liu J, Bai F-Q, Zhang H-X and Prezhdo O V 2017 Hole trapping by iodine interstitial defects decreases free carrier losses in perovskite solar cells: a time-domain *ab initio* study *ACS Energy Lett.* **2** 1270–8

[4] Han D, Dai C and Chen S 2017 Calculation studies on point defects in perovskite solar cells *J. Semiconduct.* **38** 011006

[5] Frolova L A, Dremova N N and Troshin P A 2015 The chemical origin of the p-type and n-type doping effects in the hybrid methylammonium lead iodide (MAPbI3) perovskite solar cells *Chem. Commun.* **51** 14917–20

[6] Zohar A, Levine I, Gupta S, Davidson O, Azulay D, Millo O, Balberg I, Hodes G and Cahen D 2017 What is the mechanism of $MAPbI_3$ p-doping by I_2? Insights from optoelectronic properties *ACS Energy Lett.* **2** 2408–14

[7] Lee J-W, Bae S-H, De Marco N, Hsieh Y-T, Dai Z and Yang Y 2018 The role of grain boundaries in perovskite solar cells *Mater. Today Energy* **7** 149–60

[8] Adhyaksa G W P *et al* 2018 Understanding detrimental and beneficial grain boundary effects in halide perovskites *Adv. Mater.* **30** 1804792

[9] Haruyama J, Sodeyama K, Han L and Tateyama Y 2016 Surface properties of CH3NH3PbI3 for perovskite solar cells *Acc. Chem. Res.* **49** 554–61

[10] Kojima A, Teshima K, Shirai Y and Miyasaka T 2009 Organometal halide perovskites as visible-light sensitizers for photovoltaic cells *J. Am. Chem. Soc.* **131** 6050–1

[11] Kitazawa N, Watanabe Y and Nakamura Y 2002 Optical properties of $CH_3NH_3PbX_3$ ($X = $ halogen) and their mixed-halide crystals *J. Mater. Sci.* **37** 3585–7

[12] Kashiwamura S and Kitazawa N 1998 Thin films of microcrystalline $(CH_3NH_3)(C_6H_5C_2H_4NH_3)_2Pb_2Br_7$ and related compounds: fabrication and optical properties *Synth. Met.* **96** 133–6

[13] Eperon G E, Burlakov V M, Docampo P, Goriely A and Snaith H J 2013 Morphological control for high performance, solution-processed planar heterojunction perovskite solar cells *Adv. Funct. Mater.* **24** 151–7

[14] Xiao M, Huang F, Huang W, Dkhissi Y, Zhu Y, Etheridge J, Gray-Weale A, Bach U, Cheng Y B and Spiccia L 2014 A fast deposition-crystallization procedure for highly efficient lead iodide perovskite thin-film solar cells *Angew. Chem. Int. Ed.* **53** 9898–903

[15] Jeon N J, Noh J H, Kim Y C, Yang W S, Ryu S and Seok S I 2014 Solvent engineering for high-performance inorganic–organic hybrid perovskite solar cells *Nat. Mater.* **13** 897–903

[16] Yu Y, Yang S, Lei L, Cao Q, Shao J, Zhang S and Liu Y 2017 Ultrasmooth perovskite film via mixed anti-solvent strategy with improved efficiency *ACS Appl. Mater. Interfaces* **9** 3667–76

[17] Taylor A D, Sun Q, Goetz K P, An Q, Schramm T, Hofstetter Y, Litterst M, Paulus F and Vaynzof Y 2021 A general approach to high-efficiency perovskite solar cells by any antisolvent *Nat. Commun.* **12** 1878–8

[18] Zhao P, Kim B J, Ren X, Lee D G, Bang G J, Jeon J B, Kim W B and Jung H S 2018 Antisolvent with an ultrawide processing window for the one-step fabrication of efficient and large-area perovskite solar cells *Adv. Mater.* **30** 1802763

[19] Noel N K, Habisreutinger S N, Wenger B, Klug M T, Hörantner M T, Johnston M B, Nicholas R J, Moore D T and Snaith H J 2017 A low viscosity, low boiling point, clean solvent system for the rapid crystallisation of highly specular perovskite films *Energy Environ. Sci.* **10** 145–52

[20] Xie L, Cho A-N, Park N-G and Kim K 2018 Efficient and reproducible $CH_3NH_3PbI_3$ perovskite layer prepared using a binary solvent containing a cyclic urea additive *ACS Appl. Mater. Interfaces* **10** 9390–7

[21] Chen J, Kim S-G, Ren X, Jung H S and Park N-G 2019 Effect of bidentate and tridentate additives on the photovoltaic performance and stability of perovskite solar cells *J. Mater. Chem.* A **7** 4977–87

[22] Han T-H *et al* 2019 Perovskite-polymer composite cross-linker approach for highly-stable and efficient perovskite solar cells *Nat. Commun.* **10** 520–0

[23] Huang F *et al* 2014 Gas-assisted preparation of lead iodide perovskite films consisting of a monolayer of single crystalline grains for high efficiency planar solar cells *Nano Energy* **10** 10–8

[24] Hu H, Ren Z, Fong P W K, Qin M, Liu D, Lei D, Lu X and Li G 2019 Room-temperature meniscus coating of >20% perovskite solar cells: a film formation mechanism investigation *Adv. Funct. Mater.* **29** 1900092

[25] Ng A *et al* 2018 Perovskite solar cells: a cryogenic process for antisolvent-free high-performance perovskite solar cells *Adv. Mater.* **30** 1870329

[26] Nie W *et al* 2015 High-efficiency solution-processed perovskite solar cells with millimeter-scale grains *Science* **347** 522–5

[27] Wang Z, Liu X, Lin Y, Liao Y, Wei Q, Chen H, Qiu J, Chen Y and Zheng Y 2019 Hot-substrate deposition of all-inorganic perovskite films for low-temperature processed high-efficiency solar cells *J. Mater. Chem.* A **7** 2773–9

[28] Deng Y, Zheng X, Bai Y, Wang Q, Zhao J and Huang J 2018 Surfactant-controlled ink drying enables high-speed deposition of perovskite films for efficient photovoltaic modules *Nat. Energy* **3** 560–6

[29] Li X, Bi D, Yi C, Décoppet J-D, Luo J, Zakeeruddin S M, Hagfeldt A and Grätzel M 2016 A vacuum flash–assisted solution process for high-efficiency large-area perovskite solar cells *Science* **353** 58–62

[30] Ding B *et al* 2016 Facile and scalable fabrication of highly efficient lead iodide perovskite thin-film solar cells in air using gas pump method *ACS Appl. Mater.Interfaces* **8** 20067–73

[31] Ding B, Li Y, Huang S-Y, Chu Q-Q, Li C-X, Li C-J and Yang G-J 2017 Material nucleation/growth competition tuning towards highly reproducible planar perovskite solar cells with efficiency exceeding 20% *J. Mater. Chem.* A **5** 6840–8

[32] Zhang G *et al* 2024 Suppressing interfacial nucleation competition through supersaturation regulation for enhanced perovskite film quality and scalability *Sci. Adv.* **10** eadl6398

[33] Wu C *et al* 2019 FAPbI3 flexible solar cells with a record efficiency of 19.38% fabricated in air via ligand and additive synergetic process *Adv. Funct. Mater.* **29** 1902974

[34] Liang K, Mitzi D B and Prikas M T 1998 Synthesis and characterization of organic-inorganic perovskite thin films prepared using a versatile two-step dipping technique *Chem. Mater.* **10** 403–11
[35] Burschka J, Pellet N, Moon S-J, Humphry-Baker R, Gao P, Nazeeruddin M K and Grätzel M 2013 Sequential deposition as a route to high-performance perovskite-sensitized solar cells *Nature* **499** 316–9
[36] Ko H-S, Lee J-W and Park N-G 2015 15.76% Efficiency perovskite solar cells prepared under high relative humidity: importance of PbI_2 morphology in two-step deposition of $CH_3NH_3PbI_3$ *J. Mater. Chem.* A **3** 8808–15
[37] Im J-H, Jang I-H, Pellet N, Grätzel M and Park N-G 2014 Growth of $CH_3NH_3PbI_3$ cuboids with controlled size for high-efficiency perovskite solar cells *Nat. Nanotechnol.* **9** 927–32
[38] Yang W S, Noh J H, Jeon N J, Kim Y C, Ryu S, Seo J and Seok S I 2015 High-performance photovoltaic perovskite layers fabricated through intramolecular exchange *Science* **348** 1234–7
[39] Ummadisingu A and Grätzel M 2018 Revealing the detailed path of sequential deposition for metal halide perovskite formation *Sci. Adv.* **4** e1701402–e12
[40] Bai Y, Xiao S, Hu C, Zhang T, Meng X, Li Q, Yang Y, Wong K S, Chen H and Yang S 2017 A pure and stable intermediate phase is key to growing aligned and vertically monolithic perovskite crystals for efficient PIN planar perovskite solar cells with high processibility and stability *Nano Energy* **34** 58–68
[41] Cao J, Jing X, Yan J, Hu C, Chen R, Yin J, Li J and Zheng N 2016 Identifying the molecular structures of intermediates for optimizing the fabrication of high-quality perovskite films *J. Am. Chem. Soc.* **138** 9919–26
[42] Lee J-W, Kim H-S and Park N-G 2016 Lewis acid–base adduct approach for high efficiency perovskite solar cells *Acc. Chem. Res.* **49** 311–9
[43] Lee J-W *et al* 2018 Tuning molecular interactions for highly reproducible and efficient formamidinium perovskite solar cells via adduct approach *J. Am. Chem. Soc.* **140** 6317–24
[44] Lee J-W, Kim S-G, Bae S-H, Lee D-K, Lin O, Yang Y and Park N-G 2017 The interplay between trap density and hysteresis in planar heterojunction perovskite solar cells *Nano Lett.* **17** 4270–6
[45] Yi C, Luo J, Meloni S, Boziki A, Ashari-Astani N, Grätzel C, Zakeeruddin S M, Röthlisberger U and Grätzel M 2016 Entropic stabilization of mixed A-cation ABX_3 metal halide perovskites for high performance perovskite solar cells *Energy Environ. Sci.* **9** 656–62
[46] Jeon N J, Noh J H, Yang W S, Kim Y C, Ryu S, Seo J and Seok S I 2015 Compositional engineering of perovskite materials for high-performance solar cells *Nature* **517** 476–80
[47] Yadavalli S K, Zhou Y and Padture N P 2017 Exceptional grain growth in formamidinium lead iodide perovskite thin films induced by the δ-to-α phase transformation *ACS Energy Lett.* **3** 63–4
[48] Zhang Y, Kim S-G, Lee D, Shin H and Park N-G 2019 Bifacial stamping for high efficiency perovskite solar cells *Energy Environ.Sci.* **12** 308–21
[49] Ito S, Tanaka S, Manabe K and Nishino H 2014 Effects of surface blocking layer of Sb2S3 on nanocrystalline TiO_2 for $CH_3NH_3PbI_3$ perovskite solar cells *J. Phys. Chem.* C **118** 16995–7000

[50] Rahman N U, Khan W U, Khan S, Chen X, Khan J, Zhao J, Yang Z, Wu M and Chi Z 2019 A promising europium-based down conversion material: organic–inorganic perovskite solar cells with high photovoltaic performance and UV-light stability *J. Mater. Chem.* A **7** 6467–74
[51] Kim H-S, Jang I-H, Ahn N, Choi M, Guerrero A, Bisquert J and Park N-G 2015 Control of *I–V* hysteresis in $CH_3NH_3PbI_3$ perovskite solar cell *J. Phys. Chem. Lett.* **6** 4633–9
[52] Kim J Y, Jung H S, No J H, Kim J-R and Hong K S 2006 Influence of anatase-rutile phase transformation on dielectric properties of sol-gel derived TiO_2 thin films *J. Electroceram.* **16** 447–51
[53] Yella A, Heiniger L-P, Gao P, Nazeeruddin M K and Grätzel M 2014 Nanocrystalline rutile electron extraction layer enables low-temperature solution processed perovskite photovoltaics with 13.7% efficiency *Nano Lett.* **14** 2591–6
[54] Lee J-W, Lee T-Y, Yoo P J, Grätzel M, Mhaisalkar S and Park N-G 2014 Rutile TiO_2-based perovskite solar cells *J. Mater. Chem.* A **2** 9251–9
[55] Shahvaranfard F, Li N, Hosseinpour S, Hejazi S, Zhang K, Altomare M, Schmuki P and Brabec C J 2021 Comparison of the sputtered TiO_2 anatase and rutile thin films as electron transporting layers in perovskite solar cells *Nano Select* **3** 990–7
[56] Tan H *et al* 2017 Efficient and stable solution-processed planar perovskite solar cells via contact passivation *Science* **355** 722–6
[57] Kim B J, Kim M C, Lee D G, Lee G, Bang G J, Jeon J B, Choi M and Jung H S 2018 Perovskite solar cells: interface design of hybrid electron extraction layer for relieving hysteresis and retarding charge recombination in perovskite solar cells *Adv. Mater. Interfaces* **5** 1870113
[58] Jiang Q, Zhang X and You J 2018 SnO_2: a wonderful electron transport layer for perovskite solar cells *Small* **14** 1801154
[59] Xiong L, Guo Y, Wen J, Liu H, Yang G, Qin P and Fang G 2018 Review on the application of SnO_2 in perovskite solar cells *Adv. Funct. Mater.* **28** 1802757
[60] Song S, Kang G, Pyeon L, Lim C, Lee G-Y, Park T and Choi J 2017 Systematically optimized bilayered electron transport layer for highly efficient planar perovskite solar cells ($\eta = 21.1\%$) *ACS Energy Lett.* **2** 2667–73
[61] Chen J, Zhao X, Kim S G and Park N G 2019 Multifunctional chemical linker imidazoleacetic acid hydrochloride for 21% efficient and stable planar perovskite solar cells *Adv. Mater.* **31** 1902902
[62] Chen P, Zheng Q, Jin Z, Wang Y, Wang S, Sun W, Pan W and Wu J 2024 Buried interface engineering-assisted defects control and crystallization manipulation enables stable perovskite solar cells with efficiency exceeding 25% *Adv. Funct. Mater.* **34** 2409497
[63] Yang D, Yang R, Wang K, Wu C, Zhu X, Feng J, Ren X, Fang G, Priya S and Liu S F 2018 High efficiency planar-type perovskite solar cells with negligible hysteresis using EDTA-complexed SnO_2 *Nat. Commun.* **9** 3239–9
[64] Liu D and Kelly T L 2013 Perovskite solar cells with a planar heterojunction structure prepared using room-temperature solution processing techniques *Nat. Photonics* **8** 133–8
[65] Mahmood K, S. Swain B and Amassian A 2014 Double-layered ZnO nanostructures for efficient perovskite solar cells *Nanoscale* **6** 14674–8
[66] Son D-Y, Bae K-H, Kim H-S and Park N-G 2015 Effects of seed layer on growth of ZnO nanorod and performance of perovskite solar cell *J. Phys. Chem.* C **119** 10321–8

[67] Wang L, Liu F, Cai X, Ma T and Jiang C 2018 Indium zinc oxide electron transport layer for high-performance planar perovskite solar cells *J. Phys. Chem.* C **122** 28491–6

[68] Cao J, Wu B, Chen R, Wu Y, Hui Y, Mao B W and Zheng N 2018 Efficient, hysteresis-free, and stable perovskite solar cells with ZnO as electron-transport layer: effect of surface passivation *Adv. Mater.* **30** 1705596

[69] Chen R, Cao J, Duan Y, Hui Y, Chuong T T, Ou D, Han F, Cheng F, Huang X, Wu B *et al* 2018 High-efficiency, hysteresis-less, UV-stable perovskite solar cells with cascade ZnO–ZnS electron transport layer *J. Am. Chem. Soc.* **141** 541–7

[70] Dkhili M, Lucarelli G, De Rossi F, Taheri B, Hammedi K, Ezzaouia H, Brunetti F and Brown T M 2022 Attributes of high-performance electron transport layers for perovskite solar cells on flexible PET versus on glass *ACS Appl. Energy Mater.* **5** 4096–107

[71] Wang K, Olthof S, Subhani W S, Jiang X, Cao Y, Duan L, Wang H, Du M and Liu S 2020 Novel inorganic electron transport layers for planar perovskite solar cells: progress and prospective *Nano Energy* **68** 104289

[72] Dou J, Zhang Y, Wang Q, Abate A, Li Y and Wei M 2019 Highly efficient Zn_2SnO_4 perovskite solar cells through band alignment engineering *Chem. Commun.* **55** 14673–6

[73] Koo D, Choi Y, Kim U, Kim J, Seo J, Son E, Min H, Kang J and Park H 2024 Mesoporous structured MoS_2 as an electron transport layer for efficient and stable perovskite solar cells *Nat. Nanotechnol.* **20** 75–82

[74] Krishna B G, Ghosh D S and Tiwari S 2023 Hole and electron transport materials: a review on recent progress in organic charge transport materials for efficient, stable, and scalable perovskite solar cells *Chem. Inorg. Mater.* **1** 100026

[75] Dai S-M *et al* 2017 Pristine fullerenes mixed by vacuum-free solution process: efficient electron transport layer for planar perovskite solar cells *J. Power Sources* **339** 27–32

[76] Zou D, Yang F, Zhuang Q, Zhu M, Chen Y, You G, Lin Z, Zhen H and Ling Q 2019 Perylene diimide-based electron-transporting material forperovskite solar cells with undoped Poly(3-hexylthiophene) as hole-transporting material *Chem. Sus. Chem.* **12** 1155–61

[77] Jung S K *et al* 2018 Homochiral asymmetric-shaped electron-transporting materials for efficient non-fullerene perovskite solar cells *Chem. Sus. Chem.* **12** 224–30

[78] Liu J, Wu Y, Qin C, Yang X, Yasuda T, Islam A, Zhang K, Peng W, Chen W and Han L 2014 A dopant-free hole-transporting material for efficient and stable perovskite solar cells *Energy Environ. Sci.* **7** 2963–7

[79] Kazim S, Ramos F J, Gao P, Nazeeruddin M K, Grätzel M and Ahmad S 2015 A dopant free linear acene derivative as a hole transport material for perovskite pigmented solar cells *Energy Environ. Sci.* **8** 1816–23

[80] Wang L *et al* 2018 Design and synthesis of dopant-free organic hole-transport materials for perovskite solar cells *Chem. Commun.* **54** 9571–4

[81] Jung E H, Jeon N J, Park E Y, Moon C S, Shin T J, Yang T-Y, Noh J H and Seo J 2019 Efficient, stable and scalable perovskite solar cells using poly(3-hexylthiophene) *Nature* **567** 511–5

[82] Jeng J Y, Chen K C, Chiang T Y, Lin P Y, Tsai T D, Chang Y C, Guo T F, Chen P, Wen T C and Hsu Y J 2014 Nickel oxide electrode interlayer in $CH_3NH_3PbI_3$ perovskite/PCBM planar-heterojunction hybrid solar cells *Adv. Mater.* **26** 4107–13

[83] Zhang H, Zhao C, Yao J and Choy W C H 2023 Dopant-free NiO_x nanocrystals: a low-cost and stable hole transport material for commercializing perovskite optoelectronics *Angew. Chem. Int. Ed.* **62** e202219307
[84] Li S, Wang X, Li H, Fang J, Wang D, Xie G, Lin D, He S and Qiu L 2023 Low-temperature chemical bath deposition of conformal and compact NiOX for scalable and efficient perovskite solar modules *Small* **19** 2301110
[85] Arora N, Dar M I, Hinderhofer A, Pellet N, Schreiber F, Zakeeruddin S M and Grätzel M 2017 Perovskite solar cells with CuSCN hole extraction layers yield stabilized efficiencies greater than 20% *Science* **358** 768–71
[86] Liu J, Pathak S K, Sakai N, Sheng R, Bai S, Wang Z and Snaith H J 2016 Identification and mitigation of a critical interfacial instability in perovskite solar cells employing copper thiocyanate hole-transporter *Adv. Mater. Interfaces* **3** 1600571
[87] Chatterjee S and Pal A J 2016 Introducing Cu_2O thin films as a hole-transport layer in efficient planar perovskite solar cell structures *J. Phys. Chem.* C **120** 1428–37
[88] Jeong S, Seo S and Shin H 2018 p-Type CuCrO(2) particulate films as the hole transporting layer for $CH_3NH_3PbI_3$ perovskite solar cells *RSC Adv.* **8** 27956–62
[89] Chen Y, Yang Z, Jia X, Wu Y, Yuan N, Ding J, Zhang W-H and Liu S 2019 Thermally stable methylammonium-free inverted perovskite solar cells with Zn^{2+}-doped $CuGaO_2$ as efficient mesoporous hole-transporting layer *Nano Energy* **61** 148–57
[90] Akin S, Liu Y, Dar M I, Zakeeruddin S M, Grätzel M, Turan S and Sonmezoglu S 2018 Hydrothermally processed $CuCrO_2$ nanoparticles as an inorganic hole transporting material for low-cost perovskite solar cells with superior stability *J. Mater. Chem.* A **6** 20327–37
[91] Yang B, Ouyang D, Huang Z, Ren X, Zhang H and Choy W C H 2019 Multifunctional synthesis approach of In:$CuCrO_2$ nanoparticles for hole transport layer in high-performance perovskite solar cells *Adv. Funct. Mater.* **29** 1902600
[92] Zhang H, Wang H, Chen W and Jen A K Y 2016 $CuGaO_2$: a promising inorganic hole-transporting material for highly efficient and stable perovskite solar cells *Adv. Mater.* **29** 1604984
[93] Noel N K *et al* 2014 Lead-free organic–inorganic tin halide perovskites for photovoltaic applications *Energy Environ. Sci.* **7** 3061–8
[94] Hao F, Stoumpos C C, Cao D H, Chang R P H and Kanatzidis M G 2014 Lead-free solid-state organic–inorganic halide perovskite solar cells *Nat. Photonics* **8** 489–94
[95] Ran C *et al* 2019 Conjugated organic cations enable efficient self-healing $FASnI_3$ solar cells *Joule* **3** 3072–87
[96] Chen J *et al* 2024 Efficient tin-based perovskite solar cells with trans-isomeric fulleropyrrolidine additives *Nat. Photonics* **18** 464–70
[97] Krishnamoorthy T *et al* 2015 Lead-free germanium iodide perovskite materials for photovoltaic applications *J. Mater. Chem.* A **3** 23829–32
[98] Liu D, Li Q, Jing H and Wu K 2019 Pressure-induced effects in the inorganic halide perovskite $CsGeI_3$ *RSC Adv.* **9** 3279–84
[99] Ray D, Clark C, Pham H Q, Borycz J, Holmes R J, Aydil E S and Gagliardi L 2018 Computational study of structural and electronic properties of lead-free $CsMI_3$ perovskites (M = Ge, Sn, Pb, Mg, Ca, Sr, and Ba) *J. Phys. Chem.* C **122** 7838–48

[100] Kopacic I, Friesenbichler B, Hoefler S F, Kunert B, Plank H, Rath T and Trimmel G 2018 Enhanced performance of germanium halide perovskite solar cells through compositional engineering *ACS Appl. Energy Mater.* **1** 343–7

[101] Ito N, Kamarudin M A, Hirotani D, Zhang Y, Shen Q, Ogomi Y, Iikubo S, Minemoto T, Yoshino K and Hayase S 2018 Mixed Sn–Ge perovskite for enhanced perovskite solar cell performance in air *J. Phys. Chem. Lett.* **9** 1682–8

[102] Ng C H *et al* 2019 Role of GeI_2 and SnF_2 additives for SnGe perovskite solar cells *Nano Energy* **58** 130–7

[103] Huang J, Wang H, Jia C, Yang H, Tang Y, Gou K, Zhou Y and Zhang D 2024 High-efficiency and ultra-stable cesiumbismuth-based lead-free perovskite solar cells without modification *J. Phys. Chem. Lett.* **15** 3383–9

[104] Jain S M, Edvinsson T and Durrant J R 2019 Green fabrication of stable lead-free bismuth based perovskite solar cells using a non-toxic solvent *Commun. Chem.* **2** 91

[105] Xu J, Castriotta L A, Di Carlo A and Brown T M 2024 Air-stable lead-free antimony-based perovskite inspired solar cells and modules *ACS Energy Lett.* **9** 671–8

[106] Mailoa J P, Bailie C D, Johlin E C, Hoke E T, Akey A J, Nguyen W H, McGehee M D and Buonassisi T 2015 A 2-terminal perovskite/silicon multijunction solar cell enabled by a silicon tunnel junction *Appl. Phys. Lett.* **106** 121105

[107] Werner J, Weng C-H, Walter A, Fesquet L, Seif J P, De Wolf S, Niesen B and Ballif C 2015 Efficient monolithic perovskite/silicon tandem solar cell with cell area > 1 cm^2 *J. Phys. Chem. Lett.* **7** 161–6

[108] Bush K A *et al* 2017 23.6%-Efficient monolithic perovskite/silicon tandem solar cells with improved stability *Nat. Energy* **2** 17009

[109] Sahli F *et al* 2018 Fully textured monolithic perovskite/silicon tandem solar cells with 25.2% power conversion efficiency *Nat. Mater.* **17** 820–6

[110] Tockhorn P *et al* 2022 Nano-optical designs for high-efficiency monolithic perovskite-silicon tandem solar cells *Nat. Nanotechnol.* **17** 1214–21

[111] Hou Y *et al* 2020 Efficient tandem solar cells with solution-processed perovskite on textured crystalline silicon *Science* **367** 1135–40

[112] Kim D *et al* 2020 Efficient, stable silicon tandem cells enabled by anion-engineered wide-bandgap perovskites *Science* **368** 155–60

[113] Xu J *et al* 2020 Triple-halide wide–band gap perovskites with suppressed phase segregation for efficient tandems *Science* **367** 1097–104

[114] Liu J *et al* 2024 Perovskite/silicon tandem solar cells with bilayer interface passivation *Nature* **635** 596–603

[115] Fu F, Feurer T, Weiss, Thomas P, Pisoni S, Avancini E, Andres C, Buecheler S, Tiwari and Ayodhya N 2016 High-efficiency inverted semi-transparent planar perovskite solar cells in substrate configuration *Nat. Energy* **2** 16190

[116] Leijtens T, Bush K A, Prasanna R and McGehee M D 2018 Opportunities and challenges for tandem solar cells using metal halide perovskite semiconductors *Nat. Energy* **3** 828–38

[117] Kim D H *et al* 2019 Bimolecular additives improve wide-bandgap perovskites for efficient tandem solar cells with CIGS *Joule* **3** 1734–45

[118] Liang H *et al* 2023 29.9%-Efficient, commercially viable perovskite/$CuInSe_2$ thin-film tandem solar cells *Joule* **7** 2859–72

[119] Eperon G E *et al* 2016 Perovskite-perovskite tandem photovoltaics with optimized band gaps *Science* **354** 861–5

[120] Tong J *et al* 2019 Carrier lifetimes of 1 μs in Sn–Pb perovskites enable efficient all-perovskite tandem solar cells *Science* **364** 475–9
[121] Lin R *et al* 2023 All-perovskite tandem solar cells with 3D/3D bilayer perovskite heterojunction *Nature* **620** 994–1000
[122] Zhang D, Li D, Hu Y, Mei A and Han H 2022 Degradation pathways in perovskite solar cells and how to meet international standards *Commun. Mater.* **3** 58
[123] Leguy A M A *et al* 2015 Reversible hydration of $CH_3NH_3PbI_3$ in films, single crystals, and solar cells *Chem. Mater.* **27** 3397–407
[124] Philippe B, Park B-W, Lindblad R, Oscarsson J, Ahmadi S, Johansson E M J and Rensmo H 2015 Chemical and electronic structure characterization of lead halide perovskites and stability behavior under different exposures—a photoelectron spectroscopy investigation *Chem. Mater.* **27** 1720–31
[125] Shi L *et al* 2020 Gas chromatography–mass spectrometry analyses of encapsulated stable perovskite solar cells *Science* **368** 6497
[126] Seo S, Jeong S, Bae C, Park N G and Shin H 2018 Perovskite solar cells: perovskite solar cells with inorganic electron- and hole-transport layers exhibiting long-term (≈500 h) stability at 85 °C under continuous 1 sun illumination in ambient air *Adv. Mater.* **30** 1870210
[127] Ni Z *et al* 2021 Evolution of defects during the degradation of metal halide perovskite solar cells under reverse bias and illumination *Nat. Energy* **7** 65–73
[128] Schoonman J 2015 Organic–inorganic lead halide perovskite solar cell materials: a possible stability problem *Chem. Phys. Lett.* **619** 193–5
[129] Dipta S S and Uddin A 2021 Stability issues of perovskite solar cells: a critical review *Energy Technol.* **9** 2100560
[130] Ji J, Liu X, Jiang H, Duan M, Liu B, Huang H, Wei D, Li Y and Li M 2020 Two-stage ultraviolet degradation of perovskite solar cells induced by the oxygen vacancy-Ti^{4+} states *Science* **23** 101013–3
[131] Bella F, Griffini G, Correa-Baena J-P, Saracco G, Grätzel M, Hagfeldt A, Turri S and Gerbaldi C 2016 Improving efficiency and stability of perovskite solar cells with photocurable fluoropolymers *Science* **354** 203–6
[132] Li S *et al* 2020 Van der Waals mixed valence tin oxides for perovskite solar cells as UV-stable electron transport materials *Nano Lett.* **20** 8178–84
[133] Abdelmageed G, Jewell L, Hellier K, Seymour L, Luo B, Bridges F, Zhang J Z and Carter S 2016 Mechanisms for light induced degradation in MAPbI3 perovskite thin films and solar cells *Appl. Phys. Lett.* **109** 233905
[134] Kato Y, Ono L K, Lee M V, Wang S, Raga S R and Qi Y 2015 Perovskite solar cells: silver iodide formation in methyl ammonium lead iodide perovskite solar cells with silver top electrodes *Adv. Mater. Interfaces* **2** n/a–n/a
[135] Domanski K, Correa-Baena J-P, Mine N, Nazeeruddin M K, Abate A, Saliba M, Tress W, Hagfeldt A and Grätzel M 2016 Not all that glitters is gold: metal-migration-induced degradation in perovskite solar cells *ACS Nano* **10** 6306–14
[136] Seetharaman S M, Nagarjuna P, Kumar P N, Singh S P, Deepa M and Namboothiry M A G 2014 Efficient organic–inorganic hybrid perovskite solar cells processed in air *Phys. Chem. Chem. Phys.* **16** 24691–6
[137] Grancini G *et al* 2017 One-year stable perovskite solar cells by 2D/3D interface engineering *Nat. Commun.* **8** 15684-15684

[138] Beal R E, Slotcavage D J, Leijtens T, Bowring A R, Belisle R A, Nguyen W H, Burkhard G F, Hoke E T and McGehee M D 2016 Cesium lead halide perovskites with improved stability for tandem solar cells *J. Phys. Chem. Lett.* **7** 746–51

[139] Bush K A, Bailie C D, Chen Y, Bowring A R, Wang W, Ma W, Leijtens T, Moghadam F and McGehee M D 2016 Thermal and environmental stability of semi-transparent perovskite solar cells for tandems enabled by a solution-processed nanoparticle buffer layer and sputtered ITO electrode *Adv. Mater.* **28** 3937–43

[140] Velilla E, Jaramillo F and Mora-Seró I 2021 High-throughput analysis of the ideality factor to evaluate the outdoor performance of perovskite solar minimodules *Nat. Energy* **6** 54–62

[141] Cai M, Wu Y, Chen H, Yang X, Qiang Y and Han L 2016 Cost-performance analysis of perovskite solar modules *Adv. Sci.* **4** 1600269–19

Chapter 8

Organic and polymer-based cells

Organic and polymer-based cells are photovoltaic devices that use small-molecule organic materials or polymers to absorb and convert sunlight into electricity. These devices utilize complex donor–acceptor (D–A) interfaces within their active layers, where charge separation and transport mechanisms are highly dependent on molecular organization and morphological control at the nanoscale. Critically, molecular engineering of the organic and polymer photo-absorber allows for tunable bandgap, increased absorption coefficient and improved optical and electronic properties. Compared to silicon-based devices, organic photovoltaic devices are fabricated by simple and cost-effective solution-processing techniques such as electrodeposition, spin-coating, spray deposition, printing technologies and roll-to-roll manufacturing. The organic photo-absorbers are usually soluble in organic solvents, which make it possible to produce liquid inks and apply them by printing them to flexible polymeric substrates. This allows these devices to be used in flexible and large-area solar cells. They are also lightweight, which is important for small autonomous sensors, potentially disposable, customizable on the molecular level and potentially have less adverse environmental impact. One of the attractive features of organic solar cells (OSCs) is that they can be made from a wide range of materials giving rise to a range of device architectures such as bilayer OSCs, bulk heterojunction (BHJ) OSCs and tandem/multijunction OSCs. In addition to their light weight, organic and polymer solar cells also have the potential to exhibit high transparency, which is particular attractive for applications in windows, walls and flexible electronics. Much of the research and development in OSCs is focussed on improving the device performance in terms of efficiency and cost, long-term stability and environmental sustainability. However, the lifespan of OSCs is far less compared to traditional solar cells. This is due to the nature and rate of degradation of the organic materials when exposed to the environment. Recent breakthroughs have seen all-polymer solar cells achieve power conversion efficiency (PCE) now over 20%. Notwithstanding these advancements, the fundamental challenge with

doi:10.1088/978-0-7503-3255-2ch8

© IOP Publishing Ltd 2025. All rights, including for text and data mining (TDM), artificial intelligence (AI) training, and similar technologies, are reserved.

optimizing device performance lies in balancing the competing requirements of efficient light absorption, charge separation, and charge transport while maintaining long-term operational stability under real-world conditions and is crucial for large-scale adoption.

8.1 Working principles of organic and polymer-based solar cells

While OSCs have the same fundamental structure as traditional or inorganic solar cells (ISCs), OSCs use conjugated small organic molecules and polymers instead, as their photoactive material. However, while the efficiencies of OSCs are still somewhat low relative to other solar cells, their advantages, such as being lightweight, flexible, having lower material and processing costs, failure resistant, and relative environmentally benign nature make them particularly attractive. Overall, OSCs can be distinguished by the production technique, the type and characteristics of the materials and the device configuration. Device architectures are generally categorized as: (1) simple; (2) conventional single-layer cell; (3) bilayer heterojunction; (4) BHJ, with the diffuse bilayer heterojunction as an intermediate between the bilayer and the BHJ types; and (5) tandem cells—some of these cell types are illustrated in figure 8.1 [1]. Though the single-layer cell comprises one photoactive material, the other architectures comprise an electron donor (D) and an electron acceptor (A) and therefore have different charge generation mechanisms. While ISCs utilize the bandgap of semiconductors, small organic molecules or polymers in OSCs do not have bandgaps. Instead, they utilize the highest occupied molecular orbital (HOMO) and lowest unoccupied energy orbital (LUMO) for photoexcitation and charge generation.

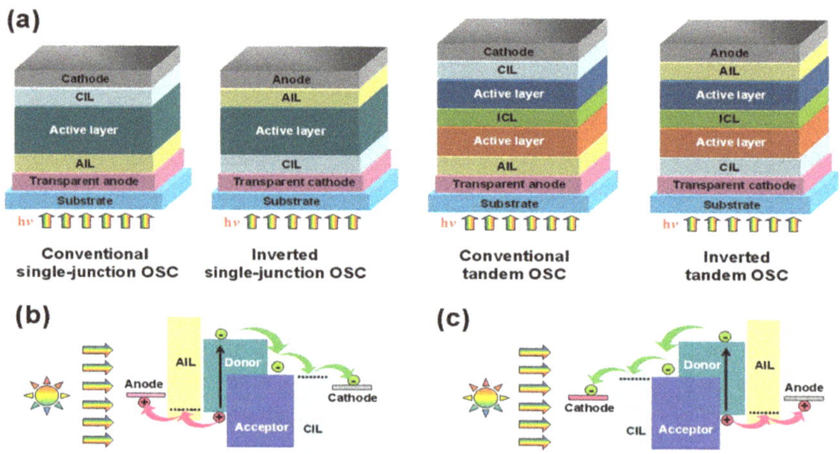

Figure 8.1. (a) Schematic device architectures of conventional and inverted single-junction/tandem OSCs. AIL: anode interface layer; CIL: cathode interface layer; ICL: interconnecting layer. Schematic illustration of the energy level diagrams and the main charge-transporting processes in (b) conventional and (c) inverted OSCs. Reprinted with permission from [1] CC BY 4.0.

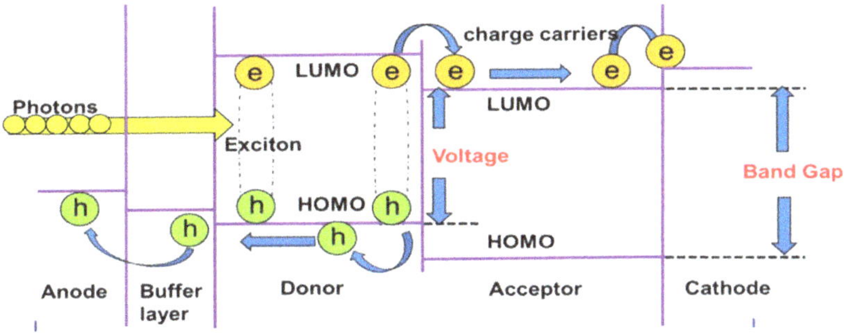

Figure 8.2. General working principle of OSCs. Reprinted from [2]. Copyright 2016, with permission from Elsevier.

As with other cell types, the primary mechanism of electrical current generation of OSCs comprises: (1) charge carrier generation through light absorption; (2) diffusion of carriers across contact layers; (3) charge separation and then extraction, as shown in figure 8.2. For a typical OSC, a photoactive layer, whether a single, BHJ or bilayer planar heterojunction is sandwiched between the cathode and anode with their corresponding interlayers [1]. The conventional single-junction device comprises a composite active layer between a modified transparent anode such as indium tin oxide (ITO), and a low work-function (WF) metal cathode (such as calcium and aluminium). Upon light irradiation of the active layer, the photogenerated excitons diffuse towards the D–A interface and separate into holes and electrons in the HOMO of the donor and the LUMO of the acceptor, respectively. Here, the LUMO and HOMO of the D–A interface of the molecular semiconductor correspond to the conduction and valance bands in inorganic semiconductors. After separation, the charge carriers transport within the respective phases of the active layer, until they are collected by the respective electrodes. In these cells, the energy level structure at the electrode interfaces plays an essential role, where an ideal one needs good Ohmic contact with minimum resistance and high charge selectivity to prevent carriers from reaching the opposite electrodes. Ideally, this can be achieved with interfacial materials of adequate WFs to match the HOMO and LUMO levels of donor and acceptor materials interfaced with the active layer and electrodes, thereby enhancing the collection efficiencies of holes and electrons on the anode and cathode, respectively. In these cells, excitons have a relatively short lifetime (<1 ns), requiring fast charge dissociation in order to prevent charge recombination [2]. Importantly, in order to produce effective free charge carriers, the distance between the generated excitons to the nearest D–A interface must be within their diffusion length.

8.2 Structures and properties of photoactive materials

One of the attractive features of OSCs is that organic photoactive semiconductors are less expensive alternatives to inorganic semiconductors. Additionally, they can have extremely high optical absorption coefficients which offer the possibility for the production of very thin solar cells, making them promising for a wide range of applications. The essential characteristic of organic photoactive semiconductors is that their electronic structure is based on conjugated π-electrons that are critical for light absorption and emission, charge generation and transport. Over time, the increasing PCEs of OSCs have benefited from the developments of new donor and acceptor photoactive materials in the active layer alongside innovations of the device structure and geometry [1]. However, a main contributor to the progress made can be attributed to the development of new light-absorbing materials. These range from the simple homopolymers, including MDMO-PPV (a dialkoxy substituted poly(p-phenylenevinylene) based conjugating polymer) and P3HT (regioregular poly(3-hexylthiophene)), as donor and fullerene derivatives as acceptors—the main materials at the genesis of this field with hundreds of more complex materials introduced with improved optoelectronic characteristics [3]. Typically, donor and acceptor materials in OSCs should possess the following properties: (1) matched absorption spectrum; (2) suitable molecular energy level alignment; (3) nanoscale phase separation; and (4) high charge carrier mobility. Figure 8.3 shows a range of typical donor and acceptor molecules used in OSCs. In tandem, significant advances have also been made in other integral components of OSCs, including electrodes and interlayers and this is particularly important in driving efficiencies, since much energy loss is associated with interfacial and recombination losses.

8.2.1 Acceptor materials

Acceptor molecules in OSCs are fundamental components that play a critical role in the charge separation process. These molecules accept electrons from donor materials after excitons are generated upon light absorption. There are a range of acceptor molecules that have been developed and incorporated into OSCs with improved performance over time (figure 8.3). However, the period of development of acceptor materials can be characterized as either fullerene dominant or non-fullerene acceptor (NFA) dominant, with the latter predicated on work by Lin and co-workers in 2015 which reported that NFA's displayed better absorption in the visible region, higher electron mobility and improved D–A miscibility [4]. Herein, is described these two categories of acceptor organic molecules for OSCs.

8.2.1.1 Fullerene derivatives as acceptors

The widely used fullerene derivatives, such as phenyl-C61-butyric acid methyl ester ($PC_{61}BM$), as electron acceptors in OSCs efficiently accept electrons and facilitate charge transport due to their high electron affinity. Fullerene derivatives are typically based on C60 or C70 fullerenes and consist of a spherical carbon cage core structure that is modified with various functional groups to enhance solubility and electronic characteristics in OSCs. They are compatible with a variety of donor materials, including both conjugated polymers and small molecules used in OSCs, allowing for effective charge separation and transport. They also exhibit good photostability compared to many non-fullerene acceptors, which contributes to device stability. A key attribute is their ability to be functionalized which improves their solubility in organic solvents, facilitating solution processing for device

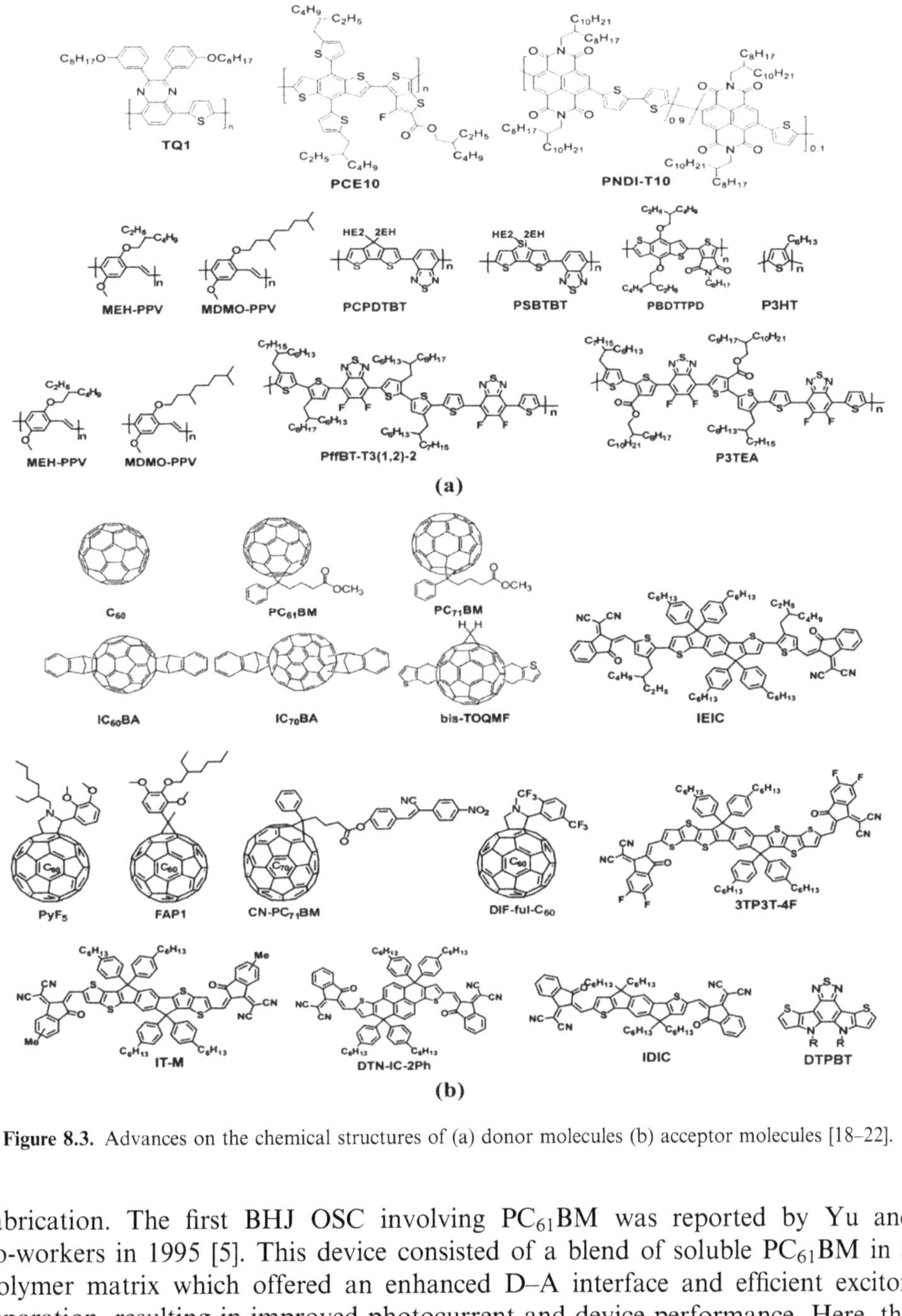

Figure 8.3. Advances on the chemical structures of (a) donor molecules (b) acceptor molecules [18–22].

fabrication. The first BHJ OSC involving $PC_{61}BM$ was reported by Yu and co-workers in 1995 [5]. This device consisted of a blend of soluble $PC_{61}BM$ in a polymer matrix which offered an enhanced D–A interface and efficient exciton separation, resulting in improved photocurrent and device performance. Here, the composite films of the donor, poly(2-methoxy-5-(2′-ethylhexyloxy)-1,4-phenylene-vinylene) (MEH-PPV) and $PC_{61}BM$ exhibited a PCE of 2.9%, more than two orders of magnitude higher than devices made with pure MEH-PPV. The efficient charge separation was attributed to photo-induced electron transfer from the MEH-PPV to $PC_{61}BM$ with high collection efficiency due to the bi-continuous network of the D–A interface. Later, a significant increase in PCE of 4.4% for an OSC device was reported in 2005 by Li and co-workers [6] involving a P3HT:$PC_{61}BM$ blend—P3HT = poly(3-hexylthiophene-2,5-diyl). This resulted in an increased carrier mobility and charge transport, with further development leading to a device performance of 5% [7]. This was attributed to improved nanoscale morphology, the increased crystallinity of the semiconducting polymer, and the improved contact to the electron-collecting electrode which facilitated charge generation, charge transport to, and charge collection at the electrodes, thereby enhancing the device efficiency by lowering the series resistance in the device.

Overall, the success of fullerenes as acceptor molecules has been due to their versatile functional chemistry which has led to a range of these derivatives with variations in the aryl group, the alkyl chain length, linkages such as the ester group, and the fullerene cage, examples of which are shown in figure 8.3 [8]. In particular, Kooistra and co-workers [9] introduced methoxy, methyl sulfur, or fluorine atoms to the benzene ring of $PC_{61}BM$ and it is found that the type and position of substituents influence the LUMO level. Furthermore, the introduction of a methoxy group to the electron group improved the LUMO level of the PCBM material, yielding a higher V_{oc}. Furthermore, Zhang and co-workers substituted triarylamine and 9,9-dimethyl fluorene for the benzene ring to synthesize TPA-$PC_{61}BM$ and MF-$PC_{61}BM$, respectively [10]. The triarylamine and 9,9-dimethyl fluorene possessing stronger electron-donating properties than the benzene ring led to an increase in the LUMO level of the molecules and V_{oc} of the cell. However, to address the drawbacks of $PC_{61}BM$, such as poor solubility and low absorption, and to further improve the PCE of fullerene-based solar cells, fullerene derivatives with higher fullerene, C_{70} were discovered. For example, work by Troshin and co-workers led to a PCE of 4.1% using P3HT:$PC_{71}BM$ as the active layer of BHJ OSCs [11]. This led to developments with a significant breakthrough in device performance achieved in 2014, where Liu and co-workers reported a new donor material, PffBT4T-2OD and a device PCE of 10.4% achieved by a PffBT4T-2OD:$PC_{71}BM$ polymer blend [12] They suggested that the high performance of this device with fill factors (FFs) up to 77% were due to a near ideal highly crystalline polymer:fullerene blend morphology that consisted of reasonably small polymer domains.

Alternative to $PC_{61}BM$ and $PC_{71}BM$, there are several other efficient fullerene derivatives acting as acceptors in OSCs, such as, bis$PC_{61}BM$, indene-C60 bisadduct (ICBA) ICBA, among others. In 2008, Lenes and co-workers reported bis$PC_{61}BM$, the bisadduct analogue of $PC_{61}BM$ which has a higher LUMO energy level of ~0.1 eV compared to $PC_{61}BM$ and resulted in a significant enhancement of V_{oc} in a P3HT:bis$PC_{61}BM$ solar cell with a PCE of 4.5% was demonstrated [13]. Similarly, He and co-workers synthesized a new soluble C_{60} derivative, indene-C60 bisadduct (ICBA), which exhibited a LUMO energy level 0.17 eV higher than that of $PC_{61}BM$. The OSCs based on P3HT:ICBA demonstrated a PCE of 5.44%, with a V_{oc} of 0.84 V under the illumination of AM1.5, 100 mW cm^{-2}, while the OSC based on P3HT/PCBM displayed a lower V_{oc} of 0.58 V and PCE of 3.88% under the same experimental conditions [14]. Later, this was further optimized to a PCE of 6.48% by Zhao and co-workers [15]. Furthermore, He and co-workers developed a cell with a soluble indene-C70 bisadduct ($IC_{70}BM$) acceptor, which showed a V_{oc} of 0.84 V and PCE of 5.79% which was further improved to 6.67% by combining $IC_{70}BM$ with low bandgap polymer (PTB7) as a donor [16, 17].

8.2.1.2 Non-fullerene acceptors (NFAs)

Although fullerene acceptors have gained tremendous attention for BHJ OSCs, they have their own intrinsic limitations. These include: (1) limited tunability of chemical structure and energy levels; (2) weak visible and near-infrared (NIR) absorption; (3) morphological instability; and (4) high synthetic cost [3]. These drawbacks have led to the development of NFAs including small molecules and polymers which possess distinct advantages in optical absorptivity, bandgap tunability, and frontier orbital energy levels, thereby yielding higher V_{oc} and J_{sc}.

Small-molecule NFAs

Non-fullerene small-molecule acceptors are characterized by a controllable energy level, convenient synthesis, easy processing and good solubility. They have received extensive attention in recent years because they have better absorption properties in the visible region than fullerenes and their derivative acceptor materials [8]. Of the small-molecule NFAs, those with the A–D–A type structure have been very effective in achieving high PCEs. These possess a conjugated 'push–pull' structure where 'A' and 'D' represent the electron withdrawing and electron-donating moieties, respectively. One of the early reports which laid the foundation for these types of acceptors was in 2015 involving the novel electron acceptor, ITIC, where OSCs fabricated using PTB7-Th as donor and ITIC as acceptor yielded a promising PCE of 6.8%, at that time [3, 4]. The structure of ITIC includes an indacenodithieno[3,2-b]thiophene (IDTT), end-capped with 2-(3-oxo-2,3-dihydroinden-1-ylidene)malononitrile (INCN) group, and four 4-hexylphenyl groups. In ITIC, the carbonyl and cyano groups of INCN down-shifted the LUMO energy level and the push–pull structure helped induce intramolecular charge transfer and extended absorption within the 500–800 nm range, and the 4-hexylphenyl groups played a role in restricting aggregation. This discovery led to the development of a range of these types of acceptors including, IDIC based on a five-ring fused core, indacenodithiophene (IDT), [23, 24] IOIC3, based on a naphtha[1,2-b:5,6-b′]dithiophene (NDT) core with alkoxy side-chains, [25] and FOIC whose core unit entailed a 2-(5/6-fluoro-3-oxo-2,3-dihydro-1H-inden-1-ylidene)malononitrile (FIC) as the electron-deficient end group [24]. Overall, these systems exhibited strong absorption within the 500–950 nm range and improved cell efficiencies up to 13% [3].

Significant improvements in PCEs were attained with devices employing Y6 NFA. This started with work by Yuan and co-workers who reported this new class of NFA involving a ladder-type electron-deficient-core-based central fused ring, dithienothiophen[3,2-b]-pyrrolobenzothiadiazole (TPBT) which displayed HOMO and LUMO energy levels of −5.65 eV and −4.10 eV, respectively [26]. Subsequent work reported by Liu and co-workers involved an efficient co-polymer donor

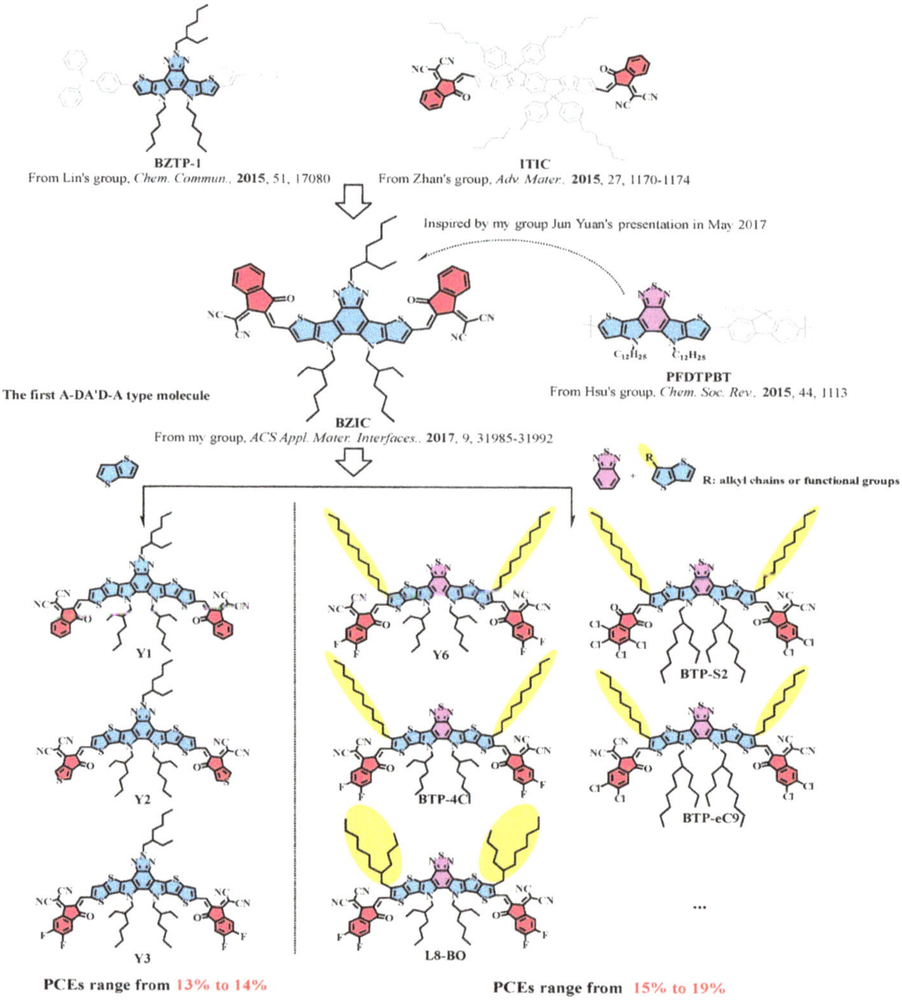

Figure 8.4. Diagram showing the progress of A–DA′D–A type non-fullerene small-molecule acceptors. Reprinted from [28]. Copyright 2022 with permission from Elsevier.

material, D18 which in combination with Y6 for a single-junction cell with a structure of ITO/PEDOT:PSS/D18:Y6/PDIN/Ag, exhibited exceptional device characteristics of PCE of 18.22% and J_{sc} of 27.70 mA cm^{-2} [27]. The Y6 class NFAs (figure 8.4) are unique molecular structures of acceptor–donor–acceptor–donor–acceptor (A–DA'D–A) characterized by a banana-shaped geometry that facilitates diverse π–π stacking in single-crystal structures, leading to a continuous and highly organized 3D network [28]. These unique arrangements enhance the effective delocalization of electron wavefunctions, diminishing Coulomb attraction between electron–hole pairs and optimizing transport channels for charge carriers. This allows Y6 class NFAs to be functionalized as bipolar materials, displaying balanced hole and electron transport properties, which, in turn, enhances exciton generation with minimal voltage loss. In particular, recent advances in the Y6 class NFAs, such as design of electron deficient cores, modification of terminal groups, incorporation of fused frameworks, and the addition of inner/outer side chains, have successfully boosted the PCEs over 20% [29].

For example, recently, Chen and co-workers [29] developed three new Y6-derived NFAs, namely, BTP-H15, BTP-H13, and BTP-H17, by introducing saturated octyl chains and unsaturated octenyl or octynyl chains to the outer positions of the DA′D core. In these, the incorporation of unsaturated chains effectively reduces the spatial steric hindrance between the outer side chains and the fused skeletons, thus enhancing their planarity. Although BTP-H13 exhibited stronger aggregation than the other two NFAs, the lack of an efficient co-oriented half-skeleton stacking mode resulted in a relatively loose 3D network structure and poor charge transport characteristics. However, BTP-H15 displayed comprehensive advantages, including optimal framework planarity, close π–π stacking distances, and a robust 3D network, leading to superior morphology and enhanced charge transport. Consequently, the additive-free OSCs based on BTP-H15 realized impressive efficiencies of 18.46% in the binary devices and 19.36% in the ternary devices.

Polymeric NFAs

Polymer acceptor materials are important alternatives to fullerene acceptors for OSCs because they more conveniently regulate HOMO and LUMO energy levels for enhanced electronic properties. Additionally, polymer acceptors have some unique advantages, including structural flexibility, morphological stability, and outstanding thermal and mechanical properties which make them more convenient for the preparation of large-area devices [8]. The first polymer acceptor, cyanogen-modified polyethylene (CN-PPV) was produced in 1995 by Yu and Heeger [30] who described an all-polymer OSC of CN-PPV with PCE values of 0.25% under light irradiation at 20 mW cm^{-2}. To date, the most common building blocks for polymeric NFAs in all-polymer solar cells include perylene diimide (PDI), naphthalene diimide (NDI), bithiophene imide (BTI), and B $\leftarrow$ N-bridged bipyridine (BN-Py) [3, 8]. In addition, adapting from the development of small-molecular NFAs, an effective approach for obtaining efficient polymer acceptors has been based on D–A co-polymer configuration where the small-molecular NFAs are utilized as acceptor (A) units with donor (D) units. Because of the D–A push–pull effect of electrons within the molecules, the effective separation of light excitons can be promoted and the HOMO and LUMO energy levels and can be effectively regulated by changing either unit.

Some of the earliest work was reported by Zhang and co-workers [31] involving novel *n*-type 2D-conjugated polymer based on bithienyl-benzodithiophene (BDT) and perylene diimide (PDI), P(PDI-BDT-T). The polymer exhibits broad absorption in the visible region and a LUMO level of −3.89 eV, which is slightly higher than that of PCBM. OSCs with P(PDI-BDT-T) as the acceptor and PTB7-Th as the donor demonstrated a PCE of 4.71% with a J_{sc} of 11.51 mA cm^{-2}, V_{oc} of 0.80 V, and FF of 51.1%. Prior to that in 2008, Guo and Watson first synthesized a new narrow bandgap polymer acceptor N2200 using naphthalimide (NDI) [32]. Since then, N2200 has been widely studied as a classic polymer acceptor material in these types of solar cells. Subsequently, Zheng and co-workers developed high-performance ternary OSCs by incorporating N2200 into a binary blend film comprising of a wide bandgap conjugated polymer PTzBI-2FP and a small-molecule non-fullerene acceptor ITIC-4F [33]. The resulting ternary devices exhibited a competitive PCE of 13.0%, which was attributable to the efficient energy transfer, enhanced charge carrier mobility, and improved morphology of the device.

Further developments in 2017 by Zhang and co-workers [34] involved the preparation of the polymerized small-molecule acceptor (PSMA), PZ1, based on the small-molecule acceptor IDIC-C16. This was prepared via the co-polymerization of idic-C16-Br and thiophenone. PZ1 displays an absorption band edge around 800 nm with a slightly higher LUMO level than that of IDIC-C16. The all-polymer OSC with PZ1 as the accepter and PBDBT-T as the donor exhibited a V_{oc} of 0.83 V, J_{sc} of 16.05 mA cm^{-2}, FF of 68.99%, and PCE of 9.19%. It was also found that PZ1 has good solubility in organic solvents generating good film-forming performance along with improved thermal stability relative to IDIC-C16. This development which involved embedding an acceptor–donor–acceptor-structured organic semiconductor building block into a polymer main chain creating narrow bandgap polymer acceptors paved the way for the design of PSMAs. In recent years, with the emergence of the class of non-fullerene acceptor, Y6, described in the foregoing, PSMA materials based on Y6 have received increased attention. Jia and co-workers [35] synthesized a novel PSMA acceptor, PJ1 by co-polymerizing Y6 with thiophene, which has a narrow bandgap of 1.4 eV and a high absorption coefficient of 1.39×10^5 cm^{-1}. When PJ1 was blended with the donor polymer PBDB-T, an all-polymer device with PCE of 14.4% was achieved, which was mainly attributed to the broad absorption, efficient charge separation and collection, and low energy loss. These cells demonstrated much better thermal stability than the control device based on a small-molecule acceptor (TTPBT-IC). These developments have proved key for improving the PCEs of all-polymer solar cells, exceeding 18% especially with respect to the molecular weight of polymer acceptors and their interactions with donors, and how they influence the morphological and mechanical properties of the device. For example, Bi and co-workers [36] systematically studied the effect of the molecular weight of polymer acceptors on the phase transition process, morphology, and photovoltaic properties by using the PYIT monomer (PYIT1), and low and high molecular weight PYIT. It was found that tuning the molecular weight effectively regulated the phase transition process of the polymer acceptor and its interaction with the polymer donor, impacting the aggregation characteristics of the polymers. With PBQx-Cl as the donor and PYIT2 as the acceptor, a high-performance binary all-polymer OSC was achieved with an efficiency of 18.39%.

8.2.2 Donor materials

Like acceptor molecules, donor molecules in OSCs are crucial components that serve as electron donors to facilitate the generation of electrical energy from sunlight. These molecules absorb photons, creating excitons that dissociate into free charges, which can then be collected as electric current. There are a range of donor molecules that have been developed and incorporated into OSCs with improved performance over time (figure 8.3) but in general can be classified into polymer or small organic molecules. In recent years, solution-processed small-molecule-based OSCs (SM-OSCs) have emerged as competitive alternatives to their polymer counterparts due to several important advantages of small molecules, such as well-defined structures and therefore less batch-to-batch variation, easier band structure control, etc [37].

8.2.2.1 Polymeric donors

In addition to their narrow dispersity and high material purity, an ideal donor should possess other essential characteristics, including high crystallinity, suitable morphology, high charge carrier mobility, frontier orbitals with energy levels matching the requirements, and enhanced photon absorption in the higher-energy area of the solar spectrum. In the early stages of the development of OSCs, the organic polymers, PPV and P3HT were the more utilized donor molecules. PPV represents a rigid-rod conducting polymer family member of poly(*p*-phenylenevinylene) with a narrow bandgap and in combination with the wide bandgap semiconducting P3HT, poly(3-hexylthiophene) can be treated into highly ordered crystalline thin films. Early developments with these polymer donors entailed OSCs with blends of the MEH-PPV donor and $PC_{61}BM$ acceptor as the active layer of OSCs which exhibited PCEs of 2.5% [38] and OSCs with P3HT exhibited PCEs of 3%–5% [7, 15, 39] with optimized cells involving indene-C_{60} bisadduct (ICBA) as acceptor yielding PCEs as high as 6.48% with an open-circuit voltage of 0.84 V, a short-circuit current of 10.61 mA cm^{-2}, and a FF of 72.7% [15].

Major breakthroughs in polymer donor developments involved thiophene derivatives such as the narrow bandgap semiconductor, PTBI reported by Liang and co-workers which constituted thieno[3,4-b]thiophene (TT) and benzodithiophene (BDT) units [40]. Polymer solar cells based on PTB1 and methanofullerene [6,6]-phenyl-C_{71}-butyric acid methyl esters ($PC_{71}BM$) exhibited a PCE of 5.6%, an external quantum efficiency of 67% and FF of 65%. This polymer which has a more rigid backbone and higher mobility than P3HT, displays a fairly narrow bandgap of ~1.63 eV, with the HOMO and LUMO energy levels being −4.90 and −3.20 eV, respectively. This development led to a series of PTB-based narrow bandgap polymers with improved solubility and modified LUMO and HOMO energy levels through various n-octyl, 2-butyloctyl, 2-ethylhexyloxy and side chain substituents [41]. In particular, introduction of electron withdrawing fluorine into the polymer backbone results in the fluorinated polymer, PTB4, which lowers the LUMO and HOMO energy levels to −3.31 and −5.12 eV, respectively, and PTB7 with lower HOMO levels. At that time, a high PCE up to 7.40% was achieved from PTB7:$PC_{71}BM$-based OSCs, which were the first polymer solar cells exhibiting a PCE over 7% [42].

Further developments were achieved with the series of donor polymers, PffBT4T, such as PffBT4T-2OD in relation to PTB7-based materials, which demonstrated an optimum morphology containing highly crystalline and sufficiently pure yet reasonably small polymer domains in the OSCs [12]. For these donors, the

2-octyldodecyl alkyl chains connected to the quaterthiophene was the key structural characteristic that led to the polymer's highly temperature-dependent aggregation behaviour, which then enabled the controlled aggregation and strong crystallization of PffBT4T-2OD during film formation. In addition, PffBT4T-2OD exhibited a bandgap of 1.65 eV with enhanced hole mobility in the polymer:fullerene blend film. These donors enabled at that time a record high PCE of 10.8%, high FF of 75%, and thick films over 250 nm in the devices containing TC71BM as acceptor and later on, an OSC with high PCE of 11.7% based on the PffBT4T-C9C13:PC71BM system [43].

Similar to acceptor molecules, significant advancements in donor molecular capabilities were made with all-conjugated alternating co-polymers adopting a D–π–A structure with primary ones including PBDB-T, PM6, and PM7 [44–47]. They consist of an electron-donating unit, benzodithiophene (BDT), and an electron-accepting unit, benzodithiophenedione (BDD). In particular, Hou and co-workers carried out various modifications to improve the optical and electrical properties of PBDB-T analogues such as the fluorinated polymer, PBDB-T-SF which displayed down-shifted HOMO/LUMO energy levels and improved absorption compared with its nonfluorinated counterpart [48]. Attributed to the suitable energy level alignment and enhanced absorption, the devices based on ITO/ZnO/PBDB-T-SF:IT-4F/MoO3/Al yielded at that time a champion PCE of 13.10%. In 2019, Yuan and co-workers enhanced the performance of these devices by incorporating PM7, a chlorinated derivative of PM6 in conjunction with the aforementioned highly effective Y6 acceptor molecule [26, 47]. The key difference between PM7 (also described as PBDB-T-2Cl) and PM6 (also described as PBDB-T-2F) lies in the nature of their BDT units. PM7 features chlorinated thienyl side chains, while PM6 incorporates fluorinated thienyl side chains. Consequently, PM7 exhibited a slightly lower HOMO energy level at −5.52 eV, in contrast to PM6's HOMO level at −5.50 eV. Devices comprising PM7:Y6 combination achieved a higher PCE of 17.04% with the J_{sc} of 25.644 mA cm^{-2} compared to the corresponding PM6:Y6 OSC device, which had a PCE of 15.94% [49]. Notably, the variance in HOMO energy levels had an effect on the V_{oc} of the devices. The PM7:Y6 device achieved a V_{oc} of 0.882 V, surpassing the 0.939 V of the PM6:Y6 device. Importantly, this shift in energy levels allowed for efficient charge transfer while maintaining an adequate energy level offset between the donor and acceptor components for both PM6 or PM7.

Recent developments of efficient polymer donors involved wide bandgap co-polymers based on a fused-ring aromatic lactone (FRAL) building blocks pioneered by Xiong and co-workers [3, 50]. For these, the strong electron withdrawing capability and extended molecular plane allows these donors to exhibit deep HOMO energy levels and high hole mobility. For example, the co-polymer, D16 with bandgap of 1.95 eV consists of a 5H-dithieno[3,2-b:2′,3′-d]thiopyran-5-one (DTTP) unit and has HOMO and LUMO energy levels of −5.48 and −2.83 eV, respectively and enhanced π–π stacking resulting in high hole mobility [50]. By optimizing the D/A ratio, active layer thickness, and additives, OSCs of the structure, ITO/PEDOT:PSS/D16:Y6/PDIN/Al displayed a PCE of 16.72%. Later, a more efficient co-polymer, D18 with a wide bandgap of 1.98 eV, was developed where another fused-ring acceptor unit, dithieno[3′,2′:3,4;2″,3″:5,6]benzo[1,2-c][1,2,5]thiadiazole (DTBT) with a larger molecular plane and a higher hole mobility than that of D16, was utilized [27]. OSCs with D18:Y6 blend film yielded a PCE of 18.22%, with a J_{sc} as high as 27.70 mA cm^{-2}. Additionally, a device consisting of the acceptor, N3 to create a D18:N3 film displayed a PCE as high as 18.56% [51], and another device with an efficient chlorinated co-polymer donor, D18-Cl, which displayed slightly deeper HOMO and LUMO energy levels had a high PCE of 18.13% [52].

8.2.3 Small-molecule donors

Small-molecule donors for OSCs include a range of molecular systems such as oligothiophene, benzodithiophene-, naphthodithiophene-, and porphyrin-based materials [53]. Relative to polymer donors, small-molecule donors possess the advantages of well-defined molecular weight, easy purification and small batch-to-batch variations for solution-processed OSCs [54]. Furthermore, enhanced crystallization, as well as their tendency to obtain high phase purity and specific crystal orientations enable fabrication of OSCs with high charge mobility and low energy losses. However, the biggest challenge with small-molecule donors is to control the interpenetrating networks in SM-OSCs, since the inefficient charge transport pathways would lead to excessive exciton recombination, decreased charge carrier mobility and unbalanced charge transport. This therefore requires small-molecule donors that are matched well with the acceptor materials for optimal morphology towards efficient devices. Accordingly, in recent years it has been demonstrated that the development of novel small-molecule donors is crucial for achieving highly efficient SM-OSCs.

Tremendous progress in the development donors for OSCs has been made with benzo[1,2-b:4,5-b′]dithiophene (BDT)—and this has been the most attractive electron-donating unit [37]. The BDT unit contains a symmetric and planar conjugated structure that causes a strong intermolecular orbital overlap, enhances the electron delocalization and π–π stacking in solid-state thin films, and leads to efficient charge transport in the device. The engineering approach for small-molecular donors based on the BDT structure includes, backbone modulation, side chain optimization, end-group engineering, and functional substitutions. The symmetrical and planar conjugated molecular backbone of the A–π–D–π–A-types with a central BDT unit could facilitate π–π stacking. In particular, Zhou and co-workers [54] reported the development of a DTBDT-based small-molecule donor named ZR1 with an A–π–D–π–A architecture. The electron-rich DTBDT, dithieno[2,3-d:2′,3′-d′]benzo [1,2-b:4,5-b′] dithiophene is an analogue of benzodithiophene (BDT), and possesses a larger coplanar core as the donor (D) which is used to extend the conjugated plane and improve the planarity of the molecule with the bithiophene as a π-bridge to deepen the HOMO level and increase the rigidity. Accordingly, ZR1:Y6-based devices were optimized to the highest PCE of 14.34% due to the improved conjugated planarity and broad absorption. Overall, BDT-based molecules display properties such as the charge carrier transport, reduction in the conformational disorder of the backbone, and increase in the molecular planarity to facilitate electron delocalization in the solid-state rendering more crystalline films with more ordered morphology [54]. Notably, previous work involved a DTBDT-based polymer PDBT-T1 reported by Huo and co-workers [55] incorporated in a device that exhibited a PCE of 9.75% and high FF of 75% due to the formation of optimized fibril network morphology, which was beneficial for efficient photogenerated exciton dissociation and charge collection —showing the potential of the BDT donor component for good charge transfer and underscores their incorporation in small-molecule donor structures. Additionally, Zhou and co-workers [56] achieved a PCE of 13.2% for ternary solar cells by using a combination of small molecules with both fullerene and non-fullerene acceptors, which form a hierarchical morphology consisting of a PCBM transporting channel and an intricate non-fullerene phase-separated pathway network. Here, it was demonstrated that for this device involving this donor component, carrier generation and transport were optimized and voltage loss was simultaneously reduced. Furthermore, in 2022, Cai and co-workers [57] reported two BDT-based molecules, C-F and C-2F, using side chain engineering with a symmetrically difluorinated benzene ring on the BDT donor core. In particular, the C-2F:N3-based devices achieved a PCE of 14.64% with a J_{sc} of 24.87 mA cm^{-2}, a V_{oc} of 0.85 V, and an FF of 69.33%. The enhancement in the PCE of C-2F:N3 was due to compact molecular packing, increased crystallinity, enhanced phase separation, and increased and more balanced charge carrier mobility.

One approach that has been used is to improve the morphology of the active layer by structural engineering. In one case reported by Feng and co-workers, [58] a new BDT-based molecule, CNS-6–8, was incorporated in a host PM6:Y6:$PC_{71}BM$ layer to further tune the morphology of the active layer. Due to the favourable miscibility of CNS-6-8 in PM6, Y6, and $PC_{71}BM$, the quaternary system achieved good phase separation morphology, enhanced crystallinity, exciton separation, and charge transport carrier, which led to competitive PCE of 18.07%, a high FF of 78.8%, and a very low voltage loss of 0.54 eV compared to ternary blends (PM6:Y6: PC71BM, 0.56 eV). Ma and co-workers in 2023, [59] reported three molecules, BO-1, HD-1, and OD-1, with alkylated thiazole side groups which differ only in the alkyl side chain length on the thiazole unit. The length of alkyl side chain significantly influenced the crystallinity and blend morphology of the active layer. Owing to good BHJ active layer morphology, the device with the HD-1 based molecule achieved a competitive PCE of 17.19% with the BTP-eC9 acceptor. More recently, Zhang and co-workers developed highly crystalline small-molecule donors, B3TR and B2 that enable over 17% efficiency for SM-OSCs [60]. In comparison with B3TR, the skeleton of B2 includes an additional benzo[1,2-*b*:4,5-*b*′]dithiophene (BDT) unit, but lacks two alkylated thiophene units. B2 when blended with the non-fullerene acceptor BTP-eC9 forms a film that exhibits a smaller π–π separation and a more favourable BHJ morphology. The device exhibited a PCE of 9.8% with a V_{oc} of 0.876 V which was enhanced to a PCE of 17.1% after solvent annealing. On the other hand, the B3TR:BTP-eC9-based device achieved a PCE of 2.36%, with an V_{oc} of 0.870 V but after annealing, demonstrated an enhanced PCE of 14.8% with a significantly reduced V_{oc} of 0.809 V.

Overall, small-molecule donors have low-lying HOMO levels with efficient absorptions in the visible range, good coplanar structures with ordered molecular packing and suitable crystallinity, and high hole mobilities [53]. These characteristics are beneficial for realizing high V_{oc}, J_{sc}, and FF values in OSCs. In particular, the BDT-based A–D–A small molecules demonstrate their superior performance, and further modifying their side-chains and π-bridges is a promising method to further improve the efficiency. These outcomes underscore the idea that the development of highly crystalline donor materials is one of the promising strategies for achieving high-performance SM-OSCs.

8.3 Cell architecture, device physics and performance

The device architecture of OSCs plays a significant role in determining their efficiency, stability, and performance. OSCs typically consist of several distinct layers, each serving a specific function and their choice and characteristics individually and in combination are critical in optimizing device performance. The basic architecture of an OSC generally includes the following components: (1) substrate—typically made of glass or flexible plastic, important for light transmission; anode—also important for light transmission and is usually a transparent conductive oxide (TCO) such as indium tin oxide (ITO) and conducts holes; (3) the active photoconductive layer of donor and acceptor which generates excitons on light absorption; and (4) the cathode—typically comprises a low WF metal, such as aluminum which collects electrons from the active layer and must be chosen to ensure proper energy level alignment with the acceptor material/layer. As outlined in the foregoing, the development of OSCs has involved several device configurations resulting in an increase in PCEs over the years—some of the major advancements in types are discussed in the following.

8.3.1 Single-layer OSCs

The most basic type of OSC is a Schottky diode which is a single-layer cell, composed of a single thin polymer layer sandwiched in between two electrodes; one of which gives an ohmic contact (e.g., TCO in the case of a *p*-type OSC) and the other one gives a rectifying contact. Upon photon absorption in the photoactive polymer layer, excitons are created with the electrons promoted into the LUMO energy level. The potential difference created by the electrodes aids the separation process of the excitons, allowing electrons to migrate to the cathode and holes to migrate the anode. Unfortunately, single-layer OSCs are typically inefficient because the potential difference across the active layer is not high enough to separate the photoexcited charge carries which travel through the same material of short diffusion length, and resulting in high recombination losses. Consequently, the current density is low, yielding low power density and efficiency. Some types of materials used for the single polymer layer include phthalocyanine, polyflourenes, polypyrenes, and polythiophenes.

Notwithstanding the inherent limitations of single-layer OSCs, developments involving various active polymer layers and solution fabrications methods which have improved device performance over the years. For example, Yuan and co-workers [26] reported a single-layer OSC with the non-fullerene acceptor, Y6, which exhibited a high efficiency of 15.7%. The device was fabricated in the configuration of the traditional sandwich structure with an ITO glass positive electrode and a PDINO/Al negative electrode in ICCAS. A thin layer (150 nm) of PEDOT:PSS (poly(3,4-ethylene dioxythiophene): poly(styrene sulfonate)) was prepared by spin-coating the PEDOT:PSS solution filtered through a 0.45 mm poly(tetrafluoro- ethylene) (PTFE) filter at 3000 rpm for 40 s on the cleaned ITO substrate. The PEDOT:PSS film (40 nm) was achieved by baking at 150 °C for 15 min in the air. The polymer PM6:Y6 (D:A = 1:1.2, 16 mg ml^{-1} in total) was dissolved in chloroform with the solvent additive of 1-chloronaphthalene (0.5%, v/v) and spin-casted at 3000 rpm for 30 s onto the PEDOT:PSS layer. Also, recently Chong and co-workers [61] reported single-junction OSCs of 19.05% PCE with improved charge extraction and suppressed charge recombination through the combination of side-chain engineering of new non-fullerene acceptors, ternary blends, and volatilizable solid additives. The 2D side chains on the BTP-Th active layer induced a certain steric hindrance for molecular packing and phase separation, which was mitigated by fluorination of side chains on BTP-FTh. Moreover, by introducing two highly crystalline molecules as the second acceptor and volatilizable solid additive, respectively, into the BTP-FTh-based host blend, the molecular crystallinity was significantly improved and the blend morphology optimized. Similarly, Sun and co-workers [62] showed single-junction OSCs with a high PCE of 19.17%. These cells incorporated an asymmetric guest acceptor BTP-2F2Cl into a PM1:L8-BO host blend. Compared with the L8-BO neat film, the L8-BO:BTP-2F2Cl blend film showed higher photoluminescence quantum yield and larger exciton diffusion length. Introducing BTP-2F2Cl into the host blend extended its absorption spectrum, improved the molecular packing of the host materials, and suppressed the non-radiative charge recombination of the OSCs.

8.3.2 Bilayer OSCs

Bilayer OSCs especially in the early years of development were a significant improvement to single-layer OSCs with the inclusion of an electron-accepting layer creating a D–A interface between the electrodes. In these, the introduction of an electron acceptor layer between the photoactive polymer material and the cathode electrode, improves the exciton diffusion length. The advantage of this configuration is that the disparate electric potential of the donor and acceptor ensures that there is more efficient migration and charge separation at the interface, increasing the exciton diffusion length. This is due to the offset between the HOMO and LUMO levels of both materials which are sufficient to separate the charge more effectively allowing for better transport to the respective electrode and thereby decreasing recombination probability of the carriers. However, one issue with these cells is that the D–A interface is still very small and only excitons near the depletion layer can reach it and become dissociated. In particular, if the LUMO level of the acceptor is sufficiently higher than the LUMO level of the donor, the excited electron will relax into the acceptor LUMO and separate from the hole. For sufficient LUMO level disparity, the charge separation is much more efficient at the D–A interface than at the electrode interface. Additionally, whilst bilayer OSCs are fundamentally more efficient than single-layer OSCs, their efficiencies are still relatively unimpressive. This primarily stems from excitons requiring more energy to cross the interface boundary to eventually reach the electrode. Additionally, the thickness of the polymer layer needs to be sufficient in order to minimize transmittance of photons. Another drawback is that since photons can only supply excitons with a limited amount of energy, excitons rarely are able to travel to the electrode, as the interface and thick polymer layers do not allow for a long diffusion length.

One of the first bilayer OSCs was that reported by Tang in 1986 [63]. The cell was comprised of a copper phthalocyanine and a perylene derivative bilayer, sandwiched between an ITO and silver (Ag) contact. The then novel cell had a FF of 0.65 and efficiency of 1% which were attributed to the interface between the two organic materials, rather than the electrode/organic contacts. However, the D–A interface still limited the efficient exciton diffusion and separation, thus not yielding high PCEs. However, an effective way to eliminate the necessity to maintain the complex nanoscale D–A network, which compromised the charge transport in these OSCs over time, was by adopting a layer-by-layer (LBL) architecture fabricated through a sequential processing approach, developed in recent years [64–66]. This alongside developments of highly effective non-fullerene acceptors, enabled the use of the LBL method to process bilayer architectures of donor and acceptor layers independently without involving the time-consuming optimization procedures to fine-tune the nanoscale morphology. In the LBL architecture, each layer has only one type of organic semiconductor and, hence, a more condensed packing and thermodynamically stable morphology for each layer can be obtained, which contrasts with the intrinsically unstable nanoscale mixed domains in the BHJ architecture (described below). As such, Kumari and co-workers [66] developed a high-performance PM6/Y7 lBL OSC by carefully optimizing its structure resulting in a conversion efficiency of 16.21%, surpassing the BHJ counterpart. Notably, this device also outperformed the traditional BHJ device in terms of long-term photostability and thermal stability under continuous illumination and temperature (85 °C) for approximately 1000 h, with similar results obtained for eight other non-fullerene acceptor systems. The improved long-term photostability and thermal stability in these LBL systems was ascribed to a mitigation of the strong phase aggregation seen in the BHJ films.

Also in recent years, quasi-planar heterojunction (Q-PHJ) OSCs with a bilayer-type D–A phase distribution and a nanoscale BHJ region at the D–A interface have attracted significant attention for their efficient exciton dissociation and charge transfer towards more efficient and stable devices [65, 67, 68]. A Q-PHJ device combines the advantages of a nanoscale BHJ film in the D–A interface and the PHJ film in the main region of the device, forming a gradient structure and optimizing the vertical distribution of carriers [65]. Here, the carrier concentration presents a distribution ordered by a vertical gradient, and constructs an effective separation and transport channel for electrons and holes. In general, orthogonal solvents are usually used to cast the donor and acceptor layer sequentially to prepare the film. In this case, the vertical phase separation of donor and acceptor in the bilayer film is easier to manipulate than the random morphology of a BHJ film. To date, the PCE of Q-PHJ OSCs has approached the level of BHJ OSCs [69]. As an example, Lai and co-workers [67] prepared a co-acceptor-based Q-PHJ (*co*-Q-PHJ) OSC with enhanced crystallinity achieved by using two structurally compatible acceptors (BTIC-BO4Cl-βδ and ITCC). With a higher LUMO energy level and a wider bandgap than BTIC-BO4Cl-βδ, ITCC broadens the absorption range and supports charge transfer for *co*-Q-PHJ devices. When combined with D18 donor, the *co*-Q-PHJ OSCs fabricated with separately depositing donor and acceptor layers by using orthogonal solvents achieved a remarkable efficiency of 17.3%, much higher than that of the control ternary BHJ OSCs (16.0%). Furthermore, Wei and co-workers [70] used sequential deposition of a D18 donor and an L8-BO acceptor film to fabricate devices which showed a very high PCE of 19.05%, exceeding that of the corresponding BHJ device.

8.3.3 Bulk heterojunction OSCs

To solve the issue of geminate recombination of carriers and short diffusion length of the D–A interface, BHJ cells were developed which have accounted for most of the reported high-efficiency OSCs in their early years of development. In the past decades, a *p*-type conjugated polymer and an *n*-type fullerene derivative have been the most commonly used acceptor and donor, respectively [71]. The tremendous progress was made in PCEs owing to the rapid development of non-fullerene acceptors [72]. In particular, the successful breakthroughs in the Y6 series NFAs boosted the single-junction BHJ OSCs efficiencies over 18% [73]. BHJ OSCs combine advantages of single-layer with bilayer OSCs by removing the rough interface of the bilayer by blending small domains of electron donors and acceptors into a single layer. The BHJ active layer, which is prepared by mixing the donor and acceptor materials together and then spin-coating or blade-coating the blends, provides a sufficient D–A interface on the nanoscale leading to more efficient exciton dissociation [74]. By introducing islands of electron donors in the electron acceptor region, the diffusion length of the exciton is shortened, increasing charge carrier migration to the electrodes before recombination. Additionally, the very small and numerous donor and acceptor domains result in a smoother interface compared to the bilayer OSCs. As such, BHJ OSCs are able to supply more electric current with the same amount of photon energy, thereby improving efficiency. However, the BHJ active layer has its inherent disadvantages. Its nanoscale active layer morphology is dependent on the solution deposition and drying conditions and it is very common that the thermodynamically favourable morphology does not yield the best device performance. In some cases, the inhomogeneous matrix consists of isolated donor or acceptor islands which confine charges hindering their transport to the corresponding cathode and anode, resulting in charge loss.

To circumvent these challenges, a graded BHJ (G-BHJ) strategy realized by sequential deposition (SD) or the LBL method has been developed [73, 75]. In this method, the D and A layers are deposited sequentially, enabling optimization and regulation of the D and A individually with well-maintained crystallinity. Compared with BHJ OSCs, LBL solution-processed OSCs separately adjust different layers, making the components distribute ideally in the vertical direction that is beneficial for exciton dissociation, charge transport, and charge collection [75]. Moreover, the LBL approach has better potential in the preparation of large-area devices, which is key for the commercialization of OSCs. In particular, Zhang and co-workers [73] developed a series of G-BHJ OSCs based on PM6 donor and BTP-eC9 NFA via non-halogenated solvent sequential deposition (figure 8.5). Spin-coated G-BHJ OSCs demonstrated a high PCE of 17.48% which was attributed to the polymer/NFA composition of high crystallinity gradient distributions. These devices were compared to non-halogenated solvent enabled G-BHJ OSC via open-air blade-coating which had a PCE 16.77%. The blade coated G-BHJ was characterized with drastically different D–A crystallization kinetics, which suppressed the excessive aggregation and induced unfavourable phase separation in the BHJ. As an alternative approach, Cheng and co-workers [76] prepared OSCs featuring an active layer comprising double-BHJ structures, featuring binary blends of a polymer donor and concentration gradients of two small-molecule acceptors. After forming the first BHJ structure by spin-coating, the second BHJ layer was transfer-printed onto the first using polydimethylsiloxane stamps. A specially designed selenium heterocyclic small-molecule acceptor (Y6-Se-4Cl) was employed as the second acceptor in the BHJ—both acceptors formed a gradient concentration profile across the active layer, thereby facilitating charge transportation. PCEs for the double-BHJ-structured devices incorporating PM6:Y6-Se-4Cl/PM6:Y6 and PM6:Y6-Se-4Cl/PM6:IT-4Cl were 16.4% and 15.8%, respectively, which were higher than those of devices having one-BHJ structures based on PM6:Y6-Se-4Cl (15.0%), PM6:Y6 (15.4%), and PM6:IT-4Cl (11.6%), presumably because of the favourable vertical concentration

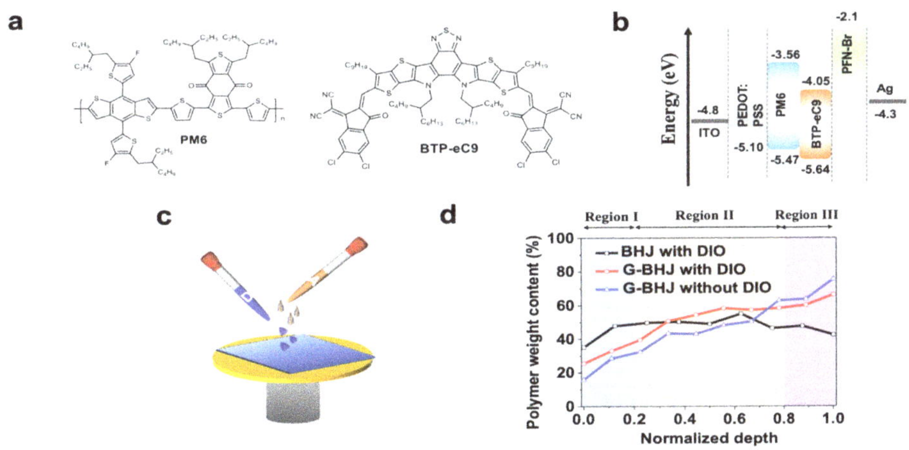

Figure 8.5. Materials, optoelectronic properties and vertical composition distribution in BHJ and G-BHJ films. (a) Chemical structure of PM6 and BTP-eC9. (b) Schematic illustration of the conventional device structure of OSCs. (c) Schematic diagram of the sequential deposition spin coating procedure. (d) Variation of polymer weight content of BHJ with DIO, G-BHJ without DIO and G-BHJ with DIO films throughout the whole film. Reprinted from [73] CC BY 4.0.

gradient of the selenium-containing small-molecule Y6-Se-4Cl in the active layer, as well as some complementary light absorption. They showed that combining two BHJ structures with a concentration gradient of the two small-molecule acceptors can be an effective approach for enhancing the PCEs of OSCs. Overall, these outcomes demonstrate G-BHJ and double-BHJ as feasible and promising strategies towards highly efficient, eco-friendly and commercializable OSCs in terms of manufacturing.

8.3.4 Pseudo-bilayer OSCs

Pseudo-bilayer (PB) OSCs represent an architecture that combines the advantages of both bilayer and BHJ designs. This configuration aims to optimize charge generation and transport efficiency—solution for the trade-off between exciton dissociation and charge collection, while maintaining a relatively simple fabrication process. The PB architecture is composed of three layers: a pure donor (or acceptor) layer at the bottom, a D–A mixture layer in between, and a pure acceptor (or donor) layer on top. Such developments have been spurred by efforts involving the NFA derivatives that have enhanced charge transport. Some of these developments involve work by Sun and co-workers [72] in which PB OSCs based on NFAs were fabricated through LBL solution processing which exhibited better performance than the conventional BHJ architecture. Additionally, sequential deposition in the FTAZ:IT-M binary and FTAZ:DPP-3T:PEBM ternary blends in PB layered OSCs demonstrated improved molecular packing and domain purity with higher performance than those of the reference BHJ devices [77]. Based on these work, Jiang and co-workers [65] developed high-efficiency OSCs with a PB architecture that outperformed conventional BHJ devices containing the donor, PM6 and a mixture of acceptors, N3 and $PC_{71}BM$. These cells possessed longer exciton diffusion length due to higher film crystallinity enabling efficient exciton dissociation and charge transport, higher short-circuit current density, FF, and a PCE of 17.42% compared to their BHJ counterpart (16.44%). Also, PB planar heterojunction OSC devices prepared by LBL sequential deposition exhibited an ideal vertical phase separation morphology with the structure of higher donor concentration near the anode and higher acceptor composition near the cathode, and the vertical graded D–A mixing in the intermediate region [74]. The active layer provided a relatively large D–A interface for exciton dissociation, and charge carriers transport to electrodes through their respective channels. These devices outperformed traditional BHJ OSCs with a PCE of 19.05%.

Recently, Liu and co-workers [78] constructed high-performance sequential deposition (SD) fabricated PB OSCs based on D18 and Y6 using chloroform as the single solvent. Benefiting from the finely adjusting of the vertical phase distribution, the SD low molecular weight D18/Y6 PB device realized high short-circuit current density of 27.24 mA cm^{-2} and a PCE of 17.94%. This work demonstrated a single solvent strategy for enabling high-performance SD OSCs and revealed the reasons that accounted for the distinct performances of these OSCs fabricated by different approaches. Also, Jo and co-workers [79] constructed a

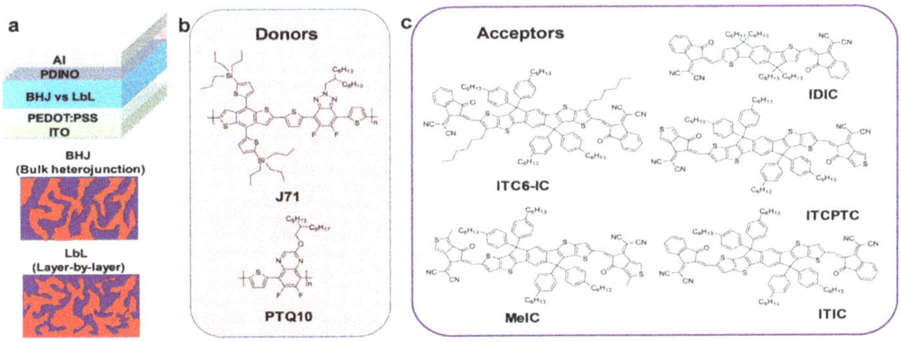

Figure 8.6. (a) Schematic device architecture of OSCs and the common active layer configurations. Chemical structures of the: (b) donor and (c) acceptor materials. Reprinted from [72] with permission from Royal Society of Chemistry.

bilayer structure using a small molecule, Y7-BO, as a bottom layer and a polymer (PM6) as a top layer. The PM6 layer was formed on water via the Marangoni phenomenon and directly stamped onto the Y7-BO layer. They developed a PB architecture incorporating a low volatility solvent additive into the polymer solution and allowed the additive to reside within the bilayer. To optimize intermixing at the D–A interface, an appropriately low amount of solvent additive (chloronaphthalene) was added, which reinforced the face-on molecular orientation of the active layers. The resulting inverted OSC achieved a PCE of 17.12%, higher than that achieved by its BHJ counterparts (figure 8.6) [72].

8.3.5 Tandem/multijunction OSCs

In general, multijunction solar cells are designed as a series device structure monolithically connecting two wide bandgap and narrow bandgap sub-cells with complementary absorption spectra. The first layer features a wide bandgap material to reduce the thermalization loss for high-energy photons thereby utilizing the spectrum more efficiently. The second layer of lower bandgap absorbs the low-energy photons that transmits the first layer. In such a configuration, a tandem cell providing improved overlap with the solar spectrum results in reduced thermalization and transmission losses than each of the corresponding single-junction cells, [80–82] and is key to overcoming the Shockley–Queisser limit of single-junction cells [83]. The device configuration is such that the two complementary absorber layers are stacked optically and electrically (figure 8.7) [80]. The interconnecting layer (ICL) between the two sub-cells must allow light to pass to sustain the photocurrent by providing an optically transparent electrical contact for recombination of electrons and holes from the adjacent photoactive layers. The Fermi level of the hole transport layer (HTL) and the electron transport layer (ETL) that jointly form the ICL must match the HOMO and LUMO levels in the adjacent photoactive layers (figure 8.7). Additionally, the ICL should not cause voltage losses and have low resistance.

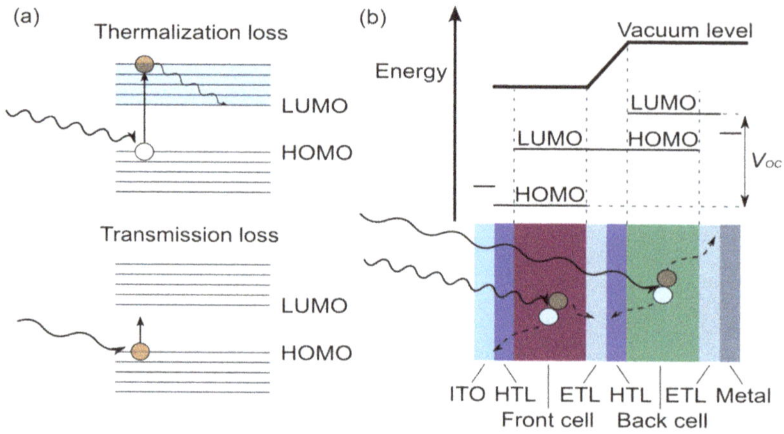

Figure 8.7. (a) Thermalization and transmission losses. (b) Arrangement of functional layers and energy levels in an organic tandem solar cell. The ETL/ HTL stack between the sub-cells forms the interconnecting layer. Reprinted with permission from [80]. Copyright 2019 Wiley-VCH Verlag GmbH & Co. KGaA.

The first reported fully solution-processed tandem polymer solar cell by Gilot and co-workers [84] in 2007 showed the potential of such devices. Following this work Kim and co-workers [85] developed similar cells which achieved a PCE of 6.5%. Both cells featured an ICL layer of poly(3,4-ethylenedioxythiophene):polystyrene sulfonate (PEDOT:PSS) as HTL, stacked on top of either a zinc oxide or a titanium oxide layer as ETL. Despite the critical nature of the other components, such as the photoactive blends and low bandgap absorbers, the properties of tandem OSCs significantly depend on the ICL, for series connection. ICLs which consist of combinations of polymeric and metal oxide materials must display key characteristics of optical transparency, uniformity, mechanical robustness, solvent orthogonality during processing, matching with the relevant HOMO and LUMO levels, and ohmic character. Although tandem OSCs based on ICLs such as molybdenum oxide (MoOx)/Ag/zinc oxide (ZnO), ZnO/n-PEDOT:PSS, and m-PEDOT:PSS/ZnO have demonstrated superior performance relative to single-junction OSCs, limitations still existed in the effectiveness of the ICLs. However, advances have been reported by Zheng and co-workers, [86] who developed an ICL composed of electron beam evaporated TiOx and PEDOT:PSS to obtain a sharp, smooth, and dense interface. This approach improved the charge recombination between two sub-cells deriving a tandem cell with 20.27% efficiency. Later, Wang and co-workers [87] fabricated another type of tandem OSC with a PCE of 20.6%. They started with a ternary OSC by introducing an asymmetric small-molecule acceptor AITC into PBDB-TCl:BTP-eC9 system and demonstrated its effectiveness in simultaneously decreasing energy disorder and non-radiative voltage losses. Here, introduction of AITC was found to modify domain size and increase the degree of crystallinity, which enhanced open-circuit voltage and PCE (19.1%). The tandem cell was constructed by selecting the wide bandgap polymer PFBCPZ to blend with AITC as a bottom sub-cell active layer material to ensure the ideal match of current with top sub-cell PM6:AITC:BTP-eC9 top active layer—the sub-cells stacked inverted. By optimizing the thickness of the active layer, the tandem OSCs with active layer thicknesses of 90 and 130 nm in the bottom and top sub-cells, respectively, exhibit a V_{oc} of 2.02 V, J_{sc} of 13.3 mA cm^{-2}, FF of 76.6% and a PCE of 20.6%.

In recent years, with the development and performance of perovskite solar cells, the development of perovskite–organic tandem solar cell is a promising way to surpass the Shockley–Queisser limit [88]. These cells combine the advantages of both perovskite and organic materials to achieve higher efficiency [89]. For example, Brinkmann and co-workers [83] demonstrated perovskite–organic tandem cells with PCE of 24.0% and a high V_{oc} of 2.15 V. These devices had the *p–i–n*-type architecture comprising a narrow bandgap sub-cell with the polymers PM6 and NFA Y6 with molybdenum oxide (MoO_x) as the hole extraction layer and a bilayer of C_{60}/2,9-dimethyl-4,7-diphenyl-1,10-phenanthroline (BCP) for efficient electron extraction. The sub-cells were connected by an ultrathin (1.5 nm) metal-like indium oxide layer of low optical/electrical losses. At that time, this work represented a milestone for perovskite–organic tandems, which outperformed the best *p–i–n* perovskite single junctions on a par with perovskite–CIGS and all-perovskite multi-junction cells. Also, over the past few years, the field of perovskite–organic tandem solar cell has developed tremendously fuelled by the emergence of narrow bandgap *Y*-derivative acceptors for OSCs [89]. Recently, a record PCE of 25.82% was been achieved for perovskite–organic tandem inverted cells based on combining with a wide bandgap organic–inorganic hybrid perovskite sub-cell. The tandem solar cells prepared by Li and co-workers [88] were fabricated with a structure of ITO/NiOx/CbzNaph/i-PVK/C60/SnOx/Au/MoOx/PM6:BTP-eC9:PCBM/PNDIT-F3N/Ag.

8.4 Cell degradation and failure

Despite the developments made with improving the PCE of OSCs, their limited stability impedes their commercialization. As such, there has been much focus on the degradation mechanisms and stability of OSCs. Overall, the degradation process can be divided into three stages: (1) burn-in stage with fast decay, (2) long-term stage which takes up most of the degradation period; and (3) failure stage leading to the failure of the device. Typically, the devices after the burn-in stage suffer a severe PCE loss of 30%–50% in several hundred hours. The long-term stage leads to another 20% PCE decrease and lasts tens of thousands of hours. The failure stage is extremely fast, resulting in the complete failure of the device in just tens of hours [90].

8.4.1 Cell stability

The overall degradation of OSCs is determined by the degradation of each layer and their interfaces, where the main mechanism by which they lose their operation lifetime may be different across devices [91]. Overall, degradation in OSCs can either be intrinsic, observed as spontaneous changes in the active layer, electrodes, or interfaces between the layers, or extrinsic, caused by external stimuli such as light, water, oxygen, elevated temperatures, and/or mechanical stress [91, 92]. Intrinsic degradation would thus be relevant for OSCs that are thermally isolated and kept in an inert atmosphere, free from mechanical stresses. It is important to note that this division between external and internal degradation factors does not imply mutually exclusive factors; instead, the different degradation mechanisms often promote one another or are in tandem. Accordingly, degradation mechanism can be categorized as outlined below.

8.4.1.1 Extrinsic factors of degradation

Physico-chemical degradation

In OSCs, conjugated polymers and small molecules are susceptible to reaction with moisture, oxygen and ozone leading to degradation of the electrodes, active layers or the interfaces, either individually or collectively [92]. The degradation of donors and acceptors occurs through changes in their chemical structures, electronic HOMO/LUMO levels, absorption and emission energies and electron mobilities [93]. Oxygen and moisture typically penetrate unencapsulated OSC devices through defects in the layers and interfaces [91]. In particular, water molecules can penetrate the active layer, forming electronic trap states that reduce charge photogeneration accompanied by formation of OH^- ions in the polymer backbone, causing the polymer to swell. For example, Bao and co-workers [94] showed that water exposure by PCBM films lead to irreversible shifting of the HOMO level and oxygen exposure lowered the WF by $\sim$0.15 eV compared to pristine PCBM. In another case, Bastos and co-workers [95] showed for C60 films and their devices that oxygen degradation was mainly responsible for the decrease in J_{sc} by inducing exciton quenching with the degradation of the ETL and C60/ETL interface being partially responsible. Also, the reduction of V_{oc} was due to an increase of saturation current in combination with a decrease of generated charges due to exciton quenching.

Pinhole defects and electrode edges have been implicated as pathways for the penetration of water and oxygen leading to corrosion of the active material and metal electrode/active layer interface [91]. Typically, degradation occurs primarily via direct oxidation or through interaction with the structurally modified polymer, as previously stated [96]. In particular, low-WF metal electrodes such as aluminium, calcium and silver which are typical in OSCs are susceptible to corrosion degradation through the formation of insulating metal oxide layers which act as charge extraction barriers, suppressing charge injection and extraction, and lowering the PCE [91, 96]. For example, Reese and co-workers investigated PCBM-based OSCs with Ca/Al electrodes and observed that oxide formation was faster due to the presence of Ca arising from exposure to air and considerable changes at the metal/organic interface [97]. Similar degradations have been reported for OSCs employing the PEDOT:PSS blend HTL, where the presence of water promotes irreversible structural transformation of its networks due to its highly hygroscopic nature, resulting in a shifting of the WF, reduction in conductivity, diminished charge extraction, lowered PCE, and a lowering of device lifetime [96, 98]. For these cells, this is also accompanied by oxidation of the electrode (typically aluminium) due to water vapour diffusion via the PEDOT:PSS layer, reducing the active area. To mitigate this issue, approaches such as decreasing the wettability of PEDOT:PSS circumvents the hygroscopic character, limiting water diffusion. It has also been shown that additives such as anionic perfluorinated ionomers were found to be effective in improving both PCE and stability of these devices [99].

Another feature of oxidation-related degradation in OSCs is *p*-type doping of oxygen on polymer donors, even in the dark. The result is excess holes, which act as deep electron traps—limiting their extraction, thereby reducing the FF and open-circuit voltage [91, 93] For example, Weu and co-workers [100] reported that degradation of PffBT4T-2OD:$PC_{71}BM$ devices was caused by oxygen-induced *p*-type doping of the active layer was exacerbated by light-induced photo-oxidation of the materials. They demonstrated that oxygen *p*-type doping was reversible in vacuum or inert atmospheres without chemical oxidation of the materials. However, in both light and oxygen, permanent photobleaching occurred due to molecular changes including the loss of chromophores via free radical diffusion and side-chain degradation [91]. The degradation resulted in a decrease in the short-circuit current because of the defect states, which increased recombination and trap density and affected overall performance.

Photo-induced degradation

Light-induced degradation or photo-degradation is one of the primary degradation mechanisms in OSCs and varies depending on the nature of the active layer and the transport layer [93]. Typically, photo-degradation of the active layer by way of photo-oxidation involves direct reaction in its excited state with molecular oxygen and proceeds through one of two pathways—*Type 1* and *Type 2* reactions [91]. The type 1 pathway involves electron transfer from the polymer to molecular oxygen to form a polymer cation and superoxide anion radical which initiates radical reactions with the donors or acceptors, especially with radicals on the polymer chain, leading to their degradation. The type 2 pathway involves energy transfer from the polymer's excited triplet state to molecular oxygen in its ground triplet state, which immediately decays to its singlet state, and then reacts with the active layer leading to its degradation. Early work in elucidating the photo-degradation mechanism was reported by Soon and co-workers involving the donor polymer, PTB7 which showed complete loss of fluorescence after being illuminated for 250 h, characterized through generation of singlet oxygen from triplet excitons [101]. Also, Inasaridze and co-workers [102] showed that degradation of OSCs based on PCDTBT blends with fullerene derivatives stemmed from accumulation of photogenerated free radicals in the fullerene moiety of the photoactive layer, and that the fullerene acceptor $PC_{60}BM$ could undergo photo-dimerization via radical species.

In OSCs comprising polymer:fullerene blends, degradation is predominantly driven by the Type 2 pathway [103]. It was demonstrated that the efficiencies of these processes not only depended on the individual properties of the donor and acceptor in the blend, but also on their ratio and the layer morphology. In other work, Blazinic and co-workers [104] showed that exposure of TQ1:$PC_{70}BM$ active layer blends in OSCs to simulated sunlight and ambient air resulted in their photo-degradation due to breakage of the fullerene conjugated bonds and formation of epoxide, diol and carbonyl compounds. These photo-oxidation products shift the positions of the LUMO and HOMO through trap states, which affected the transport and collection of charge carriers, diminishing the performance of the device of structure, ITO/MoO_3/TQ1:$PC_{70}BM$/LiF/Al. In another report, it was shown that photo-degradation of donor polymer PBDB-T, small-molecule acceptor Y5, and the co-polymer acceptors PF5-Y5 and PYT thin films under white light illumination (AM 1.5) yielded degradation products [105]. Insights into these kinds of photo-degradation mechanism were elucidated by Kim and co-workers [106]. They showed that PTB7-Th polymer films exposed to sunlight in air underwent degradation through breaking of the π-conjugated backbone and the intermolecular π–π interactions. This involved initial photo-oxidation of the thiophene ring in the benzo[1,2-*b*;3,3-*b*]dithiophene (BDT) unit, followed by the ring-opening and breaking of the π-conjugated system. Here, the thiophene ring unit reacts with oxygen to generate a BDT–O_2 adduct, which then produces thioester and carboxylic acid degradation products. Furthermore, it has also been shown that photo-degradation of polymers such as MEH PPV, P3HT and PCPDTBT in the presence of light and oxygen also involves singlet state oxygen leading to reaction with polymers and their subsequent degradation [107].

In addition to the ageing of encapsulation polymers in OSCs [108], high-energy UV and gamma radiation can induce photo-degradation of active materials in the absence of oxygen by severing chemical bonds as well as form free radicals that can further decompose the polymer [91]. In these cases, structural change of the active layer materials affect their light absorption, shift their HOMO/LUMO energy levels, limit their charge carrier generation, create sub-bandgap trap states and induce recombination and energy losses in the device. For example, it was reported that the blue-shifted absorption of P3HT under illumination in air was attributed to the reduction of the conjugated structure and lower crystallinity, and that lower optical absorption and undesirable polar groups may act as carrier quenchers that reduce the emission intensity [109]. In another study by Patel and co-workers [110], it was found that PCBM in a multilayer BHJ OSC exposed to UV light underwent oligomerization through breaking of the C–H bonds and the formation of dangling bonds which created new sub-bandgap trap states at the donor/acceptor interface. This caused a significant decrease in the photocurrent due to the reduction of the charge extraction and a change in the micromorphology of the BHJ material, leading to a reduction in photocurrent.

Beyond the photo-oxidation of active layers, carrier transport layers also play a role in the photo-degradation process of OSCs. One of the most important ETLs, ZnO has been shown to undergo photo-degradation to different extents based on the device architecture and degree of light exposure. For example, it has been reported that ZnO forms shunting channels at the ZnO/polymer interface following photo-ageing under concentrated simulated sunlight [111]. These channels consisted of surface oxygen impurities which under illumination become excited and act as electron traps, thereby reducing the open-circuit voltage and FF. It was found that doping the ZnO with aluminium impurities passivated the surface oxygen impurities and reduced electron trapping. In another study involving ZnO, Chapel and co-workers [112] reported that an OSC comprising of P3HT/ZnO-nanoparticle blend demonstrated photo-degradation of the P3HT because of photo-oxidation of the ZnO nanoparticles which was reduced in the absence of UV radiation and oxygen. Oh and co-workers [113] showed that changes in the interface between the active layer and carrier transport layers comprising ZnO-nanoparticle ETL contributed to the photo-degradation. To improve the cell performance, they replaced the ZnO ETL with TiO_2 nanoparticles (TNPs) and found that the TNPs which bonded through π–π interactions of 3-phenylpentane-2,4-dione (TNP–Ph) formed more robust ETLs than those bonded with van der Waals interactions of 3-methyl-2,4-pentanedione TNP (TNP–Me) and the ZnO-photoactive layer.

Thermally-induced degradation

OSCs operating under continuous irradiation above ambient temperatures are prone to thermally-induced degradation. The degradation is characterized by morphological changes in the active layer and other material components in the device and are related to their glass transition temperature (T_g), crystallization, and phase transition properties [91, 92]. In particular, the T_g of polymers dictates their molecular organization during solidification, which is mainly dependent on their backbone structure, geometry and side chains, which overall determines their degree of coarsening and crystallization [91]. Consequently, higher temperatures often create pinholes, fibril structures, which increases roughness of the active layer that disrupts the interpenetrating network leading to delamination of interfacial layers which ultimately affects the charge carrier transport, collection and extraction [93]. In particular, the active-layer morphology which is not usually at thermodynamic equilibrium after device fabrication undergoes structural and miscibility changes of their components due to additional thermal energy at higher temperatures [91, 93]. When the miscibility between donor and acceptor materials is low, especially at higher temperatures, they tend to aggregate and form larger-sized clusters or domains beyond the exciton diffusion length or charge transport length. For BHJ devices, it has been reported that the active layer blend, which is often metastable, is sensitive to higher temperature and transitions towards the thermodynamic equilibrium state, thus forming large phase separation which significantly reduces the D–A interfacial area, charge generation, carrier transport and carrier collection, and device efficiency [114]. For example, Wang and co-workers found that coarse phase separation occured in the PCDTBT:PC_{71}BM-based active layer above the T_g of the PCDTBT [115]. In this case, the T_g of the blend was reduced on annealing due to disruption of π–π stacking between PCDTBT molecules resulting in reduced hole mobility. In another study, Mohammed and co-workers [116] reported that the stability of P3HT-based inverted OSCs could be tuned by the choice of acceptor materials (PC_{60}BM, PC_{70}BM, and P3HT:ITIC) which was dependent on their thermal stability. They showed that the P3HT:PC_{60}BM device retained more than 50% of its initial PCE while the PC_{70}BM and ITIC-based devices retained only 22%

and 32% of their initial PCEs, respectively. They suggested that the main cause of degradation in the $PC_{60}BM$ and $PC_{70}BM$ devices was the thermally induced crystallization that led to segregation in the active layer. Recently, Li and co-workers [117] elucidated the thermally induced structural and morphological changes in the active layers in BTP-4F-12-based solar cells. The OSCs exhibited significant V_{oc} loss with increasing temperature, which recovered upon cooling. They attributed this behaviour to the charge carrier recombination, π–π stacking distances, and aggregated domains at various temperatures. Also, the irreversible loss of FF and J_{sc} during aging was due to changes in crystallinity and π–π stacking. Devices with a PBDBTCl-DTBT:BTP-4F-12 blend displayed more severe thermal degradation compared with PBDB-TF-T1:BTP-4F-12 blend due to poor miscibility of the donor and acceptor.

Another factor that degrades the device is thermally-induced chemical changes whether in the active layer or across the interfaces. For example, in the active layer, various chemical changes involve cyclic chemical reactions such as chain scission, thermo-oxidation of active-layer materials, and chain breakage due to high temperature [118]. In one study, Steinberger and co-workers [119] showed that thermo-oxidation of the active materials occurred upon annealing the active layer or during solution processing of the active materials in air at elevated temperatures. By comparing various blends of donor polymers with fullerene or non-fullerene acceptors, fullerene was identified to be the most susceptible component to thermal oxidation in both film and solution, while the polymer remained unaffected. Other thermally-induced changes such as diffusion of impurity species can lead to degradation in the device by altering the charge transport properties and introducing traps within the active layer or across the polymer/electrode interface, ultimately reducing the efficiency of the solar cell. In a recent study by Xi and co-workers [120], it was shown that OSCs of an inverted structure, ITO/ZnO/PM6/L8-BO/TCTA/MoO_3/Ag displayed reduction in J_{sc}, slow but continuous V_{oc}, and FF decay upon 150 °C thermal annealing. This thermal degradation of device characteristics involved formation of MoO_3^- impurities at the MoO_3/PM6/L8-BO interface and their diffusion through the photoactive layer and accumulation at the cathode interface, which act as acceptors leading to *p*-doping and increased charge recombination.

Mechanical degradation

In order for OSCs to be used in flexible technology requiring fabrication on flexible and transparent substrates such as thin foils and plastic, they must be made with excellent mechanical stability [91, 92]. This is dependent on their ability to limit degradation due to large mechanical strains such as stretching, compressing, twisting, and bending. Of these, tensile stress is a more dominant cause for mechanical degradation than compression stress [93]. The mechanical response to stress differs due to the intermolecular and adhesive interactions in the active layer and interfaces and depend on the structure of the polymer, their blends and the inclusion of additives [91]. Overall, structural and morphological features are directly related to the nature and extent of mechanical degradation of devices.

In particular, the degradation can be characterized as: (1) strains of organic materials in the active layer and its morphology due to linear mechanical stress; (2) delamination of the active layer, carrier transport layers, and electrodes; (3) punctures and cracks in the layers, facilitating the ingress of oxygen and water; and (4) the lateral strains caused by differences in the strains in each component [121].

In addition to material properties, the processing of materials and devices also influence the mechanical degradation. In the active layer, mechanical degradation results from weak D–A interface with reduced polymer entanglement and density, resulting in poor resistance against crack development and phase separation between the two components [91]. Hence, the mechanical robustness of the active layer materials of OSCs should be carefully considered using various criteria or approaches including processing conditions, additive components and defect engineering. For example, it has been shown that OSCs consisting of small-molecular acceptor, mPh4F-TS (TS) and polymer acceptor, PYSe2F-T (PA), introduced as ternary guests into the binary blend, PM6/mPh4F-TT (PM6/TT) fabricated via sequential deposition displayed distinct differences in mechanical performance between the PA and TS under bending stress [122]. In particular, the incorporation of small-molecule acceptor, mPh4F-TS into the flexible device compromises its mechanical stability under bending. In contrast, the integration of the polymer acceptor, PYSe2F-T significantly bolsters mechanical stability, enhancing the flexural strength and effectively mitigating crack propagation. Additionally, the rigid and flexible ternary devices, achieved superior efficiencies (19.6%/17.7% for PM6/TT+TS, and 19.2%/17.4% for PM6/TT + PA), outperforming their binary counterparts (18.3%/16.4%).

The mechanical stability of flexible OSCs still lags behind those of rigid devices and this is partly due to the inferiority of flexible transparent electrodes (FTEs). To address this, Chen and co-workers [123] developed a so-called ‘welding’ concept for the design of an Al-doped ZnO (AZO) and silver nanowire (AgNW) network FTE with tight binding of the upper electrode and the underlying substrate through strong capillary forces and enhanced adhesion of the electrode to the substrate, resulting in good mechanical and optoelectronic properties. The single-junction flexible OSCs based on this welded FTE achieved a high PCE of 15.21%. In addition, the PCEs of the flexible OSCs were less influenced by the device area and displayed robust bending durability even under extreme test conditions. Further developments in this regard were demonstrated in several reports. For example, Zeng and co-workers [124] developed FTE OCSs which employed an ionic liquid-type reducing agent containing Cl^- and a dihydroxyl group to control the reduction process of silver (Ag) in AgNW-based FTEs which facilitated an atomic-level contact between the AgNWs and the reduced Ag. This decreased the sheet resistance, and enhanced the mechanical stability of the FTEs. As a result, the single-junction flexible OSCs which achieved a PCE of 17.52% also displayed robust bending and peeling durability even under extreme test conditions.

The influence of temperature on mechanical degradation is another critical factor for OSCs since high temperatures can significantly affect the long-term reliability of solar cell packaging. For example, Cao and co-workers [125] demonstrated the effect of elevated temperatures on the mechanical properties of materials in flexible solar cell panels. In their experiments, they stretched the flexible solar cell panels at certain strain rates and found that the Young's modulus of the materials decreased sharply with increasing temperature and that interfacial delamination occurred prior to material fracture at higher temperatures. They concluded that high temperature not only reduced the mechanical stiffness of the materials but also weakened their tensile strength, which eventually undermined the reliability of the solar cell packaging and long-term stability of the solar cell.

8.4.1.2 Intrinsic factors of degradation

Intrinsic degradation in OSCs unlike extrinsic factors of degradation described in the foregoing relate to inherent molecular and structural changes that occur in the active layer and device interfaces without the direct influence of external stimuli. Usually upon solution processing, the active layer components of an OSC are metastable, depending on kinetic factors of layering and are not in thermodynamic equilibrium [91]. Consequently, the morphology evolves/transitions toward its thermodynamically stable state over time, and can lead to undesirable phase separation of the layer or diffusion of interfaces into the photoactive layer, which often does not support efficient charge generation. Furthermore, as the operating temperature of OSCs increase upon continuous illumination, such thermal stress can detrimentally accelerate the evolution of the morphology of the active layer, thereby limiting PCEs and long-term stability [126]. For example, OSCs based on NFAs, Y6 and its derivatives, evaluated under various conditions, exhibited faster degradation between 200 and 2400 h, even though few systems could be relatively stable under the dark condition without external stress [127]. It has been established that due to the long-range disorder and short-range order properties of organic semiconductors as well as the complex mixed phase regions at the interface between polymer donor and NFAs, finely regulating the film morphology of the active layer is important to improving exciton dissociation and charge transportation. For example, An and co-workers [126] demonstrated stable devices by integrating a wide bandgap polymer donor, PTzBI-dF) and two acceptors, L8BO and Y6 that feature similar structures yet different thermal and morphological properties. The OSC based on the PTzBI-dF:L8BO:Y6 blend, exhibited a PCE of 18.26% in the conventional device structure and the inverted structure demonstrated excellent long-term thermal stability over 1400 h under 85 °C continuous heating. The improved performance was ascribed to suppressed charge recombination along with appropriate charge transport due to prohibition of clustering of the amorphous phase. Importantly, strategies to limit the effect of morphological degradation and to improve efficiency have involved including additives to the active layers to improve the film morphology. For example, You and co-workers [128] reported that the amorphous small molecule SJ-IC-M was used as an efficient solid additive in the OSCs based on PM6:Y6 to enhance the PCE and long-term stability of the device. After the addition of 0.5 wt% SJ-IC-M into the active layer blends, the PCE was increased to 16.2% compared to the non-additive OSC of PCE of 15.0% along with excellent long-term stability. The PCE maintained over 90% of its initial value when the unencapsulated device was preserved in a nitrogen atmosphere for a month. This improvement was attributed to optimized crystallinity of Y6 which stabilized blend morphology, improved charge transport and enhanced device performance.

Another feature of phase related intrinsic degradation involves topological changes due to phase separation, leading to vertically separated components within the active layer, and is particularly problematic for cells with thick active layers. For example, Cai and co-workers [129] fabricated efficient thick-film OSCs with an active layer consisting of PM6 donor and BTP-eC9 and L8-BO-F acceptors. The two acceptors were found to possess enlarged exciton diffusion length in the mixed phase, which was beneficial to exciton generation and dissociation. Additionally, LBL deposition [130] was employed to optimize the vertical phase separation. Benefiting from the synergetic effects of enlarged exciton diffusion length and graded vertical phase separation, an efficiency of 17.31% was obtained for the 300nm-thick OSC, with a short-circuit current density of 28.36 mA cm^{-2}, and a high FF of 73.0%. Moreover, the device with a 500 nm active layer also showed an efficiency of 15.21%. In general, polymers with good solubility and controlled molecular aggregation allow acceptors to permeate the polymer layer, thereby increasing the D–A interface and achieving optimized vertical phase separation. Indeed, controlling vertical phase separation of the active layer to enable efficient exciton dissociation and charge carrier transport is crucial to improve PCEs. Recently, one strategy developed by He and co-workers [131] involved a ternary polymerization strategy to develop a series of polymer donors, DL1–DL4, and regulate their solubility, molecular aggregation, molecular orientation, and miscibility, thus efficiently manipulating vertical phase separation in PPHJ OSCs. In particular, the donor, DL1 not only enhanced solubility, inhibited molecular aggregation and partial edge-on orientation to facilitate acceptor molecules, Y6, to permeate into the polymer layer and increase D–A interfaces, but also sustained high crystallinity and appropriate miscibility to acquire ordered molecular packing, achieving optimized vertical phase separation to improve exciton dissociation and charge transport in the devices. These DL1/Y6-based PPHJ OSCs demonstrated excellent exciton dissociation probability, highest charge carrier mobilities and weakest charge recombination, and impressive PCEs up to 19.10%.

In addition to the morphology of the active layer, another intrinsic instability arises from the diffusion of materials at the interfaces into the photoactive layer. This highlights the importance of controlling the thermodynamic instability of the transport layers and electrodes to ensure the overall stability of the device [91]. For example, Duan and co-workers [132] investigated the mechanism of the burn-in process in the high-efficiency PM6:N3-based non-fullerene OSCs. The PM6:N3-based device achieved a PCE of 14.10% but also showed a significant performance loss after the burn-in degradation. They reported that the stable active layer experienced instability due to the diffusion of atoms from the charge transport layers and electrodes into the active layer and recommended that overcoming the instability of these interfaces and electrodes were important for long-term stability of the device. It is noteworthy that atom diffusion from the electrode and buffer layers can alter the energy levels of the buffer layers and electrodes, leading to the formation of electronic traps that promote charge recombination.

8.5 Conclusion, challenges and future prospects

Solution processible OSCs have attracted much attention as one of the most promising candidates for sustainable energy use over the past two decades. They have undergone tremendous progress in recent years, where most of the research efforts have been devoted to the development of devices with higher PCE, longer lifetime and at low cost solution processing. To date, the PCE of OSCs has reached over 20% and increasingly closed the gap with inorganic and hybrid solar cells, which exceeds the requirement in efficiency towards commercial application. In particular, much of the development has been premised on the discovery and introduction of new donor and acceptor materials including fullerene and its derivatives, polymers, and non-fullerene small-molecule materials with concomitant optimization of device architectures. With the continuous improvement of fullerene derivative acceptor materials, the performance of solar cells based on these has been improved to some extent. However, there are still some problems in fullerene materials, such as the low LUMO energy level, narrow absorption spectrum, and small electron mobility. For small-molecule acceptor materials, the best performance is demonstrated from the thick ring acceptor materials. Adjusting the optical and morphological characteristics of the thick ring acceptors through chemical modification allows the thick ring acceptors to have unique advantages in device performance. Notwithstanding the range of developments in OSCs, to achieve large-scale commercial manufacturing it is necessary to address green-solvent treatment and understand the degradation mechanisms under various environmental conditions to achieve devices based on small-molecule acceptor materials, in particular that can compete with other solar cell technologies, such as perovskite devices.

For OSCs, the attractiveness of low-cost fabrication using scalable methodologies such as roll-to-roll techniques has expanded new opportunities for this emergent technology to rival the mature ISCs sector. Despite the developments in OSCs over the years, the technology still faces limitations in terms of lower PCE compared to inorganic and perovskite solar cells, durability and stability issues, sensitivity to temperature, and limited lifespan. Of note is that low cost is generally identified as one of the potential advantages, compared with other photovoltaic technologies. However, currently most of the high-performance materials need tedious synthetic steps and/or suffer from relatively high cost. In addition, halogenated solvents which are highly toxic and expensive are widely used during processing. They are not environmentally friendly, and therefore would add to the overall cost of manufacturing in terms of their management and disposal. As such, the development of low-cost high-performance materials, less expensive and 'green' processing for OSCs is an imperative towards industrial-scale production. Notwithstanding, there are opportunities to improve the technology, including the development of new materials and improved device engineering. In particular, research and development efforts especially with respect to low cost and efficiency should be focussed on: (1) optimizing the synthetic route and reducing synthetic complexity of high-efficiency materials; (2) developing low-cost but high-performance interfacial materials and for transparent electrodes; and (3) finely adjusting the molecular structure of photoactive materials to suit green-solvent processing. Equally important are efficient encapsulation and use of more stable materials to extend device lifetime—a serious liability that has limited their large-scale commercialization. These along with the growing demand for renewable energy sources, favourable government policies and incentives to develop and deploy the technology, growing consumer awareness and demand for sustainable and environmentally friendly products, emerging markets and industries, and collaboration and investment from major companies, underscore the tremendous promise for the OSCs sector.

Overall, OSCs relative to other types of devices have the potential to revolutionize the solar energy industry due to their compatibility with printing technologies and the ability to produce thin, flexible solar cells. However, challenges such as preventing recombination energy losses, improving absorption in the visible to near-infrared part of the solar spectrum, and optimizing morphological characteristics for charge transport must be overcome. Despite these challenges, researchers are making steady progress, and the tunability and versatility of organic materials offer promise for future success. The potential applications for thin, flexible OSCs are exciting, including powering remote or underdeveloped areas and charging internal devices. The tremendous progress of OSCs in terms of PCE, solution processing and stability have shown that there is much work to be done but also that the promise of OSCs as a major player in the solar energy sector is sustainable.

References

[1] Yin Z, Wei J and Zheng Q 2016 Interfacial materials for organic solar cells: recent advances and perspectives *Adv. Sci.* **3** 1500362

[2] Wang Q, Xie Y, Soltani-Kordshuli F and Eslamian M 2016 Progress in emerging solution-processed thin film solar cells—part I: polymer solar cells *Renew. Sustain. Energy Rev.* **56** 347–61

[3] Li Y, Huang W, Zhao D, Wang L, Jiao Z, Huang Q, Wang P, Sun M and Yuan G 2022 Recent progress in organic solar cells: a review on materials from acceptor to donor *Molecules* **27** 1800

[4] Lin Y, Wang J, Zhang Z-G, Bai H, Li Y, Zhu D and Zhan X 2015 An electron acceptor challenging fullerenes for efficient polymer solar cells *Adv. Mater.* **27** 1170–4

[5] Yu G, Gao J, Hummelen J C, Wudl F and Heeger A J 1995 Polymer photovoltaic cells: enhanced efficiencies via a network of internal donor–acceptor heterojunctions *Science* **270** 1789–91

[6] Li G, Shrotriya V, Huang J, Yao Y, Moriarty T, Emery K and Yang Y 2005 High-efficiency solution processable polymer photovoltaic cells by self-organization of polymer blends *Nat. Mater.* **4** 864–8

[7] Ma W, Yang C, Gong X, Lee K and Heeger A J 2005 Thermally stable, efficient polymer solar cells with nanoscale control of the interpenetrating network morphology *Adv. Funct. Mater.* **15** 1617–22

[8] Yuan S, Luo W, Xie M and Peng H 2025 Progress in research on organic photovoltaic acceptor materials *RSC Adv.* **15** 2470–89

[9] Kooistra F B, Knol J, Kastenberg F, Popescu L M, Verhees W J H, Kroon J M and Hummelen J C 2007 Increasing the open circuit voltage of bulk-heterojunction solar cells by raising the LUMO level of the acceptor *Org. Lett.* **9** 551–4

[10] Zhang Y, Yip H-L, Acton O, Hau S K, Huang F and Jen A K Y 2009 A simple and effective way of achieving highly efficient and thermally stable bulkheterojunction polymer solar cells using amorphous fullerene derivatives as electron acceptor *Chem. Mater.* **21** 2598–600

[11] Troshin P A *et al* 2009 Material solubility-photovoltaic performance relationship in the design of novel fullerene derivatives for bulk heterojunction solar cells *Adv. Funct. Mater.* **19** 779–88

[12] Liu Y, Zhao J, Li Z, Mu C, Ma W, Hu H, Jiang K, Lin H, Ade H and Yan H 2014 Aggregation and morphology control enables multiple cases of high-efficiency polymer solar cells *Nat. Commun.* **5** 5293

[13] Lenes M, Wetzelaer G-J A H, Kooistra F B, Veenstra S C, Hummelen J C and Blom P W M 2008 Fullerene bisadducts for enhanced open-circuit voltages and efficiencies in polymer solar cells *Adv. Mater.* **20** 2116–9

[14] He Y, Chen H-Y, Hou J and Li Y 2010 Indene–C60 bisadduct: a new acceptor for high-performance polymer solar cells *J. Am. Chem. Soc.* **132** 1377–82

[15] Zhao G, He Y and Li Y 2010 6.5% Efficiency of polymer solar cells based on poly(3-hexylthiophene) and indene-C60 bisadduct by device optimization *Adv. Mater.* **22** 4355–8

[16] He Y, Zhao G, Peng B and Li Y 2010 High-yield synthesis and electrochemical and photovoltaic properties of indene–C70 bisadduct *Adv. Funct. Mater.* **20** 3383–9

[17] He Y, Shao M, Xiao K, Smith S C and Hong K 2013 High-performance polymer photovoltaics based on rationally designed fullerene acceptors *Sol. Energy Mater. Sol. Cells* **118** 171–8

[18] Song J and Bo Z 2023 Asymmetric molecular engineering in recent nonfullerene acceptors for efficient organic solar cells *Chin. Chem. Lett.* **34** 108163

[19] Liang X, Wang J, Miao R, Zhao Q, Huang L, Wen S and Tang J 2022 The evolution of small molecular acceptors for organic solar cells: advances, challenges and prospects *Dyes Pigm.* **198** 109963

[20] Al-Azzawi A G S, Aziz S B, Dannoun E M A, Iraqi A, Nofal M M, Murad A R and M. Hussein A 2023 A mini review on the development of conjugated polymers: steps towards the commercialization of organic solar cells *Polymers 2023* **15** 164

[21] Camaioni N, Carbonera C, Ciammaruchi L, Corso G, Mwaura J, Po R and Tinti F 2023 Polymer solar cells with active layer thickness compatible with scalable fabrication processes: a meta-analysis *Adv. Mater.* **35** 2210146

[22] Solak E K and Irmak E 2023 Advances in organic photovoltaic cells: a comprehensive review of materials, technologies, and performance *RSC Adv.* **13** 12244–69

[23] Lin Y *et al* 2016 A facile planar fused-ring electron acceptor for as-cast polymer solar cells with 8.71% efficiency *J. Am. Chem. Soc.* **138** 2973–6

[24] Lin Y *et al* 2018 Balanced partnership between donor and acceptor components in nonfullerene organic solar cells with >12% efficiency *Adv. Mater.* **30** 1706363

[25] Zhu J, Xiao Y, Wang J, Liu K, Jiang H, Lin Y, Lu X and Zhan X 2018 Alkoxy-induced near-infrared sensitive electron acceptor for high-performance organic solar cells *Chem. Mater.* **30** 4150–6

[26] Yuan J *et al* 2019 Single-junction organic solar cell with over 15% efficiency using fused-ring acceptor with electron-deficient core *Joule* **3** 1140–51

[27] Liu Q *et al* 2020 18% Efficiency organic solar cells *Sci. Bull.* **65** 272–5

[28] Yuan J and Zou Y 2022 The history and development of Y6 *Org. Electron.* **102** 106436

[29] Chen Y, Li Y, Zhou W, Liao C, Xu X, Yu L, Li R and Peng Q 2024 Y6-derived non-fullerene acceptors with unsaturated alkyl side chains enabling improved molecular packing for highly efficient additive-free organic solar cells *Chem. Mater.* **36** 11606–17
[30] Yu G and Heeger A J 1995 Charge separation and photovoltaic conversion in polymer composites with internal donor/acceptor heterojunctions *J. Appl. Phys.* **78** 4510–5
[31] Zhang Y, Wan Q, Guo X, Li W, Guo B, Zhang M and Li Y 2015 Synthesis and photovoltaic properties of an n-type two-dimension-conjugated polymer based on perylene diimide and benzodithiophene with thiophene conjugated side chains *J. Mater. Chem.* A **3** 18442–9
[32] Guo X and Watson M D 2008 Conjugated polymers from naphthalene bisimide *Org. Lett.* **10** 5333–6
[33] Zheng N *et al* 2019 Improving the efficiency and stability of non-fullerene polymer solar cells by using N2200 as the additive *Nano Energy* **58** 724–31
[34] Zhang Z-G, Yang Y, Yao J, Xue L, Chen S, Li X, Morrison W, Yang C and Li Y 2017 Constructing a strongly absorbing low-bandgap polymer acceptor for high-performance all-polymer solar cells *Angew. Chem. Int. Ed.* **56** 13503–7
[35] Jia T *et al* 2020 14.4% Efficiency all-polymer solar cell with broad absorption and low energy loss enabled by a novel polymer acceptor *Nano Energy* **72** 104718
[36] Bi P *et al* 2023 High-performance binary all-polymer solar cells with efficiency over 18.3% enabled by tuning phase transition kinetics *Adv. Energy Mater.* **13** 2302252
[37] Alam S and Lee J 2023 Progress and future potential of all-small-molecule organic solar cells based on the benzodithiophene donor material *Molecules* **28** 3171
[38] Shaheen S E, Brabec C J, Sariciftci N S, Padinger F, Fromherz T and Hummelen J C 2001 2.5% Efficient organic plastic solar cells *Appl. Phys. Lett.* **78** 841–3
[39] Tan Z A, Li S, Wang F, Qian D, Lin J, Hou J and Li Y 2014 High performance polymer solar cells with as-prepared zirconium acetylacetonate film as cathode buffer layer *Sci. Rep.* **4** 4691
[40] Liang Y, Wu Y, Feng D, Tsai S-T, Son H-J, Li G and Yu L 2009 Development of new semiconducting polymers for high performance solar cells *J. Am. Chem. Soc.* **131** 56–7
[41] Liang Y, Feng D, Wu Y, Tsai S-T, Li G, Ray C and Yu L 2009 Highly efficient solar cell polymers developed via fine-tuning of structural and electronic properties *J. Am. Chem. Soc.* **131** 7792–9
[42] Liang Y, Xu Z, Xia J, Tsai S-T, Wu Y, Li G, Ray C and Yu L 2010 For the bright future—bulk heterojunction polymer solar cells with power conversion efficiency of 7.4% *Adv. Mater.* **22** E135–8
[43] Zhao J, Li Y, Yang G, Jiang K, Lin H, Ade H, Ma W and Yan H 2016 Efficient organic solar cells processed from hydrocarbon solvents *Nat. Energy* **1** 15027
[44] Qian D, Ye L, Zhang M, Liang Y, Li L, Huang Y, Guo X, Zhang S, Tan Z A and Hou J 2012 Design, application, and morphology study of a new photovoltaic polymer with strong aggregation in solution state *Macromolecules* **45** 9611–7
[45] Zhang M, Guo X, Ma W, Ade H and Hou J 2015 A large-bandgap conjugated polymer for versatile photovoltaic applications with high performance *Adv. Mater.* **27** 4655–60
[46] Zhang S, Qin Y, Zhu J and Hou J 2018 Over 14% efficiency in polymer solar cells enabled by a chlorinated polymer donor *Adv. Mater.* **30** 1800868

[47] Sharma V V, Landep A, Lee S-Y, Park S-J, Kim Y-H and Kim G-H 2024 Recent advances in polymeric and small molecule donor materials for Y6 based organic solar cells *Next Energy* **2** 100086
[48] Zhao W, Li S, Yao H, Zhang S, Zhang Y, Yang B and Hou J 2017 Molecular optimization enables over 13% efficiency in organic solar cells *J. Am. Chem. Soc.* **139** 7148–51
[49] Ma R *et al* 2020 Improving open-circuit voltage by a chlorinated polymer donor endows binary organic solar cells efficiencies over 17% *Sci. China Chem.* **63** 325–30
[50] Xiong J *et al* 2019 Thiolactone copolymer donor gifts organic solar cells a 16.72% efficiency *Sci. Bull.* **64** 1573–6
[51] Jin K, Xiao Z and Ding L 2021 D18, an eximious solar polymer! *J. Semicond.* **42** 010502
[52] Qin J *et al* 2021 A chlorinated copolymer donor demonstrates a 18.13% power conversion efficiency *J. Semicond.* **42** 010501
[53] Kan B, Kan Y, Zuo L, Shi X and Gao K 2021 Recent progress on all-small molecule organic solar cells using small-molecule nonfullerene acceptors *InfoMat* **3** 175–200
[54] Zhou R *et al* 2019 All-small-molecule organic solar cells with over 14% efficiency by optimizing hierarchical morphologies *Nat. Commun.* **10** 5393
[55] Huo L, Liu T, Sun X, Cai Y, Heeger A J and Sun Y 2015 Single-junction organic solar cells based on a novel wide-bandgap polymer with efficiency of 9.7% *Adv. Mater.* **27** 2938–44
[56] Zhou Z, Xu S, Song J, Jin Y, Yue Q, Qian Y, Liu F, Zhang F and Zhu X 2018 High-efficiency small-molecule ternary solar cells with a hierarchical morphology enabled by synergizing fullerene and non-fullerene acceptors *Nat. Energy* **3** 952–9
[57] Cai S, Huang P, Cai G, Lu X, Hu D, Hu C and Lu S 2022 Symmetrically fluorinated benzo [1,2-b:4,5-b′]dithiophene-cored donor for high-performance all-small-molecule organic solar cells with improved active layer morphology and crystallinity *ACS Appl. Mater. Interfaces* **14** 14532–40
[58] Feng W *et al* 2022 Tuning morphology of active layer by using a wide bandgap oligomer-like donor enables organic solar cells with over 18% efficiency *Adv. Energy Mater.* **12** 2104060
[59] Ma K, Feng W, Liang H, Chen H, Wang Y, Wan X, Yao Z, Li C, Kan B and Chen Y 2023 Modulation of alkyl chain length on the thiazole side group enables over 17% efficiency in all-small-molecule organic solar cells *Adv. Funct. Mater.* **33** 2214926
[60] Zhang T *et al* 2024 A highly crystalline donor enables over 17% efficiency for small-molecule organic solar cells *Energy Environ. Sci.* **17** 3927–36
[61] Chong K, Xu X, Meng H, Xue J, Yu L, Ma W and Peng Q 2022 Realizing 19.05% efficiency polymer solar cells by progressively improving charge extraction and suppressing charge recombination *Adv. Mater.* **34** 2109516
[62] Sun R *et al* 2022 Single-junction organic solar cells with 19.17% efficiency enabled by introducing one asymmetric guest acceptor *Adv. Mater.* **34** 2110147
[63] Tang C W 1986 Two-layer organic photovoltaic cell *Appl. Phys. Lett.* **48** 183–5
[64] Zhou M, Liao C, Duan Y, Xu X, Yu L, Li R and Peng Q 2023 19.10% Efficiency and 80.5% fill factor layer-by-layer organic solar cells realized by 4-bis(2-thienyl)pyrrole-2,5-dione based polymer additives for inducing vertical segregation morphology *Adv. Mater.* **35** 2208279
[65] Jiang K *et al* 2021 Pseudo-bilayer architecture enables high-performance organic solar cells with enhanced exciton diffusion length *Nat. Commun.* **12** 468

[66] Kumari T *et al* 2024 Bilayer layer-by-layer structures for enhanced efficiency and stability of organic photovoltaics beyond bulk heterojunctions *Cell Rep. Phys. Sci.* **5** 102027

[67] Lai X *et al* 2022 Bilayer quasiplanar heterojunction organic solar cells with a Co-acceptor: a synergistic approach for stability and efficiency *Chem. Mater.* **34** 7886–96

[68] Cao C *et al* 2022 Quasiplanar heterojunction all-polymer solar cells: a dual approach to stability *Adv. Funct. Mater.* **32** 2201828

[69] Zhu Y and He F 2025 Quasi-planar heterojunction: enhancing stability and practicality in organic photovoltaics *ACS Energy Lett.* **10** 935–46

[70] Wei Y, Chen Z, Lu G, Yu N, Li C, Gao J, Gu X, Hao X, Lu G and Tang Z 2022 Binary organic solar cells breaking 19% via manipulating the vertical component distribution *Adv. Mater.* **34** 2204718

[71] Zhang G, Zhao J, Chow P C Y, Jiang K, Zhang J, Zhu Z, Zhang J, Huang F and Yan H 2018 Nonfullerene acceptor molecules for bulk heterojunction organic solar cells *Chem. Rev.* **118** 3447–507

[72] Sun R *et al* 2019 A universal layer-by-layer solution-processing approach for efficient non-fullerene organic solar cells *Energy Environ. Sci.* **12** 384–95

[73] Zhang Y *et al* 2021 Graded bulk-heterojunction enables 17% binary organic solar cells via nonhalogenated open air coating *Nat. Commun.* **12** 4815

[74] Yan Y, Zhou X, Zhang F, Zhou J, lin T, Zhu Y, Xu D, Ma X, Zou Y and Li X 2022 High-performance pseudo-bilayer ternary organic solar cells with PC71BM as the third component *J. Mater. Chem.* A **10** 23124–33

[75] Li X, Du X, Zhao J, Lin H, Zheng C and Tao S 2021 Layer-by-layer solution processing method for organic solar cells *Sol. RRL* **5** 2000592

[76] Cheng H-W *et al* 2021 High-performance organic solar cells featuring double bulk heterojunction structures with vertical-gradient selenium heterocyclic nonfullerene acceptor concentrations *ACS Appl. Mater. Interfaces* **13** 27227–36

[77] Ghasemi M, Ye L, Zhang Q, Yan L, Kim J-H, Awartani O, You W, Gadisa A and Ade H 2017 Panchromatic sequentially cast ternary polymer solar cells *Adv. Mater.* **29** 1604603

[78] Liu S *et al* 2023 High-performance pseudo-bilayer organic solar cells enabled by sequential deposition of D18/Y6 chloroform solution *ACS Appl. Energy Mater.* **6** 5047–57

[79] Jo J, Jeong S, Lee D, Lee S, Kim B J, Cho S and Lee J-Y 2023 Pseudo-bilayered inverted organic solar cells using the Marangoni effect *J. Mater. Chem.* A **11** 17307–15

[80] Di Carlo Rasi D and Janssen R A J 2019 Advances in solution-processed multijunction organic solar cells *Adv. Mater.* **31** 1806499

[81] Jia Z *et al* 2021 High performance tandem organic solar cells via a strongly infrared-absorbing narrow bandgap acceptor *Nat. Commun.* **12** 178

[82] Ho C H Y, Kothari J, Fu X and So F 2021 Interconnecting layers for tandem organic solar cells *Mater. Today Energy* **21** 100707

[83] Brinkmann K O *et al* 2022 Perovskite–organic tandem solar cells with indium oxide interconnect *Nature* **604** 280–6

[84] Gilot J, Wienk M M and Janssen R A J 2007 Double and triple junction polymer solar cells processed from solution *Appl. Phys. Lett.* **90** 143512

[85] Kim J Y, Lee K, Coates N E, Moses D, Nguyen T Q, Dante M and Heeger A J 2007 Efficient tandem polymer solar cells fabricated by all-solution processing *Science* **317** 222–5

[86] Zheng Z, Wang J, Bi P, Ren J, Wang Y, Yang Y, Liu X, Zhang S and Hou J 2022 Tandem organic solar cell with 20.2% efficiency *Joule* **6** 171–84

[87] Wang J *et al* 2023 Tandem organic solar cells with 20.6% efficiency enabled by reduced voltage losses *Natl Sci. Rev.* **10** nwad085

[88] Li Y *et al* 2024 Highly durable inverted inorganic perovskite/organic tandem solar cells enabled by multifunctional additives *Angew. Chem. Int. Ed.* **63** e202412515

[89] Brinkmann K O *et al* 2024 Perovskite–organic tandem solar cells *Nat. Rev. Mater.* **9** 202–17

[90] Yang C, Zhan S, Li Q, Wu Y, Jia X, Li C, Liu K, Qu S, Wang Z and Wang Z 2022 Systematic investigation on stability influence factors for organic solar cells *Nano Energy* **98** 107299

[91] Tegegne N A, Nchinda L T and Krüger T P J 2025 Progress toward stable organic solar cells *Adv. Opt. Mater.* **13** 2402257

[92] Gupta S K, Dharmalingam K, Pali L S, Rastogi S, Singh A and Garg A 2013 Degradation of organic photovoltaic devices: a review *Nano. Energy* **2** 42–58

[93] Duan L and Uddin A 2020 Progress in stability of organic solar cells *Adv. Sci.* **7** 1903259

[94] Bao Q, Liu X, Braun S and Fahlman M 2014 Oxygen- and water-based degradation in [6,6]-phenyl-C61-butyric acid methyl ester (PCBM) films *Adv. Energy Mater.* **4** 1301272

[95] Bastos J P, Voroshazi E, Fron E, Brammertz G, Vangerven T, Van der Auweraer M, Poortmans J and Cheyns D 2016 Oxygen-induced degradation in C60-based organic solar cells: relation between film properties and device performance *ACS Appl. Mater. Interfaces* **8** 9798–805

[96] Drakonakis V M, Savva A, Kokonou M and Choulis S A 2014 Investigating electrodes degradation in organic photovoltaics through reverse engineering under accelerated humidity lifetime conditions *Sol. Energy Mater. Sol. Cells* **130** 544–50

[97] Reese M O, Morfa A J, White M S, Kopidakis N, Shaheen S E, Rumbles G and Ginley D S 2008 Short-term metal/organic interface stability investigations of organic photovoltaic devices *2008 33rd IEEE Photovoltaic Specialists Conf.* vol 2008 pp 1–3

[98] Fluhr D, Züfle S, Muhsin B, Öttking R, Seeland M, Roesch R, Schubert U S, Ruhstaller B, Krischok S and Hoppe H 2018 Aluminum electrode insulation dynamics via interface oxidation by reactant diffusion in organic layers *Phys. Status Solidi* A **215** 1800474

[99] Howells C T *et al* 2018 Influence of perfluorinated ionomer in PEDOT:PSS on the rectification and degradation of organic photovoltaic cells *J. Mater. Chem.* A **6** 16012–28

[100] Weu A, Kress J A, Paulus F, Becker-Koch D, Lami V, Bakulin A A and Vaynzof Y 2019 Oxygen-induced doping as a degradation mechanism in highly efficient organic solar cells *ACS Appl. Energy Mater.* **2** 1943–50

[101] Soon Y W, Cho H, Low J, Bronstein H, McCulloch I and Durrant J R 2013 Correlating triplet yield, singlet oxygen generation and photochemical stability in polymer/fullerene blend films *Chem. Commun.* **49** 1291–3

[102] Inasaridze L N, Shames A I, Martynov I V, Li B, Mumyatov A V, Susarova D K, Katz E A and Troshin P A 2017 Light-induced generation of free radicals by fullerene derivatives: an important degradation pathway in organic photovoltaics? *J. Mater. Chem.* A **5** 8044–50

[103] Nyga A, Blacha-Grzechnik A, Podsiadły P, Duda A, Kępska K, Krzywiecki M, Motyka R, Janssen R A J and Data P 2022 Singlet oxygen formation from photoexcited P3HT:PCBM films applied in oxidation reactions *Mater. Adv.* **3** 2063–9

[104] Blazinic V, Ericsson L K E, Levine I, Hansson R, Opitz A and Moons E 2019 Impact of intentional photo-oxidation of a donor polymer and PC70BM on solar cell performance *Phys. Chem. Chem. Phys.* **21** 22259–71

[105] Prasad S, Genene Z, Marchiori C F N, Singh S, Ericsson L K E, Wang E, Araujo C M and Moons E 2024 Effect of molecular structure on the photochemical stability of acceptor and donor polymers used in organic solar cells *Mater. Adv.* **5** 7708–20

[106] Kim S, Rashid M A M, Ko T, Ahn K, Shin Y, Nah S, Kim M H, Kim B, Kwak K and Cho M 2020 New insights into the photodegradation mechanism of the PTB7-Th film: photo-oxidation of π-conjugated backbone upon sunlight illumination *J. Phys. Chem.* C **124** 2762–70

[107] Aguirre A, Meskers S C J, Janssen R A J and Egelhaaf H J 2011 Formation of metastable charges as a first step in photoinduced degradation in π-conjugated polymer: fullerene blends for photovoltaic applications *Org. Electron.* **12** 1657–62

[108] Pinochet N, Pirot-Berson L, Couderc R and Therias S 2024 UV LED ageing of polymers for PV cell encapsulation *npj Mater. Degrad.* **8** 81

[109] Chang Y-M, Su W-F and Wang L 2008 Influence of photo-induced degradation on the optoelectronic properties of regioregular poly(3-hexylthiophene) *Sol. Energy Mater. Sol. Cells* **92** 761–5

[110] Patel J B, Tiwana P, Seidler N, Morse G E, Lozman O R, Johnston M B and Herz L M 2019 Effect of ultraviolet radiation on organic photovoltaic materials and devices *ACS Appl. Mater. Interfaces* **11** 21543–51

[111] Upama M B, Elumalai N K, Mahmud M A, Sun H, Wang D, Chan K H, Wright M, Xu C and Uddin A 2017 Organic solar cells with near 100% efficiency retention after initial burn-in loss and photo-degradation *Thin Solid Films* **636** 127–36

[112] Chapel A *et al* 2016 Effect of ZnO nanoparticles on the photochemical and electronic stability of P3HT used in polymer solar cells *Sol. Energy Mater. Sol. Cells* **155** 79–87

[113] Oh H, Sim H-B, Han S H, Kwon Y-J, Park J H, Kim M H, Kim J Y, Kim W-S and Kim K 2019 Reducing burn-in loss of organic photovoltaics by a robust electron-transporting layer *Adv. Mater. Interfaces* **6** 1900213

[114] Yang W *et al* 2020 Simultaneous enhanced efficiency and thermal stability in organic solar cells from a polymer acceptor additive *Nat. Commun.* **11** 1218

[115] Wang T, Pearson A J, Dunbar A D F, Staniec P A, Watters D C, Yi H, Ryan A J, Jones R A L, Iraqi A and Lidzey D G 2012 Correlating structure with function in thermally annealed PCDTBT:PC70BM photovoltaic blends *Adv. Funct. Mater.* **22** 1399–408

[116] Mohammed Y A, Hone F G, Mola G T and Tegegne N A 2023 The roles of acceptors in the thermal-degradation of P3HT based organic solar cells *Phys.* B **653** 414666

[117] Li Z *et al* 2025 Temperature-dependent thermal behavior of BTP-4F-12-based organic solar cells *Nano Energy* **140** 111043

[118] Grossiord N, Kroon J M, Andriessen R and Blom P W M 2012 Degradation mechanisms in organic photovoltaic devices *Org. Electron.* **13** 432–56

[119] Steinberger M, Distler A, Brabec C J and Egelhaaf H-J 2022 Improved air processability of organic photovoltaics using a stabilizing antioxidant to prevent thermal oxidation *J. Phys. Chem.* C **126** 22–9

[120] Xi Q, Qin J, Sandberg O J, Wu N, Huang R, Li Y, Saladina M, Deibel C, Österbacka R and Ma C-Q 2025 Improving the thermal stability of inverted organic solar cells by mitigating the undesired MoO3 diffusion toward cathodes with a high-ionization potential interface layer *ACS Appl. Mater. Interfaces* **17** 15456–67

[121] Savagatrup S, Printz A D, O'Connor T F, Zaretski A V, Rodriquez D, Sawyer E J, Rajan K M, Acosta R I, Root S E and Lipomi D J 2015 Mechanical degradation and stability of organic solar cells: molecular and microstructural determinants *Energy Environ Sci.* **8** 55–80

[122] Wang Y *et al* 2025 Boosting the efficiency and mechanical stability of organic solar cells through a polymer acceptor by reducing the elastic modulus *Adv. Energy Mater.* 2404499

[123] Chen X, Xu G, Zeng G, Gu H, Chen H, Xu H, Yao H, Li Y, Hou J and Li Y 2020 Realizing ultrahigh mechanical flexibility and >15% efficiency of flexible organic solar cells via a 'Welding' flexible transparent electrode *Adv. Mater.* **32** 1908478

[124] Zeng G *et al* 2022 Realizing 17.5% efficiency flexible organic solar cells via atomic-level chemical welding of silver nanowire electrodes *J. Am. Chem. Soc.* **144** 8658–68

[125] Cao C, Chen X, Wang S and Liu S 2014 High temperature induced mechanical degradation in flexible solar cell and its effect on reliability of the packaging module *15th Int. Conf. on Electronic Packaging Technology (12–15 August 2014)* pp 1568–72

[126] An K *et al* 2023 Mastering morphology of non-fullerene acceptors towards long-term stable organic solar cells *Nat. Commun.* **14** 2688

[127] Liu S, Yuan J, Deng W, Luo M, Xie Y, Liang Q, Zou Y, He Z, Wu H and Cao Y 2020 High-efficiency organic solar cells with low non-radiative recombination loss and low energetic disorder *Nat. Photonics* **14** 300–5

[128] You W, Zhou D, Hu L, Xu H, Tong Y, Hu B, Xie Y, Li Z, Li M and Chen L 2021 Adjusting the active layer morphology via an amorphous acceptor solid additive for efficient and stable nonfullerene organic solar cells *Sol. RRL* **5** 2100532

[129] Cai Y *et al* 2022 Vertically optimized phase separation with improved exciton diffusion enables efficient organic solar cells with thick active layers *Nat. Commun.* **13** 2369

[130] Sun R, Wang T, Yang X, Wu Y, Wang Y, Wu Q, Zhang M, Brabec C J, Li Y and Min J 2022 High-speed sequential deposition of photoactive layers for organic solar cell manufacturing *Nat. Energy* **7** 1087–99

[131] He D *et al* 2024 Manipulating vertical phase separation enables pseudoplanar heterojunction organic solar cells over 19% efficiency via ternary polymerization *Adv. Mater.* **36** 2308909

[132] Duan L, Zhang Y, He M, Deng R, Yi H, Wei Q, Zou Y and Uddin A 2020 Burn-in degradation mechanism identified for small molecular acceptor-based high-efficiency non-fullerene organic solar cells *ACS Appl. Mater. Interfaces* **12** 27433–42

www.ingramcontent.com/pod-product-compliance
Ingram Content Group UK Ltd.
Pitfield, Milton Keynes, MK11 3LW, UK
UKHW051900100926
8556IPUK00005B/714

9 780750 332538